SCOPE 38

IPCS JOINT SYMPOSIA 9

Ecotoxicology and Climate

Scientific Committee on Problems of the Environment

SCOPE

Executive Committee, elected 10 June 1988

Officers

Members

SCOPE 38
IPCS JOINT SYMPOSIA 9

Ecotoxicology and Climate

With Special Reference to Hot and Cold Climates

Edited by

Philippe Bourdeau
Commission of the European Communities, Brussels, Belgium

John A. Haines
International Programme on Chemical Safety,
WHO, Geneva, Switzerland

Werner Klein
Fraunhofer-Institut für Umweltchemie und Ökotoxikologie,
Schmallenberg-Grafschaft, Federal Republic of Germany

and

C. R. Krishna Murti
Chairman, Scientific Commission for Bhopal Studies,
New Delhi, Madras, India

Published on behalf of the
Scientific Committee on the Problems of the Environment (SCOPE)
of the International Council of the Scientific Unions (ICSU),
and the International Programme on Chemical Safety (IPCS)
of the World Health Organization (WHO),
the United Nations Environment Programme (UNEP),
and the International Labour Organisation (ILO)

by

JOHN WILEY & SONS
Chichester · New York · Brisbane · Toronto · Singapore

British Library Cataloguing in Publication Data:
Ecotoxicology and climate: with special reference to hot and cold climates.–(SCOPE 38)–(IPCS joint symposia: 9)
1. Environment. Pollution of chemicals. Chemical pollution of the environment effect on climate
I. Bourdeau, Philippe II. International Council of Scientific Unions. Scientific Committee on Problems of the Environment. III. International Programme on Chemical safety.
IV. Series. V. Series.
363.7′384

ISBN 0 471 91831 8

Typeset by Dobbie Typesetting Limited, Plymouth, Devon
Printed and bound in Great Britain
by St. Edmundsbury Press, Bury St Edmunds, Suffolk

SCOPE 1: Global Environmental Monitoring 1971, 68pp (out of print)

SCOPE 2: Man-Made Lakes as Modified Ecosystems, 1972, 76pp (out of print)

SCOPE 3: Global Environmental Monitoring Systems (GEMS): Action Plan for Phase 1, 1973, 132pp (out of print)

SCOPE 4: Environmental Sciences in Developing Countries, 1974, 72pp (out of print)

Environment and Development, proceedings of SCOPE/UNEP Symposium on Environmental Sciences in Developing Countries, Nairobi, February 11–23, 1974, 418pp (out of print)

SCOPE 5: Environmental Impact Assessment: Principles and Procedures, Second Edition, 1979, 208pp

SCOPE 6: Environmental Pollutants: Selected Analytical Methods, 1975, 277pp (out of print)

SCOPE 7: Nitrogen, Phosphorus, and Sulphur: Global Cycles, 1975, 192pp (out of print)

SCOPE 8: Risk Assessment of Environmental Hazard, 1978, 132pp (out of print)

SCOPE 9: Simulation Modelling of Environmental Problems, 1978, 128pp (out of print)

SCOPE 10: Environmental Issues, 1977, 242pp (out of print)

SCOPE 11: Shelter Provision in Developing Countries, 1978, 112pp (out of print)

SCOPE 12: Principles of Ecotoxicology, 1978, 372pp (out of print)

SCOPE 13: The Global Carbon Cycle, 1979, 491pp (out of print)

SCOPE 14: Saharan Dust: Mobilization, Transport, Deposition, 1979, 320pp (out of print)

SCOPE 15: Environmental Risk Assessment, 1980, 176pp

SCOPE 16: Carbon Cycle Modelling, 1981, 404pp (out of print)

SCOPE 17: Some Perspectives of the Major Biogeochemical Cycles, 1981, 175pp (out of print)

SCOPE 18: The Role of Fire in Northern Circumpolar Ecosystems, 1983, 344p

SCOPE 19: The Global Biogeochemical Sulphur Cycle, 1983, 495pp

SCOPE 20: Methods of Assessing the Effects of Chemicals on Reproductive Functions, 1983, 568pp

SCOPE 21: The Major Biogeochemical Cycles and Their Interactions, 1983, 554pp (out of print)

SCOPE 22: Effects of Pollutants at the Ecosystem Level, 1984, 443pp

SCOPE 23: The Role of Terrestrial Vegetation in the Global Carbon Cycle: Measurement by Remote Sensing, 1984, 272pp

SCOPE 24: Noise Pollution, 1986, 472pp

SCOPE 25: Appraisal of Tests to Predict the Environmental Behaviour of Chemicals, 1985, 400pp

SCOPE 26: Methods for Estimating Risks of Chemical Injury: Human and Non-human Biota and Ecosystems, 1985, 712pp

SCOPE 27: Climate Impact Assessment: Studies of the Interaction of Climate and Society, 1985, 650pp

SCOPE 28: Environmental Consequence of Nuclear War
Volume I Physical and Atmospheric Effects, 1985, 342pp
Volume II Ecological and Agricultural Effects, 1985, 562pp

SCOPE 29: The Greenhouse Effect, Climate Change and Ecosystems, 1986, 574pp

SCOPE 30: Methods for Assessing the Effects of Mixtures of Chemicals, 1987, 928pp

SCOPE 31: Lead, Mercury, Cadmium and Arsenic in the Environment, 1987, 384pp

SCOPE 32: Land Transformation in Agriculture, 1987, 384pp

SCOPE 33: Nitrogen Cycling in Coastal Marine Environments, 1988, 480pp

SCOPE 34: Practitioner's Handbook on the Modelling of Dynamic Change in Ecosystems, 1988, 176pp

SCOPE 35: Scales and Global Change: Spatial and Temporal Variability in Biospheric and Geospheric Processes, 1988, 376pp

SCOPE 36: Acidification in Tropical Countries, 1988, 424pp

SCOPE 37: Biological Invasions: A Global Perspective 1989, 528pp

SCOPE 38: Ecotoxicology and Climate, 1989, 392pp

Funds to meet SCOPE expenses are provided by contributions from SCOPE National Committees, an annual subvention from ICSU (and through ICSU, from UNESCO), an annual subvention from French Ministère de l'Environment, contracts with UN Bodies, particularly UNEP, and grants from Foundations and industrial enterprises.

International Council of Scientific Unions (ICSU)
Scientific Committee on Problems of the Environment (SCOPE)

SCOPE is one of a number of committees established by a non-governmental group of scientific organizations, the International Council of Scientific Unions (ICSU). The membership of ICSU includes representatives from 74 National Academies of Science, 20 International Unions and 26 other bodies called Scientific Associates. To cover multidisciplinary activities which include the interests of several unions, ICSU has established 10 scientific committees, of which SCOPE is one. Currently, representatives of 35 member countries and 20 international scientific bodies participate in the work of SCOPE, which directs particular attention to the needs of developing countries. SCOPE was established in 1969 in response to the environmental concerns emerging at that time; ICSU recognized that many of these concerns required scientific inputs spanning several disciplines and ICSU Unions. SCOPE's first task was to prepare a report on Global Environmental Monitoring (SCOPE 1, 1971) for the UN Stockholm Conference on the Human Environment.

The mandate of SCOPE is to assemble, review, and assess the information available on man-made environmental changes and the effects of these changes on man; to assess and evaluate the methodologies of measurement of environmental parameters; to provide an intelligence service on current research; and by the recruitment of the best available scientific information and constructive thinking to establish itself as a corpus of informed advice for the benefit of centres of fundamental research and of organizations and agencies operationally engaged in studies of the environment.

SCOPE is governed by a General Assembly, which meets every three years. Between such meetings its activities are directed by the Executive Committee.

R. E. Munn
Editor-in-Chief
SCOPE Publications

Executive Secretary: V. Plocq

Secretariat: 51 Bld de Montmorency
75016 PARIS

Foreword

by J. W. M. la Rivière

Many environmental problems have first become evident in the industrialized regions of the world, largely situated in the temperate zones. Our understanding of these problems and the remedial and preventive measures that have been developed stem largely from studies carried out in these temperate climates. While some problems and their solutions are obviously universal, for others extrapolation from one climatic zone to another is not feasible, if only because there are organisms and ecosystems which are unique to each specific climatic zone.

Ecotoxicological effects of chemicals are clearly a case in point, not only for the reasons mentioned above but also because of such factors as temperature dependence of biodegradation and of mobility of chemicals.

This book on 'Ecotoxicology and Climate' (SCOPE 38) has evolved logically from previous SCOPE studies on ecotoxicology as well as on environmental problems in developing countries. It is a pioneering effort, charting a course in a subject area which until now has constituted almost entirely unknown territory.

Although this study necessarily leaves many areas still unexplored, it clearly demonstrates the need for specific follow-up research and provides the rationale for such work. Furthermore, it opens up yet another stage for co-operation between environmental scientists from developing countries and from the industrialized areas of the world. This is a high priority objective of SCOPE in view of its global mandate.

J. W. M. la Rivière
President, SCOPE

Foreword

by Michel J. Mercier

Evaluation of the effects of chemicals on the environment as well as on human health is an important part of the mandate of the International Programme on Chemical Safety (IPCS), which is a collaborative venture of the United Nations Environmental Programme, the International Labour Organization, and the World Health Organization.

The IPCS Programme Advisory Committee at its fourth meeting in October 1984 recommended that effects on the environment should be emphasized in all aspects of the Programme and that IPCS should contribute to and encourage the definition of fundamental principles for the assessment of effects on the environment. More and more of the activities of IPCS are being directed towards the needs of developing countries for assessment of effects of chemicals. In particular those chemicals widely used in these countries are given priority for evaluation in the Environmental Health Criteria documents and practical guidance is being provided through the Health and Safety Guides. It is important for this work to know whether ecotoxicological principles, established mainly on the basis of experience in the industrialized countries of the temperate zones of the world, are also valid in other climatic regions of the globe where most of the developing countries are situated.

For these reasons the IPCS has been very active in its intellectual and financial support of this project jointly undertaken with SCOPE.

Michel J. Mercier
Manager
International Programme on Chemical Safety
WHO

Preface

This publication is based mainly on a Workshop held at the Fraunhofer Institute for Environmental Chemistry and Ecotoxicology in Schmallenberg-Grafschaft, Federal Republic of Germany, from 1 to 5 July, 1985. It was produced under the guidance of the Scientific Advisory Committee on Ecotoxicology, established by the Scientific Committee on Problems of the Environment (SCOPE) in 1984 under the chairmanship of Philippe Bourdeau. Werner Klein and C. R. Krishna Murti were appointed co-chairmen for a project on 'Ecotoxicology and climate with special reference to tropical, arid, and subpolar environments'.

The International Programme on Chemical Safety (IPCS), established in 1980 as a joint venture of the World Health Organization (WHO), the United Nations Environment Programme (UNEP), and the International Labour Organization (ILO), has among its objectives the evaluation of the risks to the environment of chemicals. It joined SCOPE in organizing this project, and John A. Haines of the IPCS Central Unit was appointed to the Scientific Advisory Committee for the project.

Considerable experience has become available on ecotoxicological principles of environmental chemicals, but mainly as a result of data from temperate regions. Far less information (particularly quantified data) is available from other regions, such as tropical, arid, subpolar, and high mountain regions. Different physical and chemical conditions as well as different interactions between chemicals and various physical and biological factors throughout the world could significantly influence effects of environmental chemicals. Furthermore, on the one hand, use patterns of chemicals may vary from those in temperate zones, and on the other hand many of the ecosystems found outside the temperate zones may be more vulnerable.

With the growing recognition that prevention of adverse effects of environmental chemicals is essential for health and sustainable development, as well as that chemicals used or generated in one region of the world may affect ecosystems in other regions, it was important to check the extent to which those ecotoxicological principles now being developed are universally applicable.

With a view to establishing a better scientific basis for the control of environmental chemicals in different parts of the world, the project has endeavoured to examine those ecotoxicological principles applicable to climatic conditions found in regions other than the temperate zones. Through the preparation of 21 papers and case studies by internationally recognized experts in their field, the project has examined the environmental fate and effects of chemicals. The project has concentrated on data available from the tropical, arid, subpolar, and high mountain regions on the fate of environmental chemicals and the response of ecosystems to chemical stress.

Most of the papers and case studies were available before the Workshop, at which contributors and other invited participants reviewed the data and developed the report which forms the basis of this publication. This was edited by a small editorial group of the Scientific Advisory Committee who prepared the Conclusions and Recommendations.

This publication represents, we believe, the first systematic attempt to examine the applicability of ecotoxicological principles in non-temperate climatic zones. The report highlights areas where data are particularly lacking and recommends fields for further research and monitoring.

We are much indebted to the contributions of the various authors and their constructive comments on each other's chapters. In particular, Arthur H. Westing has provided valuable support to the work of editing various sections of the publication. Final editing of the manuscript would not have been possible without the painstaking and careful work of Margaret E. Donohoe, whose contribution is acknowledged with much appreciation.

We wish to express our gratitude for the considerable financial support which has been provided to this project by the International Programme on Chemical Safety, as well as by the Scientific Committee on the Problems of the Environment (SCOPE) and the Commission of the European Communities.

The Editors

Contributors

Bourdeau, P. Directorate General for Science, Research and Development, Commission of the European Communities, 200 rue de la Loi, 1049 Brussels, Belgium.

Bruenig, E. F. Institut für Weltforstwirtschaft und Ökologie, Universität Hamburg, Leuschnerstrasse 91, 2050 Hamburg 80, Federal Republic of Germany.

El-Sebae, A. H. Division of Pesticide Chemistry, Faculty of Agriculture, Alexandria University, Chatby, Alexandria, Egypt.

Goldberg, E. D. Scripps Institution of Oceanography, Ocean Research Division, University of California, La Jolla, California 92093, USA.

Haines, J. A. International Programme on Chemical Safety, Division of Environmental Health, World Health Organization, 1211 Geneva 27, Switzerland.

Klein, W. Fraunhofer-Institut für Umweltchemie und Ökotoxikologie, Grafschaft/Hochsauerland, 5948 Schmallenberg, Federal Republic of Germany.

Krishna Murti, C. R. Government of India Scientific Commission for Continuing Studies on Effects of Bhopal Gas Leakage on Life Systems, Cabinet Secretariat, Sardar Patel Bhavan, New Delhi 110 001, India.

Lay, J. P. Gesellschaft für Strahlen- und Umweltforschung mbH, München Institut für Bodenökologie, Ingolstädter Landstrasse 1, 8042 Neuherberg, Federal Republic of Germany.

Lévêque, C.	ORSTOM, Institut Français de Recherche Scientifique pour le Développment en Coopération, 213 rue Lafayette, 75480 Paris Cedex, France.
Magallona, E. D.	University of the Philippines at Los Banos, College, Laguna 3720, The Philippines.
Matsumura, F.	Department of Environmental Toxicology and Toxic Substances Program, LEHR Facility, University of California, Davis, CA 95616, USA.
McKay, G. A.	Environment Canada, Atmospheric Environment Service, 4905 Dufferin Street, Downsview, Ontario, Canada M3H 5T4.
Miller, D. R.	Institute for Research in Construction, National Research Council, Ottowa, Canada K1A 0R6.
Naik, Sugandhini	National Institute of Oceanography, Dona Paula, Goa 403 004, India.
Perry, A. S.	Tel Aviv University, George S. Wise Faculty of Life Sciences, Institute for Nature Conservation Research, Ramat-Aviv, 69 978 Tel Aviv, Israel.
Perry, R. Y.	Tel Aviv University, George S. Wise Faculty of Life Sciences, Institute for Nature Conservation Research, Ramat-Aviv, 69 978 Tel Aviv, Israel
Ringer, R. K.	Michigan State University, Pesticide Research Center, East Lansing, MI 48824-1311, USA.
Robinson, C. A.	International Institute for Applied Systems Analysis, Laxenberg, Austria.
Scheunert, I.	Gesellschaft für Strahlen- und Umweltforschung mbH, München Institut für Bodenökologie, Ingolstädter Landstrasse 1, 8042 Neuherberg, Federal Republic of Germany.
Sen Gupta, R.	National Institute of Oceanography, Dona Paula, Goa 403 004, India.
Sethunathan, N.	Central Rice Research Institute, Cuttack 753 006, India.
Thomas, M. K.	Environment Canada, Atmospheric Environment Service, 4905 Dufferin Street, Downsview, Ontario, Canada M3H 5T4.

Varadachari, V. V. R. National Institute of Oceanography, Dona Paula, Goa 403 004, India.

Viswanathan, P. N. Ecotoxicology Project, Industrial Toxicology Research Centre, Mahatma Gandhi Marg, Post Box 80, Lucknow 226001, India.

Westing, A. H. International Peace Research Institute, Oslo, (PRIO), Fuglehauggata 11, 0260 Oslo 2, Norway.

Zsolnay, A. Gesellschaft für Strahlen- und Umweltforschung mbH, München Institut für Bodenökologie, Ingolstädter Landstrasse 1, 8042 Neuherberg, Federal Republic of Germany.

Contents

Chapter 1

Introduction, Conclusions, and Recommendations

Ecotoxicology and Climate
Edited by P. Bourdeau, J. A. Haines, W. Klein and C. R. Krishna Murti

Introduction, Conclusions, and Recommendations

P. BOURDEAU, J. A. HAINES,
W. KLEIN AND C. R. KRISHNA MURTI

1.1 INTRODUCTION

Man-made chemicals are increasingly used throughout the world, particularly in regions outside the temperate zone, where most of the developing countries are situated. Yet, what is known of the fate and effects of these chemicals on the environment is based essentially on a body of knowledge acquired in the temperate zone. The need was felt, therefore, to review and assess the information available on exposure to chemicals and on their environmental behaviour and effects in warm (hot, wet, and dry) and cold climates and to verify the applicability of basic ecotoxicological principles under these conditions. This is the object of the present report.

The rapid development process whereby humanity aspires to attain a better quality of life is a mixed blessing in that a certain degree of consequent environmental degradation becomes almost unavoidable. Chemical pollution is the most visible cause of environmental degradation, with its potential to impair human health and to produce undesirable disturbances in ecological equilibria. It is also recognized that some chemicals which are essential to sustain life and improve its quality from the viewpoint of today's mode and style of living can indeed have a negative impact on human health and the environment.

Ecotoxicology, a relatively young subject, represents a multidisciplinary approach to the study of the adverse effects of environmental chemicals on individuals, populations, biocoenoses, and whole ecosystems, as opposed to toxicology, a much older subject, which deals with the harmful effects of chemicals on a given species of the living system. The need for rapid test systems to assess the toxic outcome, if any, of chemicals designed as therapeutic agents has led to a much wider understanding of the comparative effects of chemicals on a variety of laboratory animals, mostly rodents. Indeed, the cornerstone of the safety evaluation of chemicals on human health consists in the meaningful extrapolation of information derived from their effects under controlled conditions on a number of target species for both short-term and long-term effects. Obviously, the plethora of acute, chronic, and sublethal effects that can be expected to be elicited on ecosystems under natural conditions will be more diverse and not bear any direct resemblance to health effects seen on a single species by the same chemical. Furthermore, under natural conditions one has to take into account a variety of factors: interaction among species; climate; food-web interrelationship; synergism between chemicals; comparative biochemistry; etc. Whole ecosystems may be irreversibly changed through chemical disturbance affecting directly only part of them or particular species.

During the past two decades, a growing awareness of the principles governing the behaviour and effects of environmental chemicals has been developed, partly through the activities of SCOPE and the resulting reports: SCOPE 11, *Principles of Ecotoxicology* (1978); SCOPE 20/SGOMSEC I, *Methods for Assessing the Effects of Chemicals on Reproductive Functions* (1983); SCOPE 22, *Effects of Pollutants at the Ecosystem Level* (1984); SCOPE 25, *Appraisal of Tests to Predict the Environmental Behaviour of Chemicals* (1985).

This work was done mainly on the basis of the experience of industrialized countries, mostly located in temperate climate zones and with histories of much longer exposure to anthropogenic chemicals. During the recent past, particularly since the late sixties, the chemical industry has grown rapidly in many developing countries located in non-temperate climatic zones. It has thus become important to ask whether the ecotoxicological principles evolved as a result of retrospective study of conditions prevalent in industrialized countries of the temperate zone of Europe and North America are applicable universally to the present situation in rapidly developing countries in the tropical zone or the semi-arid zone. Moreover, are these principles relevant to the fate and impact of toxic chemicals in the arctic and antarctic zones or in high elevation ecosystems?

These questions have been addressed in the present project, while the somewhat related issue of acidification was the object of another SCOPE study, to be published under the title 'Acidification in Tropical Countries' (H. Rohde, editor), SCOPE 36.

1.1.1 Quantity and Range of Chemicals

It has been estimated that there are at present over 60 000 man-made chemical substances in use today, of which some 4000, accounting for 99% of total production volume, are in common use globally. In addition, many naturally occurring chemicals and those produced by the metabolic activity of many microbial organisms, insects and other pests, snakes, etc. are known to be extremely toxic to animal life. Risks of exposure to chemicals faced by ecosystems arise in many ways, but mainly through their production, storage, transport, use and disposal. Some hazardous chemicals, even when their use is restricted, can escape into the environment by accidental release. Others widely used in industry, agriculture, food, commerce or the home are readily dispersed in the environment. Thus many anthropogenic chemicals will ultimately appear as pollutants in air, water, soil and food, either as metabolites or residues, or as wastes. Furthermore, human activities related to mining, dredging, land development, and coastal management release naturally occurring chemicals at a rate more rapid than that attained by normal geochemical processes. The consequent disturbances of chemical dynamics can have serious repercussions on various ecosystems.

Other dimensions of pollution by chemicals are their spread, which can transcend frontiers from the region where they are produced to a region where they are diffused and eventually deposited in the sinks, and their fate at the interfaces between land and water, particularly in the oceans, as well as between air and land, as when chemical clouds move over large tracts of vegetation. The impact of transfrontier export of air pollutants on terrestrial (e.g. forests) and aquatic ecosystems has attracted much attention.

There is growing recognition of the fact that for sustainable development one must ensure prevention of the adverse effects of environmental chemicals. At the same time a balance has to be struck between the essential needs of chemicals for attaining development goals and the urgency of recycling and conservation of non-renewable resources of the earth.

It thus becomes imperative to determine whether the toxic effects of chemicals on ecosystems manifest themselves in varied forms in different climatic zones. It is equally important to establish a more scientific basis for the control of environmental chemicals in different parts of the world.

1.1.2 Content

The background material presented in Chapter 2 includes overviews on climates of the world from an ecotoxicological angle (2.1), the diversity of ecosystems (2.2), atmospheric transport of chemicals (2.3) and aquatic transport of chemicals (2.4). The environmental fate of chemicals determined by abiotic degradation and by biotic degradation is described in Chapter 3. Also in Chapter 3, existing knowledge of exposure of non-temperate ecosystems to environmental chemicals is synthesized for cold environments (3.3) and for tropical and arid regions (3.4). Chapter 4 deals with the role of temperature and humidity on comparative toxic effects on diverse living systems (4.1), toxic effects seen in arid regions (4.2), tropical marine ecosystems (4.3), arctic and subarctic ecosystems (4.4) and on domestic animals (4.5). Chapter 5 contains eight case studies: coastal pollution in the Indian subcontinent (5.1), biodegradation of pesticides in rice ecosystems (5.2), effects of insecticides on rice ecosystems (5.3), fate and effects of aldrin/dieldrin in terrestrial ecosystems in hot climates (5.4), blackfly control in West Africa (5.5), herbicides in warfare and effects on coastal ecosystems (5.6), effects of pesticides in Egypt (5.7), and mercury in Canadian rivers (5.8).

The assessment of the above-mentioned papers led to the conclusions and recommendations for future research listed in the following sections.

1.2 CONCLUSIONS

1.2.1 Exposure of Ecosystems to Chemicals

In carrying out this study, it soon became evident that few data were available on the kinds and amounts of chemicals used in non-temperate regions. Information on residue levels, environmental concentration, effects on health and on the environment, is even scarcer.

With world trade in chemicals and chemical products burgeoning in an unprecedented manner and with their ever increasing use in developing countries in every sector, the potential for chemical pollution has expanded by leaps and bounds. Most of the data discussed in this study came from non-temperate climatic zones of developed countries and rapidly industrializing countries in the Third World: Egypt, India, Indonesia, Israel, the Philippines. Some data from Vietnam and West Africa were also available.

A wider coverage, however desirable, was not possible within the constraints of the present study. No uniform system is followed in the different countries for maintaining statistics of production, import, export, and use of chemicals. Except for the FAO statistical reports there appears to be no consistent international effort to compile data on production of classes of chemicals. FAO reports deal primarily with agrochemicals, viz. fertilizers and pesticides.

Statistics of production and use of bulk chemicals and fine chemicals are maintained by some countries of the industrial world and are accessible for survey and analysis. Data are also collected for internal use by industrial associations of chemical manufacturers. This source of information is not readily available. There is a paucity of information on the internal trade of chemicals among the Eastern European countries. In regard to developing countries with programmes of rapid growth of chemical industries, this study could cover only the data for industrial chemicals for India and the data for production, import, and use of pesticides globally. Thus information on chemical production and usage pattern in China, Indonesia, Malaysia, and the Latin American countries could not be obtained.

Note has been taken of the exceedingly useful work being done by the International Registry of Potentially Toxic Chemicals (IRPTC) as part of the UNEP activities to compile information as a register of a select group of toxic chemicals. The network of national correspondents of IRPTC should enable the organization to widen the scope of the Register in the coming years. The elaborate list prepared by the Environmental Protection Agency, USA, in connection with the Toxic Chemicals Control Act, is a useful compendium for assessing chemical exposure in the USA. Similar lists are available in the European

Communities (EINECS). Thus the framework for establishing registers is already in existence and can be used in compiling and collecting information from developing countries.

Some of the tropical countries, like India, have sizeable industrial chemical complexes and processing plants with a high potential for pollution, being located on the banks of inland water bodies, including rivers, in coastal areas, or in the proximity of pristine forests. Emissions into air as smoke, particulates, and gaseous vapours are on the increase, as is evident from the data being generated by the national network of air monitoring stations, particularly in industrial and urban zones. Chemicals discharged as effluents into inland water bodies and coastal waters also show an increasing trend of aquatic pollution. By and large, the impact of biological pollution on human health has been of primary concern from the point of view of control.

Even for industry, the problems of dealing with BOD due to discharge of biodegradable effluents have received far greater attention than related problems resulting from discharge of toxic chemicals.

Another uncharted problem area is the size and nature of chemical wastes generated or deposited in non-temperate ecosystems. There is a growing trade in chemical wastes generated in industrialized countries in the temperate zone being disposed of in developing countries in non-temperate regions. Stricter environmental controls exercised in industrialized countries in the developed world have also encouraged the manufacture and export of hazardous chemicals from developing countries in the non-temperate regions. As of today, no concerted efforts seem to have been made to locate toxic waste dumps in developing countries and to assess their impact on the environment. However, even the limited information available from India, indicates that the problems of diffusion of toxic chemicals from waste dumps may assume serious proportions if appropriate checks are not introduced.

Although data are sparse, there is a growing use of chemicals in all non-temperate climate zones and, where data are available, levels in the environment of the persistent chemicals appear to be on the increase. In particular, there is evidence of an increasing use of chemical pesticides in tropical countries, a trend which is not likely to be reversed in the foreseeable future.

The problem is compounded in some places by the inadequacy of administrative and legislative structures for implementing pollution control.

It is not easy to quantify the increase, region by region, nor is it feasible to identify the chemical species involved as the use patterns exhibit wide changes. While the use of persistent pesticides, particularly the chlorinated ones, is being phased out in some countries, there is no evidence that the use and diffusion of some of them on a global scale is decreasing. It may be recalled that the assay of levels of DDT, HCH, and PCB residues in human blood and human milk fat executed under the GEMS programme of WHO/UNEP revealed a several-fold difference between the values reported from China, India, and

Mexico on the one hand and Sweden, France and Yugoslavia on the other. Obviously this high tissue level reflects an unexpectedly high level in the environment and a consequently greater uptake by living systems. Pesticide residue levels in food crops, milk and dairy products, poultry and animal food are available from some tropical areas, such as India.

1.2.2 Environmental Fate

Basic climatic and oceanographic patterns are sufficiently well-known on a global basis for the purpose of this study. Long-range transport of chemicals by air and water occurs between various climatic zones. Tropical storms accelerate the rate of removal of chemicals from the atmosphere by rain and eventually lead them to the seas. There is perceptible evidence of accumulation of toxic chemicals by coastal leaching in continental shelves and eventual drift to colder regions. Natural transport of chemicals between hemispheres accounts for 25% per annum of global transport.

Disappearance of chemicals from the environment appears to be greater in hot-humid rather than in cold-dry or high-altitude regions. In hot climatic zones, movement of chemicals through the media of soil, water or air may be enhanced by vaporization. Solubility in water increases also with temperature. Under hot-dry conditions microbial degradation of chemicals is, however, decreased. Information as to whether this is compensated for by thermal degradation or photodegradation is not adequate. One expects significant differences in the ecology of soil microflora between hot-humid, hot-dry, and temperate zones and hence divergences in the pathways of microbial degradation.

As far as cold regions are concerned, it appears that chemicals tend to concentrate in various ecosystem compartments, probably because of slower reaction rates.

Mention should also be made here of the effects of fluorinated and chlorinated hydrocarbons on the ozone layer, especially in the arctic and antarctic regions (the so-called antarctic ozone 'hole').

1.2.3 Ecosystem Effects

Information on the presence of chemical hazards in the environment and their adverse effects on the ecosystems is fragmentary. This is particularly true of the possible chronic or synergistic effects or the effects elicited in highly exposed or relatively more susceptible populations. Adverse effects on ecology of toxic chemicals documented so far range from description of the effects on a single species to those on whole ecosystems. The synergistic effects of naturally occurring chemicals and man-made chemicals could present an altogether different picture. The impact of physical agents, such as heat, humidity or radiation, may exacerbate ecotoxic effects. Furthermore, metabolic disturbances elicited by a toxic chemical in one species of an ecosystem may lead to irreversible changes in the ecosystem as a whole.

Against the background of ecotoxicological principles derived from observations on the temperate zone, the present study showed that ecosystems with low diversity, such as mangroves or tundras of the arctic zone, have a greater potential sensitivity to environmental chemicals than those with high diversity. In general, warmer climates tend to favour an acceleration of the ecotoxic effects of chemicals. In arid systems there is a slower degradation of chemicals due presumably to lower microbial activity. However, this does not seem to modify the spectrum of effects on the ecosystem.

The accumulation of chemicals in subpolar and arctic ecosystems may be due not only to the fact that reaction rates are slow but also that lipophilic chemicals can accumulate in fat, which is abundantly present in the tissues of the organisms inhabiting cold regions. In these ecosystems the melting snow releases certain chemicals and induces an additional seasonal stress.

Episodic soil flushing has been reported from tropical rain forests, and pulse emissions of dissolved organic matter into the drainage can reach levels toxic to aquatic life.

In general, there appears to be little basic difference between ecotoxic effects seen in terrestrial tropical ecosystems and those seen in temperate ecosystems. The same is not true in the marine environment. However, there are differences among species in their sensitivity to chemicals. The difference is thus not qualitative but quantitative. It is likely that a species' sensitivity might result in overall deterioration of the ecosystem if the species is critical in maintaining ecological balance. As regards effects on managed systems, no data are available on the corresponding effects on the related natural ecosystems. For a managed ecosystem like the rice ecosystem, ecotoxic effects of a given chemical appear to be the same whether the managed ecosystem is located in the temperate or the tropical zone.

1.3 RECOMMENDATIONS FOR RESEARCH AND MONITORING

1.3.1 Exposure Assessment and Monitoring of Ecotoxic Effects

More quantitative information is required on the persistence of chemicals in non-temperate zone ecosystems to enable assessment of exposures in both the short term and the long term. The tropical zones support and sustain a variety of managed ecosystems: rice, coconut, cocoa, sugar-cane, and monoculture-based forest systems for biomass and fuel. The help of the chemical industry and of agriculture and silviculture must be sought to accomplish this task.

There is an urgent need to build and update data bases on the types of chemicals to which non-temperate ecosystems are presently exposed or are likely to be exposed in the coming years. This could go hand in hand with sustained monitoring programmes on selected managed ecosystems to unravel subtle changes in the fragility and resilience of such systems.

1.3.2 Modulation of Ecotoxic Effects by Physical Factors

So far, degradation and mobility studies on environmental chemicals have been undertaken for the range of temperature, humidity and light intensity found in temperate zones. It is recommended that carefully designed studies should be conducted under conditions found in tropical and polar zones.

Current understanding of the interrelations between toxicity and temperature, toxicity and humidity, and toxicity and light intensity is inadequate. There is scope for further research in this area using selected specific species representing diverse ecosystems from different climatic zones.

1.3.3 Soil and Sediment Parameters

There are known differences in soil parameters between climatic zones. It is recommended that studies be undertaken to assess the extent to which such differences modulate rate of degradation and toxicity.

The role of sediments in absorption, storage and release of pesticides in tropical aquatic environments is not well known and should be further investigated.

1.3.4 Transport of Chemicals

Transport of chemicals within the temperate zone is relatively well understood. Using that experience it is desirable to establish transport models in and between other zones, by field observations as well as by experimental studies.

1.3.5 Ecotoxic Effects Produced by Degradation Products

The nature of degradation products formed by the combined action of microbes and light remains to be explored, especially from the point of view of exacerbating toxic effects or producing new toxic manifestations. The potential of degradation products to produce long-term effects, particularly mutagenic effects, on sensitive species remains to be explored.

1.3.6 Experimental Ecotoxicology

Experimental toxicological work on non-human species is normally undertaken at an ambient temperature of 25°C. It may be worthwhile to carry out a series of experiments over a range of temperatures, e.g. 10°C, 25°C and 40°C (but not more than 30°C for aquatic organisms), using a range of species more representative of different ecosystems.

1.3.7 Methodological Problems

Several problems related to methodologies of assessing toxic effects of chemicals on the ecosystem as a whole were highlighted in SCOPE 22. These have not been resolved and further systematic work is required towards devising suitable ecotoxicological models, microcosms, etc. The need for more precisely defined end-points of toxic effects is also apparent. So far productivity has been the main criterion in assessing effects on whole ecosystems.

A more specific recommendation can be made regarding pesticide research in arid zones. Parameters such as temperature, humidity, solar radiation and soil type cannot be simulated properly in the laboratory. Experiments *in situ* under field conditions would yield more accurate information and thus should be preferred to laboratory tests.

Chapter 2

Background: World's Climate and Ecosystems; Global Transport of Chemicals

Ecotoxicology and Climate
Edited by P. Bourdeau, J. A. Haines, W. Klein and C. R. Krishna Murti

2.1 Climates of the World Seen from an Ecotoxicological Perspective

G. A. McKAY AND M. K. THOMAS

2.1.1 INTRODUCTION

The influence of weather on the environment is very apparent in growth processes, which are controlled by the availability of light, heat, and water. However, climate and weather are a continuum and climate is often defined as the sum of weather. It is an integral part of the natural environment, and is recognized as a major biological control. The effects of climate are evident in occurrences of skin cancer and the extent of biomes. From a toxicological viewpoint, interest in climate usually is focused on the atmosphere's ability to influence the toxicity of pollutants as well as their dispersal and transport.

Study of climate information reveals the nature of commonly experienced weather as well as extremes and the risk of occurrence of harmful combinations of climatic elements. As a rule these characteristics are relatively conservative over large areas. However, climatic regions or zones display many anomalous features on the meso and micro scales, i.e. climates that differ from the regional average because of special terrain or relatively local effects of the atmosphere. For example, the Los Angeles smog results from geography, a relatively unique combination of atmospheric processes and a cold offshore ocean current. Similarly valley climates differ noticeably from those nearby at higher elevation.

Over the globe there are many different climatic regions, the number being defined by whatever classification system is employed. Broadly speaking, the climatic differences exist because of differences in geography, the latitudinal variation in the energy from the sun, the availability of moisture, and the influence of these factors on atmospheric dynamics. These differing climates have contributed to the creation of ecosystems and in turn are influenced by the vegetative cover. The differences are manifest in the major biomes of the earth: tropical rain forests, deserts, grasslands, boreal forests, etc. This chapter focuses on the large-scale characteristics of climate that identify with these features, and to a lesser degree on atmospheric processes that relate to pollution concentrations, transport, and performance within climatic zones.

2.1.2 THE ATMOSPHERE

The atmosphere can be considered as consisting of horizontal layers. Adjacent to the earth lies the troposphere, which contains the bulk of the atmospheric mass and which extends vertically to about 10 km in mid-latitudes and to 20 km near the equator. It contains the planetary boundary layer within which occurs an active exchange of heat, water vapour, and contaminants between the oceans and land surfaces. Above the troposphere lies the stratosphere, which extends to a height of above 82 km. Within the stratosphere is the ozone shield, which is vulnerable to the catalytic effects of dissociated chlorofluorocarbons.

The tropospheric wind system is highly organized, being controlled primarily by the earth's rotation and differential heating by the sun. Tropospheric winds generally increase with height. At the equator are found the relatively calm doldrums. Poleward of the doldrums the trade winds exist, and then the horse latitudes, a region of light winds near 30° latitude. Seasonal changes in the relative warmth of the ocean and land areas results in monsoons; these are pronounced in tropical and subtropical areas. Farther poleward, roughly between 30° and 60°, are stormy regions within which the winds blow mainly from the west. At high latitudes the winds tend to be weaker and more variable. Many departures from these generalizations are found, both seasonally and annually. The departures are extremely important since they are often responsible for the exchange of pollutants between zones, exposure to more critical levels of pollutants, and the occurrence of anomalous climatic environments at ground levels. Synoptic-scale analyses provide insight into their character.

The vertical movement of air is of great importance in the initial lifting, diffusion, deposition, and transport of pollutants. Large-scale ascending motions occur over mountains, weather fronts, active low-pressure areas, and in equatorial regions; however, smaller-scale turbulence and convection are of greater relevance in the vertical movement of pollutants. Turbulence induced by frictional drag and convection is the cause of the earth's boundary layer. Well-mixed, the layer has a depth of 1–3 km, varying diurnally and seasonally with the stability of the air, the roughness of the ground, and wind velocity. Land, sea, and other topographically controlled breezes or drainage of air occur within the boundary layer. Strong convection lifts air from the lower to the faster-flowing middle and high regions of the troposphere, and at times into the stratosphere. Strong convection is found daily over the tropical land areas, but elsewhere it is infrequent and the exchange of air between the troposphere and stratosphere is slow.

Horizontal tropospheric circulations are strongest in mid-latitudes and at times are concentrated into 'jet streams'. Best known is the polar jet stream, found at about 7 km height and within which winds may be in excess of 250 knots. Highly variable in character, it may be sinuous or more commonly move from west to east. The subtropical jet, found at about 28° latitude and at a height of 10 km, has maximum winds between 100 and 150 knots. Other jets are found at both high and low levels, for example the tropical easterly jet that moves from West Africa to the Caribbean.

The capacity of the winds to transport pollutants varies with the wind field. Those confined to the boundary layer have relatively short residence times and their depositions tends to be local. Exceptions occur particularly at higher latitudes where ducting is more common. Convection lifts contaminants to stronger wind-levels where they can be transported over thousands of kilometres within short time spans. For example, dense smoke from western Canada forest fires in 1950 was jetted across the North Atlantic in a day.

The somewhat zonal, relatively uniform fields of wind, along with temperature, water vapour, and radiation, are the bases of the world's climatic zones. Within the zones conditions, both at the surface and in the upper air, tend to be relatively conservative, such that general statements can be made as to how they distribute and influence the performance of contaminants.

2.1.3 CLIMATE AND ATMOSPHERIC CONTAMINANTS

The atmosphere acts primarily as a conduit by which pollutants are conveyed from a source to an ecosystem. It may also play a role in photochemical change that alters the toxicity to the emitted substances. Furthermore the behaviour of the toxics can be influenced by climatic conditions, and the pollutant acting alone or in combination might be instrumental in altering the nature of the climate.

Toxics released into the biosphere can move along a variety of pathways, for example by streamflow into oceans, by food chains, and by the atmosphere to be deposited on the land or on water surfaces. The atmosphere is a major pathway that can be used at different stages in this process. For example, the atmospheric PCB contribution to Lake Superior is estimated to be 6600–8300 kg/year, five to seven times the direct input by industry or tributary streams (Bruce, 1983). 'Re-emission' of toxics, i.e. a secondary release of already transported contaminants, allows them to be transported still farther from their original source (Ottar, 1979). The effectiveness of these processes within a zone relates to the fundamental nature of the climate (winds, stability, etc.) of the zone, which must be appreciated to understand the role of climate in the dispersal of toxics, and to the nature of the pollutant.

Gases that are relatively inert remain unchanged for long periods of time and become widely diffused throughout the atmosphere. Carbon dioxide background levels, for example, are similar from pole to pole, although seasonal and regional differences occur. More active gases, such as sulphur dioxide and oxides of nitrogen, enter readily into chemical reactions and, therefore, have a shorter residence time. For that reason they are more variable in their global and regional distribution. The same holds true for carbon monoxide. Thus while the stable gases create effects that are generally global, the reactive gases are best considered in the context of zonal climates.

Pollutants are removed from the atmosphere most effectively by precipitation. Precipitation cleanses the atmosphere, but in doing so, it can deposit large amounts of pollutants over specific areas. Droplets of about 2 mm diameter are theoretically the best collectors, so that even a very light rainfall or drizzle can deposit large quantities. Dry deposition is much slower but it, too, can be quite significant. The subsequent dissolution of the particulates by rainfall may result in relatively strong chemical concentrations in runoff and on vegetation. In the case of melting snow cover, both the dry and wet deposition chemicals are leached from the snow in the early stages of melting, thereby producing the phenomenon of 'acid shock'.

Variations in climate can severely damage existing ecosystems and may lead to the introduction of others that are highly vulnerable. The warm decade of the 1930s enabled the introduction into sub-Arctic regions of many new crops that could not survive the colder years that followed. Climate variations are marked in the semi-arid transition zones, such as the Sahel, where ecosystems may alter greatly from year to year. Low temperatures reduce metabolic processes and thereby may reduce impacts on biota. On the other hand the anomalously cold winter of 1976–77 in North America did major damage to coastal marine ecosystems. Different seasons also introduce their peculiar problems. The turbulent floods that arise with the onset of a rainy season may produce soil erosion and stir up toxin-bearing river bed sediments. Seasonal droughts may reduce surface water volumes and thereby increase the pollutant concentration, as well as raise the water temperature. Climate variations, such as drought, may act to predispose to, or incite damage within, ecosystems that are threatened by toxins. Allergens, dust storms, monsoonal rains, and the temperature/humidity combinations that favour the explosive development of rusts or blights all have a seasonal character. By examining the anomalous episodes, seasonal shifts, longer-term variations, and other characteristics of climate, much can be learned of climate-related hazards.

2.1.4 CLIMATE ZONATION

The Sahara, boreal forest, and the tundra conjure up impressions of vast regions with relatively unique, but uniform climates. Many attempts have been made to classify the Earth's climates to obtain a brief comprehensive appreciation of their characteristics, similarities and differences. Early classifications, such as that of Koeppen (Koeppen and Geiger, 1935), are based largely on values inferred from vegetative cover. More recent systems have introduced spatial and temporal variability using statistics and features such as ecoclimate, topoclimate, microclimate, and mesoclimate, etc. to introduce different scales of interest.

Global zonation, of necessity, must be highly generalized and concentrate on the major differences as indicated by temperature and moisture as well as their patterns across the continents. However, a too generalized classification scheme will not reveal ecologically important phenomena that occur on fine scales of time and space. For example, specific incidents of episodic stagnation or specific trajectories of pollutants will not be revealed by highly generalized zonation techniques. Nevertheless many processes of importance to the ecosystem must and can be inferred from the zonal characteristics. Occurrences such as periods of stagnation tend to be repetitive, and their frequency, extent, and intensity as well as the nature of extreme incidents may be revealed by reference to climate statistics for representative locations within a zone.

The zonal boundaries and characters are coarse averages and they may shift briefly or for longer intervals of time. Such upsets are often associated with the displacement of weather fronts and monsoonal systems from their mean position. The result usually is widespread regional droughts as well as areas of moisture excess and anomalous heat and cold. A semi-arid region may become arid or subhumid during a climate fluctuation. The El Nino/Southern Oscillation phenomenon that occurs every 2 to 7 years is commonly associated with regional climatic fluctuations in tropical areas, for example torrential rains in Ecuador, and drought in northeast Brazil, the Sahel, and southeast Asia in 1982–83. In arid areas most variability is interannual, but dry episodes that span several growing seasons and sequences of dry episodes may occur—a climate feature that is recurrent in the Sahel.

Five generalized zones are identified. These are: (1) Equatorial—where the climate is hot and moist all or during the larger part of the year; (2) Desert/Steppe—where water supply is insufficient to meet growth potential; the zone is generally hot all year but with a cold season at higher latitudes; (3) Mid-Latitude—a somewhat humid, variable climate; (4) Boreal/Taiga—a region with a short growing season and cold winters; and (5) Tundra/Ice-cap—a zone where mean temperatures are below freezing for most of the year and precipitation is generally light. These five zones are delineated on the accompanying map (Figure 2.1.1) and their relevant characteristics are specified in the zonal narrative that follows.

Not identified are the climates of oceanic areas, and these cover 71% of the earth's surface. Oceanic climates strongly reflect the nature of the ocean surface, as well as the characteristics of the air masses that traverse it. Precipitation is pronounced in areas of atmospheric convergence, such as along the intertropical Convergence Zone (ITCZ), and in those areas swept by the major west-to-east moving mid-latitude storms. Also worthy of note are the highland climates, such as are found in East Africa, Tibet, and Ecuador. Altitude strongly influences climate and the climates of highlands in any general zone are anomalous. The normal lapse rate of temperature with height is about $-6°C$ per 1000 m, but there are important exceptions. Precipitation and fog occurrences

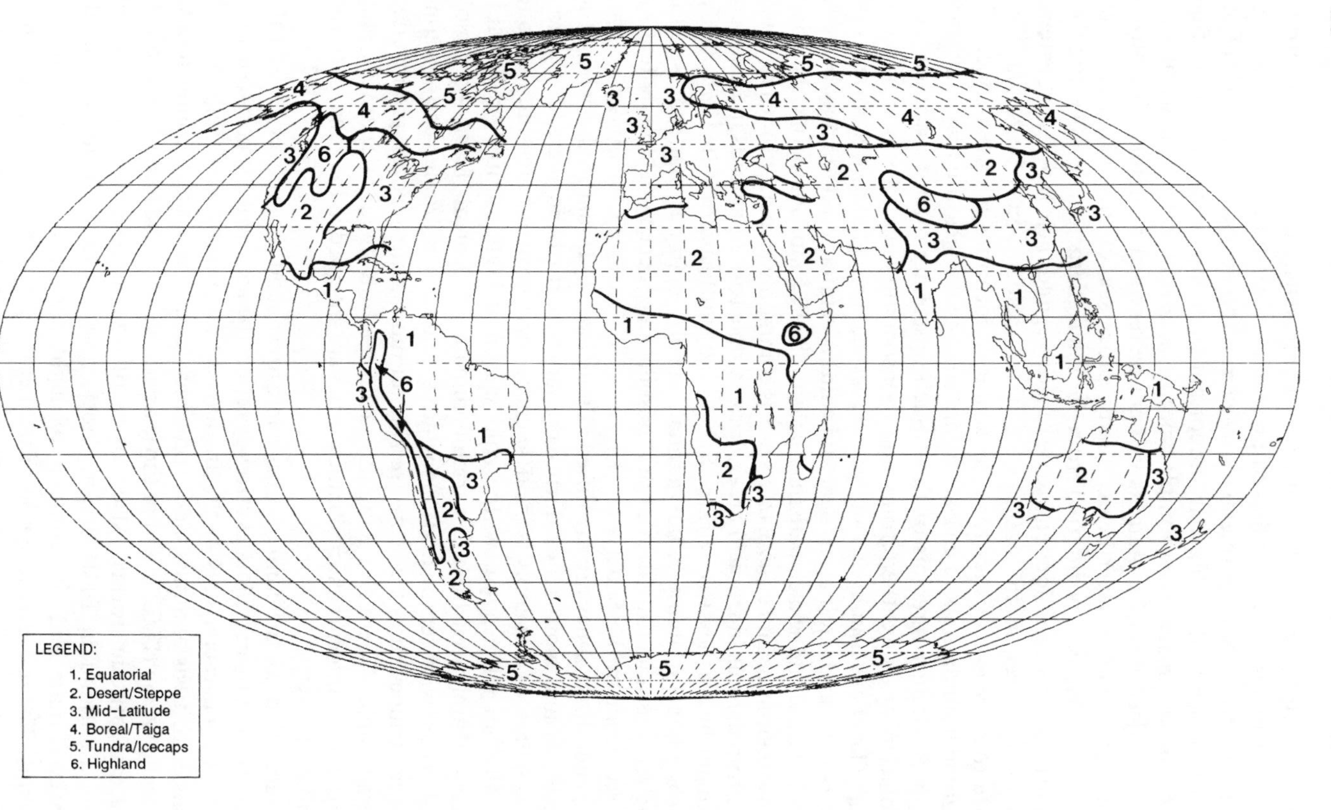

Figure 2.1.1 Major climate zones for the world's land areas

generally increased with height along a windward slope, and snowfall is general whenever precipitation occurs and air temperatures are sub-zero., Pronounced orographic precipitation can purge large volumes of pollutants from the atmosphere, as along the southwestern coast of Norway. Mountain ranges act as barriers to the inflow of moisture so that downwind regions, particularly valleys and sheltered slopes, are relatively arid. Highland climates typically are subject to large diurnal fluctuations in temperature and night frosts.

2.1.5 EQUATORIAL ZONE

This hot moist zone covers most of Central America, the Caribbean, Brazil, and the central core of Africa from Senegal eastward through Nigeria and Zaire to Mozambique and Madagascar. Most of India, Burma, and the remainder of southeast Asia along with Indonesia and northern Australia also belong in this zone. Tropical rain forest covers the core. Grassland-savannah on the poleward edges experiences both wet and dry seasons.

A major feature of the poleward side of the zone is the monsoon. That of India and southeast Asia is best known. There the wet monsoon arrives in summer from the southwest, and the dry monsoon occurs as winds blow outward from the continent. The onset of rain is strongly marked and the impact on the environment is spectacular. Other regions experiencing pronounced monsoons include Central Africa, the Philippines and Northern Australia. The summer monsoon in West Africa is wet, but both of the East African monsoons are dry. The depth of penetration inland is variable. Delays in onset or failure of the wet monsoon can have devastating effects on agriculture and water resources.

Mean daily temperatures for the zone range from 20 to 26°C. Absolute extremes are rarely above 37°C or below 15°C. The diurnal range may be as high as 10°C, but the annual range from the coldest to the warmest month is usually less than 3°C. Copious dew occurs at night under clear skies. Precipitation is heavy, usually about 2500 mm a year. Most falls from afternoon thunderstorms that are of short duration and high intensity. Hurricanes account for much of the autumnal rainfall on coastal areas at higher latitudes. Heaviest annual rainfall occurs where the trade winds impinge on mountain ranges. The core area has no dry season, but short-duration droughts occur because of the high rates of evaporation and transpiration. On both polar limits of the zone there is a definite dry season with little or no precipitation. Seasonal rainfall there is more variable and both unseasonal droughts and floods may occur with the occasional anomalous displacement of the ITCZ.

Cyclonic storms are few, but hurricanes or typhoons and tropical storms spawned over oceanic areas can devastate island and coastal areas. Examples are provided by the widespread loss of life and flooding off the Bay of Bengal in 1970 and again in 1985. On the other hand hurricanes provide areas of Mexico

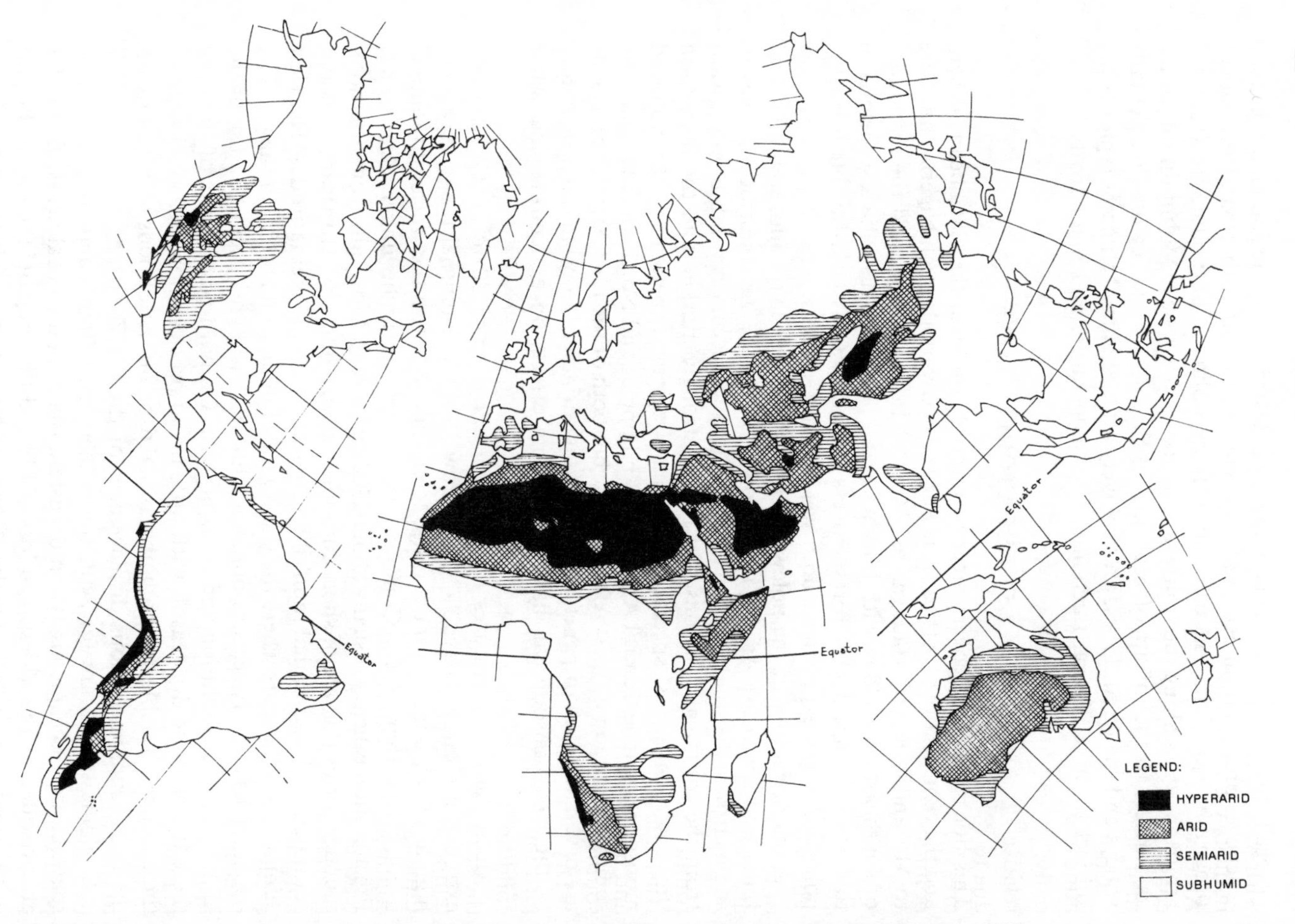

Figure 2.1.2 Map of world distribution of arid regions (UNESCO, 1977)

with water that is urgently needed for agriculture. The storms are most frequent in the northern hemisphere, approaching the southeast coasts of Asia and North America in late summer and autumn. Within the southern hemisphere the north coast of Australia and southeast Africa are most vulnerable. They quickly lose their strength inland, but the moist winds can feed rainstorms far inland.

Convection is pronounced by day, due to the strong insolation, and the atmosphere is mixed to great heights. The strong convection and intense, almost daily, precipitation along with low levels of fossil fuel combustion maintain relatively good air quality. Recognized sources of pollution are agricultural areas, due to the burning of the trash cover and the use of pesticides, and industrial areas. The erosive power of the rainfall is great and this may contribute to the pollution of water courses and lakes.

2.1.6 DESERT/STEPPE ZONE

The extent of this zone is depicted in Figure 2.1.2, which shows the distribution of arid land over the world (United Nations, 1977). Generally speaking, there are two arid belts around the globe somewhat aligned along the tropics of Capricorn and Cancer (there are notable exceptions). The major deserts are found in both North and South Africa, Arabia, south-central Asia, and Australia. The main areas of steppe climate are found between latitudes 35° and 50° in both hemispheres, but primarily in Asia, western North America, southwestern Africa, and South America.

Dry climates occur where the demand for water (for evaporation and transpiration) exceeds the supply (precipitation). Within this zone the temperature conditions are favourable but precipitation is limiting to growth. Because of the seasonal nature of the precipitation, growth, if it occurs, is seasonal. The temperature range is large. Even warmer conditions than are found in the Equatorial Zone occur at low latitudes, and quite cold temperatures occur in the continental interiors at high latitudes. In the Sahara the surface layer of the soil reaches 30°C in winter and as high as 70°C in summer at midday. The average relative humidity is about 10%, dropping at times to as low as 2%. With low humidity and often bare ground, daily air temperature ranges as great as 18°C occur. There is little rainfall over deserts and steppes and adjacent oceanic areas. Precipitation is erratic and highly localized. Over the steppes it occurs in winter in the north and in summer in the south. Over the deserts it is brief, missing in many years, and there is no seasonality. Thunderstorms are the major source in the tropics, and in some areas storms along the ITCZ provide seasonality to the precipitation. ITCZ storms provide precipitation between East Africa's two dry monsoons. Mid-latitude storms are usually the cause of winter precipitation in poleward regions. Over the steppes precipitation totals sometimes exceed 750 mm a year, but because of its seasonal nature and the hot demanding temperatures, aridity is the dominant characteristic.

Except for the small core areas, there are marked seasonal changes in climate within this zone. For example, in the northern hemisphere, the tropical rainy zone migrates into the southern edge of the dry zone in summer, and winter storms bring precipitation to the northern margins in that season. Convection is vigorous by day in the summer but limited by large-scale atmospheric stability. Winds tend to be light except in the northern margin in winter. There is little rain to scavenge atmospheric pollutants. Strong nocturnal temperature inversions and the highly stable air masses that form along coasts having cold water offshore, favour the concentration of emitted pollutants. Coastal fog, drizzle, and low temperatures reduce the soil water deficiency and favour vegetative growth along the western coasts of continents. The zone is not highly industrialized so that there are few concentrated sources of atmospheric pollution. However, where they exist there can be serious pollution problems, as in Cairo, Mexico City, and newly urbanizing areas that are located in topographic basins.

2.1.7 MID-LATITUDE ZONE

This zone includes areas in North America along the Pacific Coast, eastern United States and southeastern Canada, Western Europe extending into central USSR, and eastern USSR, China, and Japan. It also includes parts of central Peru and northern Chile, parts of eastern South America between 20° and 35°S, New Zealand, and parts of coastal South Africa and southern and eastern Australia. In a sense the climate of this zone is heterogeneous. It is most readily identifiable by what it is not. It is not tropical, arid, or frigid, but rather transitional between these types and at times displays some of the characteristics of each.

The zone exhibits a seasonal rhythm from cold winters to warm summers, and marked seasonal and interannual variations in precipitation. Summers have relatively stable weather with somewhat tropical conditions existing where the circulation is off warm oceanic areas. In the northern hemisphere, mean monthly temperatures in summer are about 27°C in the south and 20°C in the north, and daily extremes about 35°C. The incursions of cold air in winter result in large annual ranges of temperature, which reach 38°C at Winnipeg and 29°C at Moscow. Winter storms are numerous as cold and warm air masses alternate in the south, the cold dominating in the north. Coastal areas have most of their precipitation in the winter months, while interior continental areas have the most precipitation in the summer. Snowfall contributes a significant portion of annual precipitation totals, which average over 2500 mm in the rainiest maritime areas and less than 500 mm in the more continental areas. Snow cover may persist on the ground 3–5 months of the year in the more poleward areas. In maritime areas many locations have over 150 days a year with precipitation; Bahia, Chile, averages 325 precipitation days because of its oceanic exposure.

With the seasons the ground cover changes from green grassland and forest throughout the summer, to dead, brown, or leafless vegetation in winter, with snow cover for appreciable periods on the poleward side. Evaporation and transpiration demand exceeds precipitation in summer, but ample winter precipitation compensates for most deficiencies. Extended droughts occur, but they are generally of lesser consequence than in more arid regions. There is only a slight risk of ground frost on the equatorial side of the zone (except at high altitudes), but the incidence of frost increases poleward so that frost is likely on 270 days of the year near the poleward limits.

Ventilation is generally good but there are notable periods of stagnation. Prevailing winds and air-mass stability have led to extensive regional deposition of pollutants. Stagnation high pressure systems are most common in the spring and autumn, and their effects are most pronounced in long nights of autumn, especially in valley locations. Convection is more pronounced in summer; mixing is effective generally to a height of about 1000 m in winter and 1500 m in summer. Four-fifths of the world's fossil fuel consumption occurs within this zone. Frequent storms both transport emitted pollutants over long distances and provide effective precipitation scavenging mechanisms. Depositions are heaviest near and downwind of the major cities and concentrations of industry. Even longer pathways, from Australia and Asia to South America, are suggested for traces of pesticides recently found off the cost of Peru (Bruce, 1983).

2.1.8 BOREAL/TAIGA ZONE

This zone hosts the major boreal and 'scrub' subarctic forests of the world. The boreal/taiga forests extend from Alaska to Newfoundland in North America and in Eurasia from Norway across northern USSR to the Kamchatka Peninsula. Boreal forests also exist in alpine areas of other zones, but the forest is virtually absent from the southern hemisphere for want of land areas within favouring latitudes.

Winter is the dominant season. The annual range of temperature is typically about 30°C in the most continental locations. At Yakutsk, USSR, the mean monthly temperature ranges from 20°C in July to −43°C in January. Winter temperatures are extremely cold, sometimes colder than in the more polar regions. Record low temperatures of −68°C were measured at Verkhoyansk and Oimekon, USSR, and the difference between the lowest and highest recorded temperatures at Verkhoyansk is 102°C. During the short summer season mean air temperatures are conducive to growth, which is also favoured by very long day length. Summer droughts occur, while winter temperatures may fall below −30°C for weeks at a time. Diurnal temperature ranges average about 10–15 C°, and inland the temperature rises occasionally on summer days to above 30°C. The frost-free season inland ranges from 50 to 90 days, and there is risk of frost in all months.

Storms are infrequent in summer, and the weather is frequently clear and void of haze. Precipitation is light compared to that in mid-latitudes, and about one-third falls as snow. Convective showers provide the greater portion of summer precipitation, which ranges usually from 200 to 500 mm. Annual precipitation ranges from over 1000 mm in maritime areas to less than 250 mm at inland higher-latitude locations. Snow may fall in any month and persistent snow cover is present from early November to early May. The average duration of snow cover is about 120 days near the temperature limit, and up to 240 days near the poleward limit. Maximum depths occur in February and March.

The long winter season, snow cover, and relatively low storm frequency/intensity make the atmosphere in this zone less effective in purging itself of pollutants than the Mid-Latitude Zone. The zone receives a net poleward advection of pollutants from mid-latitudes and this is exacerbated by the encroachment of industry and urban development along its southern frontier. Acid deposition is highly episodic, the zone being frequented by storms, periods of stagnation, and drought as occur in the Mid-Latitude Zone.

2.1.9 TUNDRA/ICE-CAP ZONE

This zone is found in both hemispheres poleward of 68°, but tundra also occurs along coasts where sea ice prevails into the summer or where the ocean temperatures are below 10°C. In the absence of permanent ice the average monthly temperature rises above freezing for a few months of the year, and the temperature of the warmest month averages about 10°C. Winter temperatures average between −30°C and −45°C, not as severe as those found in the Boreal/Taiga Zone because of the moderating influence of the oceans.

Within this zone frost and snow occur in every month and snow covers the ground for about 10 months of the year. Annual precipitation is generally less than 350 mm and is primarily due to cyclonic storms. Areas of shallow snow cover are underlaid by permafrost, and growth occurs mainly in favoured microclimates. Snow is dry and powdery and quite variable in depth. It averages 30 cm depth inland, but a vast portion of the barren lands in arctic Canada accumulate relatively little snow cover and may be called a polar desert. The snow depths increase towards sea coasts and reach about 300 cm on exposed mountain slopes. Ice disappears but briefly in summer arctic lakes and along many coasts.

Within the ice-cap area no month of the year has mean temperatures above freezing. The mean annual range is about 20–35°C, being largest in the interior: there is no significant diurnal range of temperature. Lowest temperatures occur at the end of the dark period. The lowest mean annual temperature for a specific year was recorded at the Amundsen-Scott Station: −50°C, in 1976. Melting occurs for a few days in summer, but the highest temperature recorded at the South Pole is −15°C. In Antarctica precipitation is estimated at 50 mm in the interior plateau and 500 mm along the coast. Most precipitation is due to cyclonic storms: frost and rime also contribute to the accumulation. Insolation is exceptionally great in summer, but the surface albedo is high and the snowy environment limits the temperature rise.

The ground is snow covered much of the time. Accordingly there is little convection and ventilation is poor in the absence of storms. Very strong winds are produced orographically and by storms near the land-open water margin during the winter season, with storm wind-speeds in excess of 50 knots and gusts over 100 knots. Pronounced inversions that persist all day in valley areas coupled with the incapacity of the cold air to hold significant quantities of moisture result in locally intense concentrations of pollutants and ice-crystal fogs.

The relative absence of precipitation scavenging and strong inversions have been blamed for the net accumulation of pollutants, such as sulphates, mercury, cadmium, vanadium, and manganese in the Arctic. The sources are temperate-zone industrialized areas. The pollutants along with particles from other land and oceanic sources, form an 'arctic haze', which is widespread and well mixed. It is seasonal in concentration, the winter peak attaining levels that are 20 to 40 times greater than in summer. The haze is acidic, being about 30% sulphate, but the levels of acidity are still ten times lower than in the industrialized areas (Rahn and McCaffrey, 1979). This phenomenon is not reproduced in the southern hemisphere.

ACKNOWLEDGEMENTS

The authors surveyed many literature sources in preparing this chapter. Particular mention must be made of H.J. Critchfield's *General Climatology* (1983), which was the source of many statistics. Critchfield's zonation concepts closely conform to those of the authors, and his text is recommended for more detailed information on zonal climates.

2.1.10 REFERENCES

Bruce, J. P. (1983). Climate and water quality. WMO Technical Conference on Climate for Latin America and the Caribbean, Bogota. World Meteorological Organization, Geneva.

Critchfield, H. J. (1983). *General Climatology* (4th edn.). Prentice-Hall Inc., Englewoods Cliffs, N. J.

Koeppen, W., and Geiger, R. (1935). *Handbuch der Klimatologie*. Verlag von Gebrüder Borntraeger, Berlin.

Ottar, B. (1979). Long-range transport of air pollutants which are re-emitted to the atmosphere by sublimation. In: Proceedings of WMO Symposium on Long-Range Transport of Pollutants and its Relation to General Circulation including Stratospheric/Tropospheric Exchange Processes, pp. 125a–125b. World Meteorological Organization, Geneva.

Rahn, K. A., and McCaffrey, R. J. (1979). Long-range transport of pollution aerosols to the Arctic: a problem without borders. In: Proceedings of WMO Symposium on Long-Range Transport of Pollutants and its Relation to General Circulation including Stratospheric/Tropospheric Exchange Processes, pp. 25–36. World Meteorological Organization, Geneva.

United Nations (1977). Report of the United Nations Conference on Desertification, Nairobi, 29 August–9 September 1977

Ecotoxicology and Climate
Edited by P. Bourdeau, J. A. Haines, W. Klein and C. R. Krishna Murti

2.2 Ecosystems of the World

E. F. BRUENIG

2.2.1 DEFINITION

An ecosystem is a community of organisms and their physical and chemical environment interacting as an ecological unit. It represents all biological and abiotic components, including man, within a defined and delimited biotope, and is characterized by distinct ecological biocoenotic features of structure and functioning. The ecosystem relates to many different scales. At the lowest end of the scale, the ecosystem may comprise a rock, a fallen tree, or a certain layer in a plant community (pico-ecosystem, as described by Ellenberg, 1973). At the largest end of the scale it relates to life support media and covers oceanic, terrestrial, limnic, or urban ecosystem categories (mega-ecosystem). The present description is restricted to terrestrial ecosystems and focuses in particular on potential natural vegetation. Centred on the plant formation, it classes natural forests and woodlands in hot and cold climates at macro-ecosystem level. The description is structured by classifying the forests and woodlands at micro-ecosystem (formation group) level according to major physiognomic, phenological, and growth features in relation to the site and environmental conditions. These units correspond roughly to the bioclimatic life zones of Holdridge (1967). The description is patterned on the serial structure of the physiognomic-ecological classification of plant formations of the earth (Müller-Dombois and Ellenberg, 1974, pp. 466–488) commonly quoted as the UNESCO Classification.

2.2.2 NATURAL VEGETATIONAL ZONES

2.2.2.1 Latitudinal Zonation

The characteristic features of the zonal vegetation in each latitudinal zone are primarily determined and shaped by climate. In hot climates, vegetation physiognomy and phenology express the effects of the change from weak seasonality near the equator to pronounced radiation and rainfall seasonality around the tropics. The climatic equator is straddled by the Predominantly Evergreen Humid Tropical Forest, which extends north and south of the tropics along major paths of tropical cycles (Figure 2.2.1). This humid to supersaturated forest ecosystem belt is broken on land by considerable gaps in Africa and India. In West Africa, the gap is caused by wind circulation and barrier effects which allow relatively dry conditions to develop seasonally. Similarly, the pattern of atmospheric circulation caused by orography together with the effects of human activity combine to create semideserts, dry woodlands, and man-made savannas in East Africa and south India.

To the north and south of the Predominantly Evergreen Humid Forest lies a fragmented mosaic of Predominantly Deciduous Humid to Subhumid Forests and replacement vegetation, such as various forms of man-induced savannas, dry woodland, and scrub. Agricultural land, extensively used grazing land, and increasingly degraded wasteland occupy large portions of the Predominantly Deciduous Humid to Subhumid Forests and the semi-arid zone. The boundaries between these two zones are indistinct, interlocked, and fluctuating, generally moving towards more xeric conditions in the course of advancing desertification. The tropical and subtropical semideserts and deserts occupy relatively small areas in the neotropics, but cover larger areas elsewhere, where oceanic and atmospheric circulation and the effects of topography cause more xeric conditions.

In contrast to the tropics, the vegetation in the cold climates is primarily determined physiognomically and phenologically by the short period in which sunlight and temperatures suffice to achieve a positive net photosynthesis and assimilation balance, as well as by the severities of the long winter. The latitudinal vegetation zones in the cold climates in the northern coniferous forest and tundra biomes are of varying widths in America and Eurasia. Figure 2.2.1 shows the extent of the Boreal Coniferous Forest (close stipples) and Tree Tundra (wide stipples), the latter becoming treeless tundra and cold deserts further north. The climate diagrams in Figure 2.2.2 show typical examples for the lowland climate in the equatorial evergreen forest, the predominantly deciduous moist tropical forest, and the oceanic and continental boreal forest. The tropical and boreal forest and woodland vegetation represents 80% of the world forest and woodland area (Table 2.2.1); the tropical forests and woodlands alone represent 50% of the world's wooded land.

2.2.2.2 Altitudinal Zonation

The general features of the altitudinal zonation of the natural vegetation are shown in Figure 2.2.3, excluding hot semideserts and desert. The change of physiognomy and phenology is primarily determined by the decrease of temperature with increasing altitude, interacting with the simultaneously changing moisture and radiation regimes. In the tropics, temperature decreases by 0.5–0.6 C° per 100 m, while rainfall increases up to the cloud belt, which

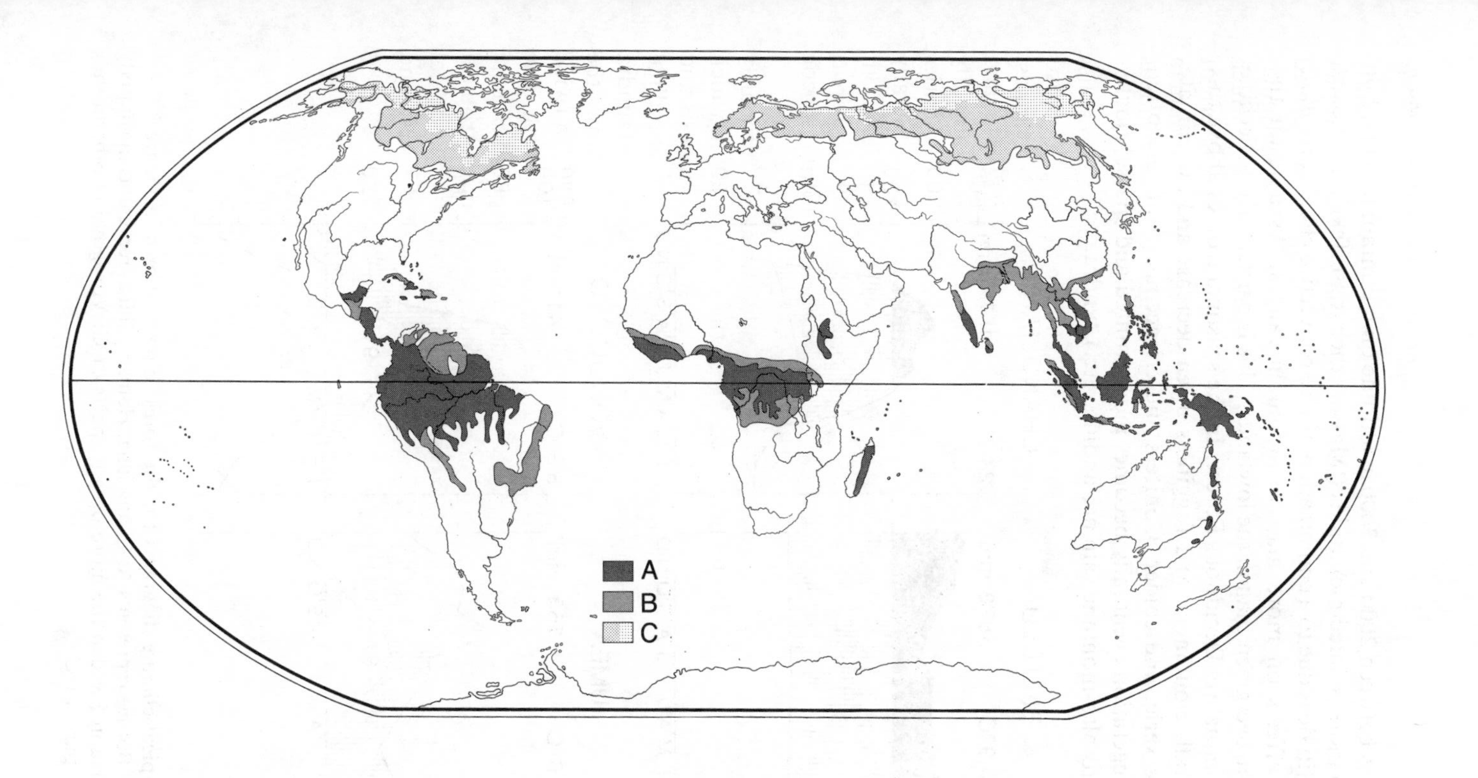

Figure 2.2.1 World Distribution of: (A) the Predominantly Evergreen Tropical Humid Forest Biome (including perhumid to saturated life zone); (B) Predominantly Deciduous Tropical Forest (including outliers in subtropical areas); (C) Boreal Coniferous Forest (taiga); and Tree Tundra

frequently centres between 2000 and 3000 m altitude on mountain massifs, but may be lower on more isolated mountains ('Massenerhebungseffekt'). Increased humidity with altitude is due to a combination of free convective cloud formation and of barrier effects on more massive mountain ranges. As a result the vegetation is more evergreen than in the lowlands. This is particularly noticeable above the Predominantly Deciduous Forest. In the seasonal tropics above this cloud belt, rainfall, cloudiness, and fogginess often decrease and the climate is sunnier, more xeric, and colder. Conifers and ericaceous plants are often common. The vegetation eventually becomes more stunted and sparse, finally degenerating into alto-montane (alpine, andine, etc.) tundra.

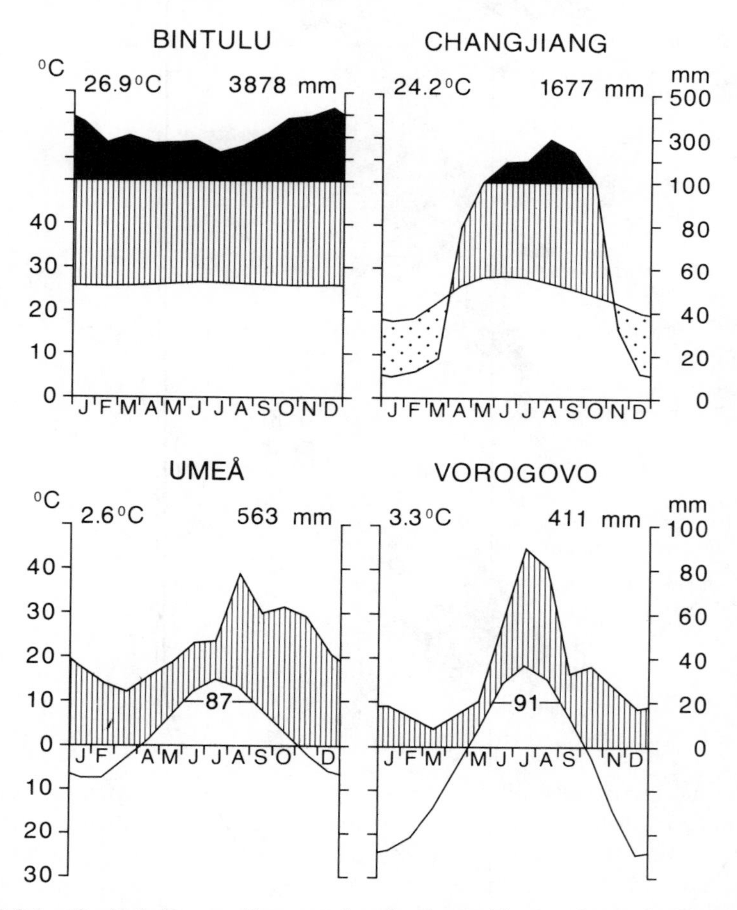

Figure 2.2.2 Typical climate diagrams for the forest biomes shown in Figure 2.2.1. Bintulu, Sarawak, Borneo represents A; Changjiang, Hainan, China, represents the humid section of B; Umeå in Sweden the more oceanic parts of C; Vorogovo in Siberia the more continental parts of C

Table 2.2.1 a = optimistic prediction; b = pessimistic prediction. Compiled from Lanly (1982) and World Resources Institute and IIED (1986)

Forest formation class	Area (10^6 km^2)				
	1965	1975	1985 a	2000 a	2050 a
1. Closed virgin and modified high forest equatorial-tropical predominantly evergreen; wet to moist; (saturated/humid)	5.5	5.0	4.4	4.0	3.0
Tropical predominantly deciduous; moist to dry; (humid to subarid)	7.5	6.5	6.0	5.5	4.5
2. Open virgin and modified natural high forest and woodland; deciduous and evergreen; dry to parched; (semi-arid to arid)	7.5	6.5	6.0	5.5	4.5
Sum natural high forest					
a	20.5	18.0	16.4	15.0	12.0
b	20.5	17.0	15.0	12.5	8.0
Tropical tree plantations					
Forestry	0.04	0.05	0.08	0.16	?0.30
Agriculture	0.20	0.25	0.27	0.28	?0.50
3. Natural and man-made forests and woodlands					
Subtropical	7.6	3.6	3.7	3.7	3.8
Warm/cool temperate	2.6	2.6	2.7	2.7	2.8
4. Cold temperate (boreal) virgin and modified					
High forest	6.0	5.9	5.7	5.5	?5.0
Open tundra woodland	> 6.0	6.1	6.3	6.5	?7.0
5. Total world forest and woodland					
a	38.2	36.0	34.5	33.7	30.6
b	38.2	35.5	33.0	29.0	26.6
World population ($\times 10^9$)					
a	–	4.0	5.0	5.8	?
b	–	4.2	5.3	6.6	>11.0

The zonation of altitudinal belts in the tropics is not analogous to the latitudinal zonation of vegetation in the lowlands. One very obvious difference is the increasing seasonality of moisture and temperature at higher latitudes. The diurnal pattern of temperature, wind, precipitation, and cloud becomes less regular and the free convective component of the moisture regimes becomes relatively less important than in the tropics. This influences the physiognomy of the various vegetation belts, although the effect is obscured by the general dominance of evergreens at higher altitudes (Figure 2.2.3).

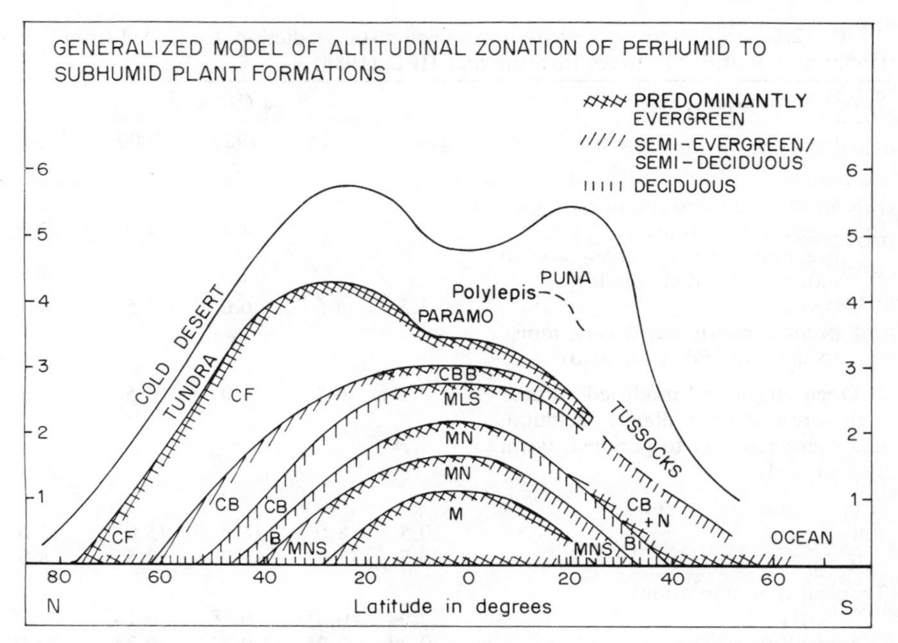

Figure 2.2.3 Generalized model of altitudinal and latitudinal zonation of perhumid to subhumid plant formations and latitudinal on big mountain massifs. The predominant physiognomic vegetation formations are: CF = conifer forest, mainly boreal and temperate-montane; CB = mixed conifer-broadleaved, mainly temperate and subtropical; CBB = mixed conifer-broadleaved with bamboo, tropical, and subtropical alto-montane; B = broadleaved forest, predominantly noto-mesophyll deciduous forest; M = mesophyll forest, mainly tropical evergreen and deciduous; MN = meso-notophyll forest, tropical montane; MNS = meso-nanophyll sclerophyll, australo-pacific forest; NMS = noto-microphyll sclerophyll forest; MLS = micro-leptophyll sclerophyll forest, tropical montane and australo-pacific forest; N = notophyll broadleaved forest, australo-pacific forest. Adapted from Troll, 1948

2.2.3 EDAPHIC ZONATION

Within each ecological vegetation zone and altitudinal belt, different physical and chemical soil conditions are associated with differences of physiognomic and functional features of plants and vegetation. These could be interpreted as adaptations along the edaphic gradient similar to the changes which occur along the latitudinal and altitudinal gradients. These edaphic gradients often have a pattern of repetitive catenas primarily determined by variations in soil moisture and nutrients, and by correlated biological activity on and in the soil. At one end of the nutrient gradient are nutrient-rich soils, while at the other end of the gradient are the oligotrophic soils. The edaphic moisture gradient runs from permanently wet and water-logged, through mesic conditions, to alternating wet and dry, and finally to permanently dry conditions. In addition the soil texture, structure, and organic matter content modify the effects of any conceivable combination of moisture and nutrient regimes. In the humid tropical climate, rates of weathering of parent materials, leaching, mass wasting, and acidification of soils, as well as the intensity of biological activity in the soil, are generally high, and in any event higher than on sites with corresponding soils in other climatic zones. But this relatively high intensity of soil-forming processes in the humid tropics has not led to any uniformity of edaphic conditions. On the contrary, even on very old land surfaces, tropical soils are at least as heterogeneous and diverse as temperate soils, over both small-scale and large-scale areas.

2.2.4 TIME-RELATED TRENDS

Structure and processes in the natural forest ecosystems are subject to changes over time. Natural ecosystems are not in a state of homeostasis, and at no stage do they function as closed systems. Long-term successional phasic development, processes of build-up of biomass, storage and release of nutrients, and change of species, are combined with short-term processes of regeneration, maturing, and mortality caused by small- to large-scale perturbations, disturbances, and, in extreme cases, full-scale catastrophies. Disturbances over small areas prevail in climatic climax tropical forests. Small gaps are created continuously and provide the mechanism by which structural complexity and species diversity are maintained.

By contrast, natural monocultures in the tropics, such as *Shorea albida* peatswamp forests in Sarawak, Borneo, and in the Boreal Coniferous Forest Zone are liable to large-scale catastrophic collapse from insect pests, windthrow, or fire. The ecosystem reverts over large areas to almost initial stages of succession. Reconstruction involves the early phases of pioneer vegetation.

While the effect of small-scale perturbance and disturbance on chemical cycles and organizational structure is small and ephemeral, large-scale destruction profoundly disrupts ecosystem functions and causes drastic changes in stocks, turnover rates, and organization. Figure 2.2.4 gives an impression of the factors and structural changes involved in the case of evergreen lowland tropical forest. This figure gives a generalized illustration of the hypothetical course of forest development as a result of successional dynamics; the cycles of regeneration, growth, and mortality; and the effects of perturbations caused by biotic and abiotic agents. To be added are the effects of inputs of gaseous, dissolved, and solid substances from the atmosphere and the discharges into it. In the example man-made or spontaneous fire (or landslide) is incorporated as a completely destructive exogenous factor to initiate a new successional sequence. Perturbations by climatic damage, pest, and disease, and by pollinating, dispersing, and consuming organisms are essential components of the ecosystem's dynamics and its capacity to adapt and survive.

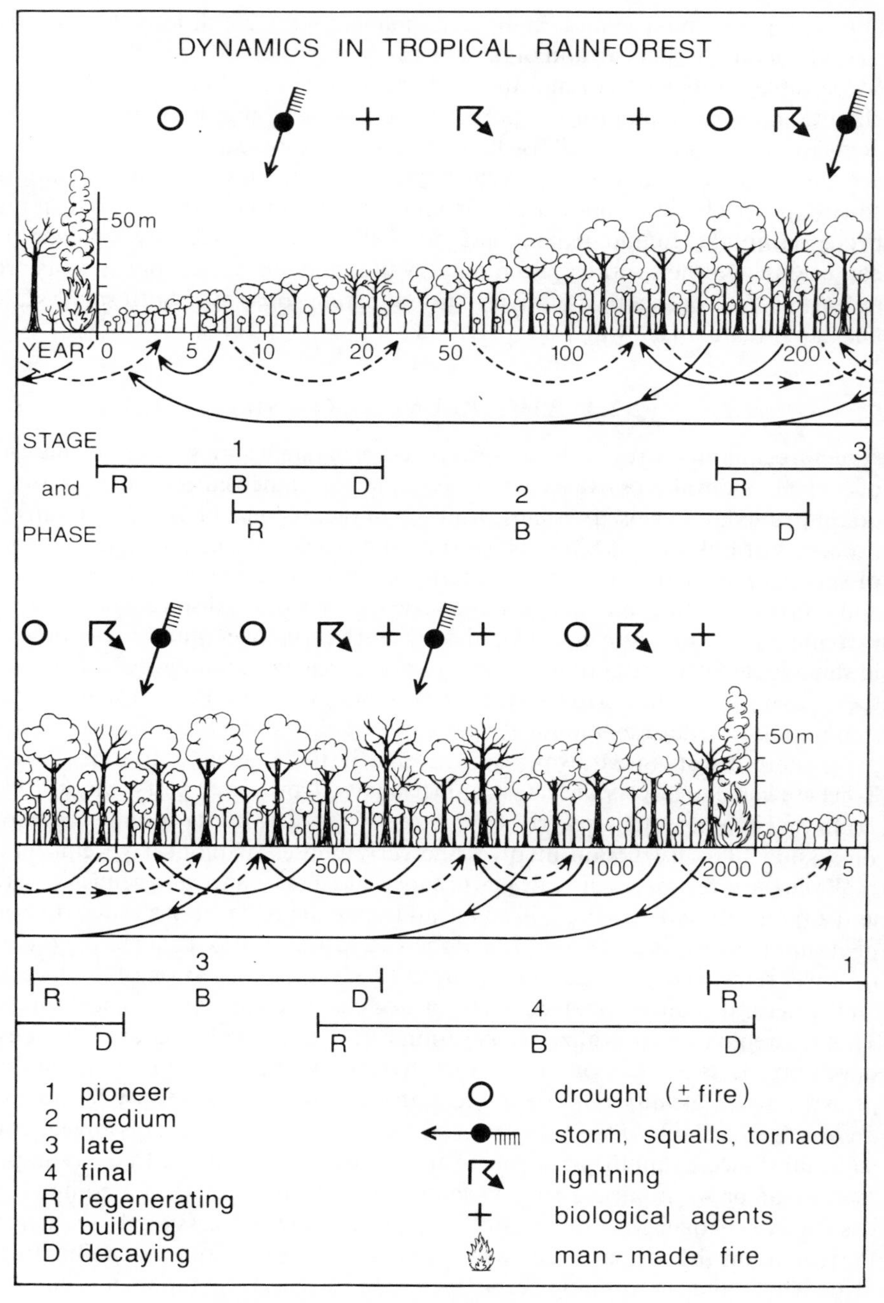

Figure 2.2.4 Dynamics in tropical rainforest

To generalize, the following processes are characteristic components of time-related trends of vegetation dynamics:

(a) energy fixation and negentropy formation;

(b) build-up of structure to the limits of the site carrying capacity;

(c) build-up of architectural and organizational complexity to the limits set by floral and faunal richness interacting with environmental growth and disturbance factors.

The natural terrestrial ecosystems are thermodynamically open systems in which dynamical change; rates of input, throughput, and output; internal cycling; and growth and decay are closely linked with the architectural structure and the floristic and faunal richness and diversity.

2.2.5 STRUCTURAL ADAPTATION TO ENVIRONMENTAL FACTORS

Prevailing minimum factors for primary productivity in the humid tropics are mainly edaphic: low absorptivity of the soils; low mineral nutrient contents; and poor aeration of the soil. In the subhumid tropics and subtropics, aeration is a critical factor only in certain hydromorphic soils. In the semi-arid tropics moisture supply is the major limiting climatic factor, with downward leaching being replaced by upward accumulation. Critical factors are water availability, high potential evapotranspiration, and soil salinization. In the boreal zone, minimum factors are the length of the growing season, low temperatures, and consequently low biological soil activity, which causes a tendency for the accumulation of dead biomass on the soil and hence soil acidification.

The structure, physiognomy, biochemistry, and dynamics of natural vegetation are designed to use site resources efficiently under the average climatic conditions, but at the same time to reduce risk of damage in the event of extreme conditions. Sporadic extreme events must be endured with resilience or elasticity. Catastrophic collapse under strain must be overcome by the capability to restore organization by structural reconstruction and build-up of biomass and self-regulating mechanisms. The architectural, biochemical, and biological features of the vegetation canopy as an active-exchange surface layer between earth and atmosphere determine the intensity of the processes of change and in fact the stability of the ecosystem itself. The illustration of the water balance in a tropical rain forest and in a temperate mixed forest demonstrates the high rates of input, turnover, and output in the complex, aerodynamically rough tropical evergreen lowland rainforest (Figure 2.2.5).

Greater aerodynamic canopy surface roughness is associated with:

(a) greater interception and absorption of radiation and matter;

(b) greater penetration of irradiance into the canopy;

(c) greater crown surface area which absorbs and re-radiates energy;

(d) greater turbulence, deviation, and vertical speed of airflow;

(e) more rapid exchange of moist air within the canopy with drier air from above;

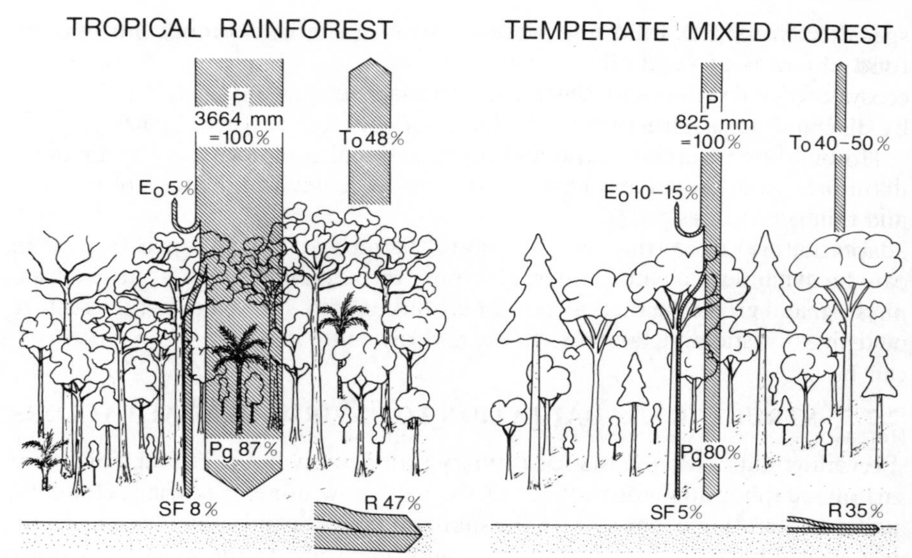

Figure 2.2.5 The water balances of the lowland rain forest in the Amazon Ecosystem Research Area at San Carlos de Rio Negro, Venezuela, and of average cool-temperature mixed forest in Central Europe. This illustrates the comparatively very high rates of input, throughput, output, and runoff of water in the tropical rain forest. Similar relations characterize the energy balance. Both are vital for the system to function effectively, but also put heavy stress on it. P = precipitation; Pg = precipitation reach ground; E_o = evaporation from overstorey; T_o = transpiration from overstorey; SF = stemflow; R = runoff

(f) greater potential and actual evapotranspiration;

(g) greater potential primary gross productivity and greater associated rates of respiration from greater leaf area and crown volume.

Lesser aerodynamic canopy surface roughness is associated with:

(a) lesser absorption of radiation and greater reflection at low sun angles; therefore greater albedo;

(b) lesser turbulent air exchange and kinetic wind energy absorption;

(c) greater or lesser penetrability and transfer resistance to light, humidity, heat, and air, depending on leaf and foliage features.

The vegetation can adjust to the opportunities and risks of the site by a suitable combination of features, such as canopy surface roughness, crown architecture, foliage density, leaf size, leaf orientation, pigmentation, leaf surface characteristics, and internal diffusion and conductance capacities (Bruenig, 1970; 1971).

Higher species richness and diversity increase differences of albedo, surface temperature, and emission of latent and sensible heat in the canopy, and thereby the tendency to free convection and lower canopy diffusion resistance. Greater species richness and diversity is associated with more complex organization of food chains, self-regulating mechanisms and functions. Consequently, the ecosystem is likely to react to impacts by change, but to maintain its functionality by diversion (Bruenig, 1973).

However, the ecological interpretation of features of the physiognomy of natural ecosystems is complicated by the fact that many site factors interact and that individual, seemingly adaptive, features are compensatory, and that adaptation to certain physical, chemical, and biotic stress and strain factors can be achieved in various ways. However, certain common features of physiognomy and structure of natural ecosystems along the ecological gradients of temperature, moisture, and, to some extent, nutrients and solar radiation, can be recognized.

Outward from the perhumid and humid tropical lowlands on medium soils (ferrasol, acrisol, oxisol, ultisol) and on medium sites the stature, complexity, and species richness and diversity of the natural ecosystem decline toward super-saturated and xeric sites in the equatorial zone and the drier, seasonal subhumid and semi-arid tropical zones. This is accompanied by a decline in the gross primary productivity, in the respiration and decomposition rates, and also in the speed of chemical processes and intensity of turnover of matter and energy.

This general trend continues beyond the arid and hyperarid subtropics as the latitude increases. Here again in each zone, vegetation stature, physiognomy, and functions follow similar trends along the gradients of moisture, temperature, and nutrients.

2.2.6 REFERENCES

Bruenig, E. F. (1970). Stand-structure, physiognomy and environmental factors in some lowland forests in Sarawak. *Bull. Int. Soc. Trop. Ecol*, **11** (1), 26–43.

Bruenig, E. F. (1971). On the ecological significance of drought in the equatorial wet evergreen (rain) forest of Sarawak (Borneo). In: Flenley, I. R. (ed.) Transactions of the First Aberdeen-Hull Symposium on Malaysian Ecology, pp. 66–97. University of Hull, Dept. of Geography Miscellaneous Series No. 11, Hull, England.

Bruenig, E. F. (1973). Species richness and stand diversity in relation to site and succession of forests in Sarawak and Brunei. Third Symposium on Biogeography and Landscape Research in South America, Max Planck-Institut, Plön, 3.5. *Amazoniana*, **4** (3), 293–320.

Bruenig, E. F. (1987). The forest ecosystem: tropical and boreal. *Ambio*, **16** (2–3), 68–79.

Ellenberg, H. (1973). Die Ökosysteme der Erde: Versuch einer Klassifikation der Ökosysteme nach funktionalen Gesichtspunkten, pp. 235–265. In: Ellenberg, H. (ed) *Ökosystemforschung* Springer-Verlag, Berlin.

Holdridge, L. R. (1967). *Life Zone Ecology*. Tropical Science Center, San José, Costa Rica.

Kellog, W. W. and Schware, R. (1983). Society, science and climatic change. *Dialogue*, **3**, 62–69.

Lanly, J. P. (1982). *Tropical Forest Resource*. FAO Forestry Paper 30. FAO, Rome.

Müller-Dombois, D. and Ellenberg, H. (1974). *Aims and Methods of Vegetation Ecology*. John Wiley, London.

Troll, C. (1948). *Der assymetrische Aufbau der Vegetationszonen und Vegetationsstufen auf der Nord- und Südhalbkugel*. Ber. Geobot. Forschungsinst. Rübel, f. 1947, 1948: 46–83.

World Resources Institute and the International Institute for Environment and Development (1986). *World Resources* 1986, Basic Books Inc., New York.

Ecotoxicology and Climate
Edited by P. Bourdeau, J. A. Haines, W. Klein and C. R. Krishna Murti

2.3 Atmospheric Transport of Chemicals

D. R. MILLER AND C. A. ROBINSON

When the atmospheric transport of polluting chemicals is considered, the problem seems naturally to divide itself into three categories, based largely on the time and distance scales involved. These categories are: (1) local transport, by which we mean distances of up to perhaps 50 km, in which the main phenomena are mixing within the atmospheric boundary layer, prevailing winds, and immediate deposition; (2) mesoscale transport, up to a few hundred kilometres, for which synoptic weather pattern information and knowledge of the rate of photodegradation processes are vital; and (3) global transport, for which major considerations are the balances between the various global sources and sinks involved, as well as understanding of particular transport routes and pathways that may lead to much higher pollutant levels in remote areas than can be understood on the basis of general diffusion phenomena. See Table 2.3.1.

All of these processes are important for the purposes of this volume. However, since in general they have been summarized in many published works, we restrict ourselves to describing each of the phenomena involved only briefly, and then addressing issues which are of particular significance to the often isolated areas considered here, particularly in northern climates. Here, a major concern is the long-range input of pollutants from built-up temperate areas into regions where the prevailing ecosystems are rather delicate. Local pollution generation is limited in magnitude but important over short distances. The main conclusions are that the problem is serious and becoming more so, and that the main source of uncertainty in our conclusions is the lack of monitoring facilities in the remote areas involved, as well as a persistent lack of fundamental understanding of the processes by which these more remote ecosystems do or do not cope with the stresses they suffer.

Table 2.3.1 Features of the three main types of atmospheric transport

	Distance scale	Transport mechanisms	Removal mechanisms	Modelling
Local	0–50 km	Prevailing wind; atmos. stability	Deposition	Wind rose, stability classes
Mesoscale	50–1000 km	Vertical mixing throughout troposphere; weather patterns	Deposition/ resuspension; photochemistry	Trajectory analysis multiple-level models
Global	Global	General circulation	Oceanic and Arctic sinks	General circulation models

2.3.1 INTRODUCTION

The rejection of the old adage that the 'solution to pollution is dilution' is a quite recent phenomenon. Until just a few years ago, once a pollutant was dispersed just enough that its presence was not immediately obvious, it was regarded as no longer any kind of a threat. The same opinion was held for a great many pollutants in the environment and for some persons, indeed, it still is.

Most people now realize, however, that dispersion of a pollutant from a concentrated point source (or a localized collection of point sources) to a level at which it is not an immediate threat to man, his structures, his domesticated plants and animals, or the ecosystem at large, is but a first step. We now know that significant ecosystem effects can be produced by pollutants generated many hundreds of kilometres away (mesoscale pollutant transport). We have even finally come to recognize, alas, that even the vast mass of the atmosphere on a global basis has a very definitely limited capacity to dilute, assimilate and ultimately dispose of the chemicals it receives. It seems clear that cessation or curtailment of input to the atmosphere is the ultimate solution. Until that can be arranged, however, it is vital that we understand the fate of what does get released.

2.3.2 CHEMICAL AND PHYSICAL PROCESSES

When a chemical, in whatever form (gas, particle, aerosol, etc.), is released into the atmosphere, it has only three possible fates in any time period: it must be chemically converted into some other form; it must be effectively removed from the atmosphere, or at least that part of the atmosphere with which we expect to come into contact in the near future; or it must remain in the atmosphere and become part of a general accumulation phenomenon. Which of these happens depends strongly, of course, on the chemical and physical nature of the substance involved; but relatively few polluting substances (mercury vapour may be a counterexample) travel very far without a chemical transformation or a (perhaps temporary) removal process being effective, and therefore it is appropriate to consider these before discussing details of transport processes. Chemical transformations, particularly photochemical degradation of organic pollutants in the atmosphere, have been extensively dealt with in a previous volume in this series (Korte, 1978).

2.3.3 DEPOSITION AND EXPOSURE CALCULATIONS

Deposition onto ground, water or vegetation is perhaps the most important sink, or removal process, for all atmospheric contaminants that are not chemically degraded in the atmosphere itself. From the point of view of land- or water-based organisms, such deposition is perhaps the most important source. Fortunately, the problem may in most cases be dealt with in a fairly simple way.

The suggestion that the rate of deposition should be proportional to the concentration in the atmosphere at any given point or time, although it may seem an obvious hypothesis today, was actually made fairly recently. Credit may be given to Gregory (1945) for the observation, and to Chamberlain (1953) for the subsequent formulation (Yamartino, 1985). In any case, if the rate of deposition, in terms of mass per unit area per second, say F, is proportional to the concentration in the atmosphere, C, in terms of mass per unit volume, so that

$$F(g/cm^2/s) = v/C(g/cm^3)$$

then the proportionality constant, v, obviously has the dimensions of a velocity. The interpretation that one may imagine is that the entire concentration of a pollutant in the atmosphere is settling at a uniform velocity onto the surface of the earth (and simultaneously being uniformly remixed in the atmosphere). Although it is clear that the actual processes at work are much more complicated than that, the term 'Deposition velocity' has gained a popularity in the literature that is not to be challenged.

Realistically, one would expect that for large particles (say, exceeding a few micrometers), the deposition velocity would be something like the terminal, sinking velocity given by Stokes's law. For very small particles (say, less than 1 μm), on the other hand, Brownian movement would maintain a uniform mixture in the lower layers of the atmosphere. Thus, a minimum deposition rate would be expected for particles around 0.1–1.0 μm. This is, indeed, observed.

2.3.4 MECHANISMS OF TRANSPORT

The mathematical equations governing the behaviour of fluids were formulated around 1850, in essentially the same form as we know them today. They are described in detail in many texts, and we see no reason in the present work to make any more than a brief mention of their general characteristics.

Briefly, the equations are fourth-order time-dependent partial differential equations which are strongly nonlinear. They describe a slightly viscous compressible fluid subject to very complicated driving forces, including differential heating, very much modified by Coriolis forces on a global basis. The boundary condition at the lower edge is quite difficult to prescribe due to the highly variable nature of the terrain over which the air flows.

Once the velocity field, temperature distribution, and so forth for the air itself are known, either by calculation or from very detailed observations in the field, the movement of a pollutant is a further problem involving convection (or advection) and diffusion, generally of a complicated form (see below), as well as deposition and chemical transformation.

Not surprisingly, exact analytical solutions for such an array of equations are not generally available. Indeed, they exist only for a small number of highly specialized and geometrically rather artificial situations. The full-fledged numerical solution in any given case is theoretically possible, but it is extremely expensive in terms of computer time.

The whole subject, therefore, becomes a matter of making approximations that allow one to obtain quantitative information but are still realistic for the particular problem under consideration. Naturally, there is a considerable literature about various possible simplifying techniques. We will leave to other sources the discussion of techniques for solving the fluid mechanical equations themselves, and start from the assumption that we have information (or will be willing to make simplifying assumptions) about the velocity field, and discuss the issue of how to calculate the pollutant distribution starting from that point.

2.3.5 TYPES OF APPROXIMATION

Generally, what kind of approximation is appropriate depends on the time and distance scale of the problem. To begin with, although the equations involve diffusion, what the theoretician regards as (molecular) diffusivity as might be measured in the laboratory, is quite inappropriate. Small- and large-scale eddies and circulation patterns caused by general turbulence, wakes behind objects, rotations induced by low- and high-pressure weather patterns, and so forth produce much more mixing in most situations. The standard response is to replace the molecular value of diffusivity with a much larger value, often itself a function of time and/or distance, determined empirically to make the solution match reality in some sense. Other approximations are made in similar ways.

Environmentally significant modelling studies of pollutant behaviour seem naturally to divide themselves into three levels, for a variety of reasons.

At the strictly local level, most pollutants are after all produced close to the ground, and their effect upon anything in the immediate vicinity will depend strongly on vertical movement of the chemical within the atmospheric layer closest to the earth, the so-called meteorological boundary layer. This is best understood in terms of stability classes, inversion layers, and the like. Lateral transport depends largely on the prevailing winds at the time, and short-range deposition is reasonably well understood.

At somewhat larger distances, however, of the order of 50 km or more, the prevailing wind is likely to have changed in direction by the time the pollutant has travelled that far, and the speed and direction of the prevailing winds at a point are no longer accurately predictive of the pollutant's fate. For such cases, real-time consideration of weather patterns and of their changes in space and time are required, and much more complex models are needed. Also, the times involved allow photochemical reactions to proceed, which must therefore be considered in considerable detail.

Finally, at the global level, the overall efficacy of mechanisms for final removal of chemicals from the atmosphere must be considered, and although general circulation models are critically important in particular cases, the overall holding and disposal capacities of the atmosphere are perhaps the dominant considerations.

2.3.6 THE GAUSSIAN PLUME

The direct analytical solution of the governing equations under a number of simplifying assumptions (uniform wind field, including close to the ground, wind uniform over time for a long period, constant diffusion coefficients, no up- or downstream diffusion, etc.) leads to what is usually called the Gaussian Plume Model, for which an explicit formula for the pollutant density as a function of position can be written. A slight generalization which allows spatially varying diffusion coefficients is called the Gradient Transport or the K-model, and is conceptually quite similar. One or the other of them is used very widely in formulating models of air pollutant transport (Hanna, 1985).

The Gaussian plume approach is a very convenient way of handling air pollution estimation problems as far as it goes. However, the assumptions are very often not valid; the most important of these, perhaps, is the idea that the air is moving past the pollutant source essentially as a solid, and that lateral and vertical diffusion are the only mechanisms by which pollutants can be distributed. Even if the numerical values are increased to reflect the existence of eddy diffusivity, the results suffer because of larger-scale instabilities connected with buoyancy, stability, and other phenomena.

To make this general approach more realistic, Pasquill (1961) devised the concept of Stability Classes. The idea was to classify meteorological situations into a number of categories that influenced the speed at which vertical movement, particularly movement from the height at which the pollutant was released down to ground level, can take place. The results of the study were called

Pasquill Stability Classes, and are now more often known as Pasquill–Gifford, or even Pasquill–Gifford–Turner Stability Classes, to reflect further work (Gifford, 1961, 1976; Turner, 1970). These classes are very widely used to evaluate and predict short-range short-term air pollution events.

2.3.7 AVERAGE EXPOSURE: THE WIND ROSE

The basic Gaussian and Pasquill approaches deal, fundamentally, with situations where the wind is in a constant direction. We know, however, that it varies in direction and in speed, and we should consider the modifications needed to deal with these cases. The first situation one must consider is the problem of predicting or evaluating the long-term average pollution levels that will result from a given source, considered first to be emitting uniform amounts of material over a long period of time. For this purpose the so-called Wind Rose was developed. This is a pictorial representation of the magnitude and direction of prevailing winds at a geographical location, information which is typically available from existing records (or which can be recorded automatically for a relatively low price).

The application of this concept is simple enough. One imagines preparing a sort of map, by superimposing a Gaussian-type plume on each of a number of direction vectors around the compass circle presented on a wind-rose diagram, noting how much pollution (on a long-term average) results from each one, and adding the results for a fairly fine grid or network of points on the resulting two-dimensional map. The process may sound laborious, but it can easily be computerized, and a readily understood result, such as a contour map, can be produced. Again, this approach is widely used for localized pollution problems.

The main drawback to the wind-rose concept is that it is severely limited in terms of distance from the source. This is especially true in directions other than that of the prevailing wind. Even for locations where the wind is almost always in one direction, there will be (perhaps brief) periods when the wind direction is quite different. A Gaussian plume approach applied to these directions will lead to the incorrect conclusion that some pollution will travel 'upstream', as it were, for unlimited distances. In fact, the time over which the wind continues in such a direction may be quite limited and the polluting material may virtually never have the opportunity to travel any great distance. Thus, the wind-rose approach will overestimate the amount of pollution that will be felt in an 'upstream' direction and, for that matter, will give incorrect results for larger distances in all directions from the source. In practice, this limits the distance over which the approach can be used to something between 30 and 50 km.

2.3.8 CHANGING WIND PATTERNS: TRAJECTORY ANALYSIS

In order to deal with distances great enough that weather and wind patterns may have changed by the time a packet of air traverses the region considered, it is necessary to have actual (or simulated typical wind-field information on at least a two-dimensional basis. Then, one can think of a plume with a 'wavy' centre line changing its position and shape with time.

With a reasonable amount of computer time available and a certain amount of care given to the formulations, we may think of integrating the resulting ground-level exposure and deposition to calculate a time-integrated realistic value for both.

Unfortunately, the analytic expression for the Gaussian formulation in any realistic sense breaks down if the centre line of the plume is not straight. What is then needed is a calculation scheme that synoptically takes into account the entire field, incorporating the wind pattern on at least a two-dimensional basis. Two approaches are typically discussed, generally referred to as the Euler and the Lagrange formulations (these have been discussed in many places; see, e.g., Venkatram, 1985). In the Euler approach, one considers dividing the region into cells or subdivisions of some shape, and calculating for each one the inflow and outflow resulting from wind transport and the source and sink (deposition) terms. One can incorporate easily such things as chemical reactions. The drawback is that many numerical problems have yet to be resolved, and inordinate amounts of computer time may be required, even on the largest of modern machines.

The Lagrangian approach is conceptually easier to deal with and, in its simplest form, Trajectory Analysis, permits a number of extremely useful, if approximate, calculations. In a Trajectory Analysis approach (Eliassen and Saltbones, 1975), one imagines a series of 'puffs' to be released from a point, and follows each one across the map using time-dependent information on wind patterns, allowing for a dispersion (which does not have to be isotropic) as it goes. For time-averaged exposure, one keeps track of the accumulated exposure from each puff at each time, a procedure which may sound computationally expensive but in fact is vastly less involved than Euler-type calculations. It is easy to show that, provided the puffs are spaced closely together, the procedure produces results identical to those of the Gaussian plume approach under conditions in which the latter is applicable, so at the very least trajectory analysis may be considered a legitimate generalization of plume-type calculations.

The approach is criticized since it is not clear what wind field should be used. Wind differs in direction and speed at different heights, and so far each such calculation has been produced using what the authors regarded as a 'typical' altitude (such as the wind-field at the 850 mb level) to follow the trajectory, although it is demonstrable that if a different altitude is used for the wind field, the appearance of the trajectory on the map will indeed be noticeably different (Venkatram, 1985). It is interesting to note that all Gaussian plume approaches are subject to the same criticism and more, since they assume the vertical variation does not exist at all, but this drawback is typically ignored.

On the positive side, there are several things one can do with trajectory analysis that are not possible using other approaches. For one thing, it is immediately obvious from short-term simulations that the deposition pattern is naturally extremely 'patchy' for relatively short-term releases (Zuker *et al.*, 1979). This observation seemed nonetheless to come as something of a surprise in, for example, observations of depositions of radioactive materials from the Chernobyl accident (Hohenemser *et al.*, 1986).

Another important application which trajectory analysis makes possible is the investigation of the source of sudden air pollution episodes. The approach allows one to use historic weather pattern information to trace air masses backwards in time to get a rough estimate of where they arose. This has been done in Scandinavian countries, for example, to trace the origins of sudden appearances to locations in other parts of Europe and, indeed, provided one of the first indications of the location, as well as characterizing the type of source, of the Chernobyl incident (Hohenemser *et al.*, 1986).

Finally, an application which is increasingly seen as important is that of tracing a polluted air mass in real time. Incidents such as the accidents at Chernobyl or at Three Mile Island, nonradioactive spills such as Bhopal and elsewhere, and others make it desirable to be able to follow the movement of the main bulk of the polluting substance geographically using an atmospheric model. There is no reason why on-line weather data cannot be combined with existing forecasts for a few hours in advance to provide a quite reasonable prediction of air pollutant movement for even as much as a day before action by regulatory officials, or those in charge of emergency evacuation procedures, for example, might be required. Such on-line modelling systems will inevitably help with decisions concerning evacuation, protective measures, and so forth in case of future accidents which, after all, seem inevitable and ought to be anticipated, if not foreseen.

2.3.9 REFERENCES

Briggs, G. A. (1973). Diffusion estimation for small emissions. ATDL Contribution File no. 79. ADTL/NOAA, Oak Ridge, Tennessee.

Chamberlain, A. C. (1953). Aspects of travel and deposition of aerosol and vapour clouds. AERE Report H.P./1261. Atomic Energy Research Establishment, Harwell, Berkshire, UK.

Eliassen, A., and Saltbones, J. (1975). Decay and transformation rates for SO_2 as estimated from emission data, trajectories and measured air concentrations. *Atmos. Environ.*, **9**, 425–429.

Gifford, F. A. (1961). Use of routine meteorological observations for estimating atmospheric dispersion. *Nucl. Saf.*, **2**, 47–57.

Gifford, F. A. (1976). Turbulent diffusion-typing schemes: a review. *Nucl. Saf.*, **17**, 71.

Gregory, P. H. (1945). The dispersion of airborne spores. *Trans. Br. Mycol. Soc.*, **28**, 26–72.

Hanna, S. R. (1985). Air quality modelling over short distances. In: Houghton, D. D. (ed.), *Handbook of Applied Meteorology*, pp. 712–743. John Wiley, New York.

Hohenemser, C., Deicher, M., Ernst, A., Hofsass, H., Lindner, G., and Recknagel, E. (1986). Chernobyl: an early report. *Environment*, **28**, No. 5, June 1986, pp. 6–13.

Korte, F. (1978). Abiotic processes. In: Butler, G.C. (ed.), *Principles of Ecotoxicology*, (SCOPE 12), pp. 11–36. John Wiley & Sons, Chichester.

Pasquill, F. (1961). The estimation of the dispersion of wind-borne materials. *Met. Mag.*, **90**, 33–49.

Turner, D. B. (1970). *Workbook of atmospheric dispersion estimates*. 999-AP-26. U.S. Department of Health, Education and Welfare, Washington, D.C.

Venkatram, A. (1985). Air quality modelling over long distances. In: Houghton, D. D. (ed.), *Handbook of Applied Meteorology*, pp. 744–753. John Wiley, New York.

Yamartino, R. J. (1985). Atmospheric pollutant deposition modelling. In: Houghton, D. D. (ed.), *Handbook of Applied Meteorology*, pp. 754–766. John Wiley, New York.

Zuker, M., Ridgeway, J. M., and Miller, D. R. (1979). A study of atmospheric radionuclide transport and exposure using trajectory analysis. In: *Biological Implications of Radionuclides Released from Nuclear Industries*, pp. 381–398. International Atomic Energy Agency, Vienna.

Ecotoxicology and Climate
Edited by P. Bourdeau, J. A. Haines, W. Klein and C. R. Krishna Murti

2.4 *Aquatic Transport of Chemicals*

E. D. GOLDBERG

2.4.1 INTRODUCTION

Continental debris, both natural and anthropogenic, is transferred to the oceans by the winds, rivers, coastal runoff, waste outfalls, and dumping from ships. The first three processes can be markedly influenced by climate, affecting both entry to and movements within the marine environment. Two climate-related factors are dominant in affecting the fates of substances: the annual precipitation, which determines the amounts and compositions of river flow and coastal runoff; and primary plant productivity in the hydrosphere. There are well-defined geographical differences in these two factors as a function of present-day climatic regimes. An understanding of these climate-controlled parameters can be used to describe and predict the fates of pollutants introduced to marine systems.

A key role is played by organic substances. The behaviour of anthropogenic chemicals during transport within water systems will be strongly affected by organic compounds, both in the dissolved and particulate states. The solid forms can take up pollutants, such as heavy metals or artificially produced radionuclides, and enhance their descent through water columns to the underlying sediments. They can also make pollutants more available to filter-feeding organisms. Similarly, some of the dissolved phases have the abilities to form complexes with inorganic cations and to alter their involvements with solids through sorption processes or their accumulation by marine organisms. Further, the reducing abilities of some of the organics can involve them in redox reactions with pollutants.

2.4.2 PRIMARY PRODUCTIVITY

The primary plant productivity in the surface waters of the oceans initiates most of the chemical reactions within the marine system. The photosynthetic fixation of carbon dioxide and water into organic matter is accompanied not only by

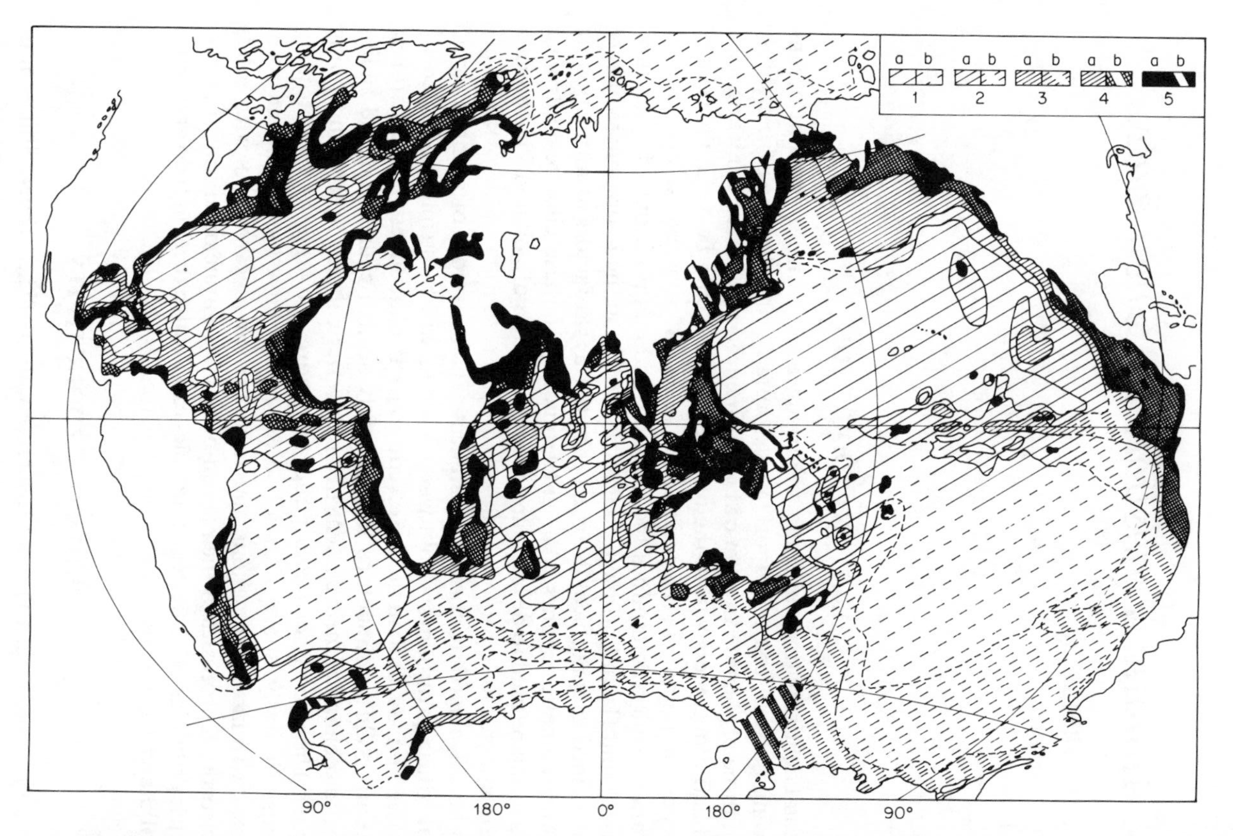

Figure 2.4.1 Distribution of primary production in the World Ocean. Units are in mg C/m^2/day. (1) Less than 100; (2) 100–150; (3) 150–250; (4) 250–500; (5) more than 500. a = data from direct ^{14}C measurements; b = data from phytoplankton biomass, hydrogen, or oxygen saturation. From Koblentz-Mishke *et al.* (1970). Reproduced with permission

the uptake of plant nutrients, such as nitrate and phosphate, but also by the incorporation of many inorganic elements, such as manganese, copper, and iron. The plants provide the food base for higher organisms. As a consequence there is an overall transfer of particulate organic phases from the upper layers of the oceans where photosynthesis occurs to depths; this occurs through gravitational settling, through uptake of materials by higher trophic levels, and through the discharge of metabolic waste products. Also, the organic phases are oxidized by dissolved oxygen with the release of various chemical species back into the dissolved states. Overall, the initial plant productivity results in the transfer of many elements from surface to deeper waters.

As an example, the particulate organic phases have been associated with the transport of naturally occurring thorium isotopes (Bacon and Anderson, 1982) and artificially produced plutonium isotopes (Toggweiler, 1983) downwards in the oceanic water column. Bacon and Anderson developed a simple scavenging model in which dissolved thorium, the principal form of the element in the oceans, is taken up by small particles which subsequently can become attached to larger particles. These larger particles during their fall through the water column can be broken up and re-formed through biological activity. Further, during this sinking, there is simultaneous adsorption/desorption of the metals from the solids.

Plutonium profiles in the Pacific Ocean have a unique character—maxima at depths around 500 m. Toggweiler (1983), utilizing the rate constant of Bacon and Anderson (1982) for the dissolved thorium uptake and loss term for the destruction of larger particles to the smaller particles and to the dissolved state, was able to model the observed plutonium distribution. Utilizing a 50 m/day sinking rate, the destruction of the large particles is found to take place daily. The loss term is related to the average lifetime of faecal pellets produced over all depths by grazing organisms. With particle destruction, the plutonium enters the dissolved state for a period of several years before encountering another sinking particle. The downward motion of plutonium takes place at about 40 m/year as derived from the model. This overall transfer of plutonium from upper to deeper waters is thus clearly dependent upon the production of primary organic materials.

Present-day primary plant productivity in the world ocean has been summarized by Koblentz-Mishke *et al.*, (1970). The distribution of values was divided into five classes: less than 100, 100–150, 150–250, 250–500, and more than 500 mg of carbon fixed per square metre per day (Figure 2.4.1).

Quantitatively, the global picture of primary productivity is described by two main variables: the supply of nutrients and the quantity of light energy available to the algae. The latter factor is of importance, for example, in the polar regions where the presence of sea ice and the brevity of daylight in winter limit the amount of light penetrating the waters. The lowest levels of primary productivity are found in such areas.

The higher levels of primary productivity occur in shallow coastal waters, i.e. within the 180 m depth contour (Ryther, 1969). Herein there is re-use of biostimulants, such as P, N, and Si. Especially productive waters are those of the continental shelf and slope, islands, and the upwelling regions of the oceans. The highest values of primary productivity are found along western continental coasts, especially at subtropical latitudes where there are prevailing offshore winds and eastern boundary currents. Surface waters are diverted offshore and are replaced by nutrient-rich deep waters—the upwelling phenomenon. The upwelling areas constitute about 0.1% of the world ocean (Ryther, 1969).

2.4.3 RIVER FLOW

The world's rivers deliver both natural and anthropogenic substances in the dissolved and particulate states to the coastal ocean. Variations in compositions of the major chemical elements (Al, Fe, Mg, Ca, Na, Si, Ti) can be understood in part through the climatic features of the river system (Martin and Meybeck, 1979). The intensity of weathering, both present and past, appears to be the controlling factor. Al, Fe, and Ti are high in the suspended materials of tropical rivers, while the concentrations of Ca and Na are low. This is a consequence of the intense leaching action of waters upon tropical soils. Over time, the readily soluble materials will have been removed. On the other hand, the temperate and Arctic rivers have higher levels of Ca, Fe, and Ti. Their suspended loads are derived from poorly weathered rock and as a consequence their compositions are quite similar to those of average crustal materials.

In addition to climatic factors, the topography of the river basin and human activities can characterize river composition. With respect to morphology, Richey (1983) indicates there are four patterns: (1) the river load is derived from mountains marginal to the flatlands, like the Amazon and Mississippi; (2) the river is marginal to a fold belt and flows parallel to it (Ganges and Parana); (3) the river flows along the strike of a mountain chain (Mekong and Magdalena); or (4) is superimposed across mountain chains (Columbia and Danube). The combination of precipitation, discharge (rainfall), basin area, and suspended load then influence the natural character of a river. For example, Asian rivers transport relatively large amounts of suspended solids as a consequence of high erosion rates and the steep relief of the continent.

The riverine export of organic materials that can interact with a variety of substances is clearly climate related. The greatest export appears to be in the sediment-rich waters of Asia and the organic-rich rivers of tropical America and tropical Africa. Meybeck (1982) indicates the linkages between the dissolved organic carbon (DOC) levels and climatic features. Taiga rivers have high values, with a median of 10 mg/l, while the wet tropics and temperate zones have 6 and 3 mg/l respectively. Lower concentrations exist in rivers draining tundra, with levels of around 1 mg/l. Maximum natural DOC values, at levels of 25 mg/l, are found in rivers passing through swamps or poorly drained soils. The lowest values, <1 mg C/l, are found in mountain rivers of the French Alps and New Zealand.

The concentration of total organic matter is highly variable, ranging from <1 mg C/l in alpine streams to >20 mg C/l in some tropical or polluted rivers (Richey, 1983). A substantial portion of this organic matter is in the form of fulvic and humic acids, which constitute the bulk of the organic matter in soils and sediments. These highly polymerized and highly aromatic compounds have molecular weights in the thousands and tens of thousands, with large numbers of functional groups, including phenolic, carboxyl, ketonic, alcoholic, and quinoidal types. As a consequence they have remarkable complexing abilities with metals, especially those characterized as hard acids. The dissolved fulvic and humic acids can significantly complex such metals as Zn(II), Cd(II), and Pb(II) if their concentrations are of the order of 1 mg/l (Raspor *et al.*, 1984). In the open sea, where these acids rarely exceed 0.2 mg/l, their associations with metals will be minimal. At 1 mg/l, around 10% of these three metals will be complexed as humates and fulvates.

The complexing of heavy metals with humic substances and with other organics in marine waters has been presumed to reduce the free metal ion concentrations, such as those of copper, which is toxic to aquatic organisms. Thus, Florence *et al.* (1984) found that many naturally occurring complexing agents, including the fulvic and humic acids, decrease the toxicity of copper. However, with synthetic ligands, lipid-soluble complexes turned out to be more toxic than the inorganic copper. Such complexes include oxine and 1-(2-pyridylazo)-2-naphthol. These investigators indicated that strong chelation with copper was essential for high toxicity. This sense that toxic responses may be a function of type and concentration of organic ligand present was also found for Ni and Cd. For example at 10 μM citric acid, about 25% of the Cd is complexed. Yet there is increased toxicity to algae at lower Cd concentrations (Laegreid *et al.*, 1984). Thus, the possibility that the organic molecules are toxic to marine organisms, either alone or in association with metals, seems reasonably high. Further, some pollutant organic chemicals carried to marine systems could in principle behave similarly.

A second important group of organic compounds entering the oceans from the rivers includes the plant building-block lignin and its degradation products, which can be oxidized to cinnamyl, syringyl, and vanillyl acids (Hedges and Mann, 1979).

The river load of particulate organic carbon (POC) is inversely related to the total amount of suspended matter (Meybeck, 1982). This relationship is apparently valid for about 99% of the world's rivers and can be coupled to the well-known increase of total suspended load with water discharge. The river POC, taken to mainly autochthonous, is diluted with increased levels of land-derived mineral matter. This phenomenon explains the relationship.

Table 2.4.1 DOC levels in rivers, range of TOC export rates compared to terrestrial primary production, and TOC budget to the ocean. From Meybeck (1982)

Environment	Total area ($10^6 km^2$)	DOC (mg/l)	TOC export rate $g/m^2/yr$	Average total TOC load (10^{12}g C/yr)	Mean net primary production (g $C/m^2/yr$)
Tundra	7.55	2	0.6	4.5	65
Taiga	15.85	10	2.5	39.6	360
Temperate	22.0	3	4.0	88	225–585
Tropical	37.3	6	6.5	241	315–900
Semi-arid	13.5	3	0.3	4.6	32
Desert	1.7		0	0	1.5
Total of the exoreic runoff	99.9		3.8	378	480
Tropical rain	15.9		10	159	900

The dissolved organic carbon loads in rivers and the export is given in Table 2.4.1. The relationship between the export of total organic carbon (TOC) and river runoff is evident. Maximum rates are found in the rain forests of temperate and tropical regions. Lowest rates are noted in semi-arid regions (Meybeck, 1982).

Of importance to the fate of some pollutants is the organic carbon entry into the oceans, both in the dissolved state (DOC) and particulate state (POC). For example, high levels of components of the dissolved load, such as the humic and fulvic acids, can chelate metals and reduce their availability to marine organisms and their sorption onto solid phases. The greatest input of organic carbon appears to be in the tropical rivers of South America and Africa (Richey, 1983). Further, there is evidence that a large amount of the organic carbon within rivers is respired or stored. Richey and Salati (cited as an article to be published in Richey, 1983) suggest that only 30% of the exported organic matter from the Amazon river actually enters the ocean.

One of the widespread impacts of human activity upon the oceans is the increased mobilization of nutrients to the coastal zone and the consequential increase in plant productivity. Richey (1983) estimates that the river fluxes of phosphorus and nitrogen have increased from prehistoric times to the present by a factor of two: N from 2–5 to 11 Tg/year and P from 0.5 to 1 Tg/year. He indicates that the current consumption of phosphate fertilizer is greater than the natural river export. Nitrogen is also mobilized by human society as a consequence of the high-temperature fixation of nitrogen to oxides of nitrogen during the burning of fossil fuels and of its use in fertilizers.

Nutrient supplies within riverine systems have been related to increased plant productivity in the lower reaches of the river (Richey, 1983). The entry of these biostimulants has been related to the productivity of the Amazon, the flood plains of Shatt el Arab, and the Zaire and Nile rivers. Further, high fish populations have been found in the flood plain rivers of southeast Asia, Africa, and South America. The relationship of natural nutrient levels in rivers and climate is not yet clearly defined and is in need of further investigation.

The entry of excessive levels of plant nutrients can lead to the high production of organic matter, which may be capable of consuming all of the dissolved oxygen in the waters and in the sediments. This leads to a state of anoxia, where sulphate becomes the primary oxidizing agent with the production of hydrogen sulphide. When all of the sulphate is utilized in oxidative processes, fermentation and carbon dioxide reduction result, with the production of methane. In addition, the diatoms, which form the base of the marine food chain, are displaced by dinoflagellates, the weeds of the sea. This alteration of normal communities can be felt in higher trophic levels.

The trend towards anoxia may be self-sustaining. The increased flux of organic matter to the sediments may be compensated by an increased flux of nutrients out of the sediments. Following oxidation of the organic matter in the sediments, the nitrate and phosphate ions can return to the water column primarily by molecular diffusion. Coupling this flux with that of the rivers and coastal runoff to the surface waters can result in further enhanced productivity, and the formation of organic matter can thus become self-sustaining.

Biomass increases, anoxia, and alteration of the plant community structure have been observed worldwide in coastal waters and in estuaries. Chesapeake Bay, the New York Bight, the southern and northern California coasts, the Baltic Sea, Kaneohoe Bay, Hawaii, the Adriatic Sea, the Oslo Fjord, Omura Bay, Japan, and Tokyo Bay, Japan, are but a few of the zones responding to these anthropogenic stresses of increased inputs of biostimulants.

2.4.4 RIVERS AND PRIMARY PRODUCTIVITY

Changes in climate can exert an influence upon the fate of pollutants through coupling of river flow and the production of organic matter in the coastal ocean. Clearly, the importation via a river of substances that are toxic to photosynthetic activity, such as copper, halogenated hydrocarbon pesticides, etc., is well recognized. However, there are other possible impacts upon nutrient and light availability that may be significant.

The effects of rivers on primary productivity in the oceans may take several forms: (1) an alteration of nutrient concentrations of the coastal waters by dilution, enhancement, or removal on river-borne particulates; (2) changing light penetration through the introduction of particles; and (3) an effect on the stability of the water column. For example, by increasing its stability through low-density surface layers, the possibility of photosynthesizing cells being carried below the photic zone is minimized.

2.4.5 ESTUARIES

Estuaries play an important regulatory role in the transport of river-borne materials to the coastal ocean. Depending upon the morphology and hydrodynamical properties of an estuary, the materials introduced by rivers can be held or delivered to the open ocean (Wollast, 1983). The vertical stratification of estuarine waters and the amounts of mixing in part govern the movement of materials through the estuaries. The waters of rivers entering well-mixed estuaries have longer residence times than those flowing into highly stratified systems. In the latter case, the freshwater flow takes place in well-defined surface layers with little mixing with the underlying more saline waters. Such river waters rapidly reach the sea (Wollast, 1983).

Wollast (1983) points out that the particulate organic matter of terrestrial origin is often partially, if not completely, removed in the estuarine zone and carries with it many inorganic species. Similarly, colloidal species in rivers, usually negatively charged, flocculate in estuarine streams. Positively charged materials are carried down with the flocs through neutralization of the surface colloidal charges. Thus, river-borne pollutants may meet an accommodation in estuarine sediments.

The primary plant production within the estuary and the subsequent respiration of these materials and of river-borne organics determine the outflow of organic phases. In those estuaries where the waters have long residence times, bacterial degradation of the non-refractory organic matter takes place. If primary productivity and the river fluxes are high, anoxia can develop (Wollast, 1983).

The fate of nitrogen and phosphorus entering estuaries has been investigated for the Scheldt Estuary, Belgium (Wollast, 1983). Three biochemical reactions govern the speciation of nitrogen: nitrification, denitrification, and uptake during primary productivity. A mass balance for the nitrogen species has been prepared (Table 2.4.2). The total riverine nitrogen input of 34 tonnes is compensated by a loss of 20 tonnes annually to the coastal ocean, a loss of 3 tonnes to the sediments, and a loss of 11 tonnes of nitrogen gas to the atmosphere through denitrification.

Table 2.4.2 A proposed mass balance for nitrogen species in the Scheldt Estuary. The processes considered include freshwater input, denitrification, nitrification, primary productivity, and sedimentation, as well as output to the North Sea (Wollast, 1983). Units are tonnes of N/year

Species	Input	Denitrification	Nitrification	Primary Productivity	Sedimentation	Output
NH_4	23	–	−12	−6	–	5
NO_3+NO_2	9	−11	+12	–	–	10
N particulate	2	–	–	+6	−3	5
N total	34	−11	–	–	−3	20

Table 2.4.3 A proposed mass balance for phosphorus for the Scheldt Estuary in units of kilotonnes per year (Wollast, 1983)

	Input	Chemical precipitation	Plankton uptake	Sediment	Output
Dissolved P	5.6	−3.3	−0.8	0	1.5
Particulate P	1.5	+3.3	+0.8	−4.9	0.7
Total P	7.1	–	–	−4.9	2.2

Phosphorus occurs nearly completely in the phosphate form, both dissolved and particulate. The estuarine input of 7000 tonnes per year is compensated by an output of 2200 tones and a deposition of 4900 tonnes (Table 2.4.3).

Clearly, the above data are specific for the Scheldt. A greater percentage of the river input can reach the oceans for more highly stratified estuaries.

2.4.6 MONITORING

The monitoring of atmospheric and riverine fluxes of pollutants to the coastal environment takes a variety of forms depending upon the time interval to be averaged. For example, instantaneous measurements of pollutants in the air or water streams can be made by direct sampling. Such measurements usually involve time periods of days or less. At the other extreme, the sedimentary strata, where dating techniques are applicable, provide integrations of fluxes over periods of years or less. The use of sentinel organisms, such as bivalves, has been effective in monitoring pollutants where the integration periods appear to be of the order of months to years, depending upon the nature of the pollutant. Atmospheric fallout collected upon plates or into buckets can also be integrated over periods of days to months.

The sedimentary record of fluxes to the coastal zone depends upon the development of an adequate time base (usually a radioactive geochronology) and upon the absence of disturbance to the sediment, either through bioturbation or physical mixing processes. For example, the anthropogenic fluxes of Pb, Cr, Cd, Zn, Cu, Ag, V, and Mo into the coastal zone off southern California were measured in anoxic sediments (Bruland *et al.*, 1974). The contributions from different transporting agencies—the winds, sewer outfalls, storm runoff, and river runoff—could not be evaluated on the basis of the measurements. The time base was provided both by the counting of the annual varves of the sediments and by ^{210}Pb geochronologies. There was no evidence of mixing of the solid phases of the deposit. The lead fluxes were increased by a factor of four as revealed in the sedimentary record, whereas there was less than a 50% increase in vanadium fluxes over the past century.

The atmospheric fluxes of metals, including the radioactive element plutonium, to the California coastal zone were determined with buckets, which collected fallout over monthly periods (Hodge *et al.*, 1978). The flow of these aerosols to the oceans appeared to measurably govern some metal concentrations in surface waters but did not account for the overall accumulation of metals in coastal sediments. The metal-containing aerosols were removed from the atmosphere mainly by dry fallout. The lower atmospheric residence times of the metals appear to be of the order of a half day to several days.

Monitoring of sentinel organisms for their pollutant concentrations is most useful to determine measures of environmental levels and of their bioavailability at a given site. However, often the behaviour of pollutants can be understood through monitoring of extended coastal areas (Farrington *et al.*, 1983). For example, elevated Cd and $^{239+240}Pu$ concentrations in bivalves from the central California coast compared to their counterparts either to the north or south are apparently related to enrichments of these metals in intermediate depth waters of the north Pacific and to the upwelling of this water associated with the California current system. Both metals are transferred from the surface waters to deeper waters through biological activities.

Bivalves, including mussels and oysters, are especially adapted to pollutant monitoring. They are cosmopolitan, often in communities with large populations which can be sampled without fear of destroying a community. Being sedentary as adults they can effectively characterize through their body burdens some pollutants in a given area. The concentration factors of such pollutants as chlorinated hydrocarbon pesticides, heavy metals, and transuranic elements often achieve values of between 100 and 100 000. The bivalves clearly assess the biological availability of the pollutants under study. In addition, there is little evidence that they metabolize many xenobiotics. Thus, a quite accurate evaluation of the environmental levels can be made through measurements of their body contents. Moreover, they survive under pollutant concentrations that reduce or eliminate other species, and they can be transplanted from one area to another. Finally, since some bivalves are commercial seafood resources, the level of their contamination is important from a public health standpoint (Farrington *et al.*, 1983).

Primary productivity measurements over long times and over wide areas in the oceans and rivers may be effectively pursued using satellite imagery. Present measurements from ships give inadequate spatial and temporal coverage to study such problems as a trend towards anoxia which may be occurring at a very slow pace.

The Coastal Zone Color Scanner (CZCS) aboard the Nimbus 7 has been obtaining near-surface phytoplankton pigment concentrations in the world ocean since 1978. The procedure for converting the satellite pigment images into water column primary productivity has been described by Eppley *et al.*, (1985). The spatial resolution is around 1 km^2. Cloud cover often obscures the pigment image. Still a coverage of 50 images a year for a given area will provide an understanding of the short-term temporal changes in primary productivity and will allow comparisons to be made with other areas and subsequent years. A trend towards eutrophication of coastal areas will most effectively be assessed with these techniques.

2.4.7 OVERVIEW

The time scales over which climate related parameters can change and hence influence primary productivity and river flow/coastal runoff can extend over periods of days (the episodic event) through years to millenniums. The resultant impacts upon the fate of pollutants introduced to natural waters will have similar time constants.

For instance, tropical storm Agnes in June 1972 caused massive flooding of Chesapeake Bay, Maryland. The movement of suspended solids from the Susquehanna River into the estuary during a one week period exceeded that of the previous several decades (Schubel, 1974). The solids were distributed in the upper reaches of the Bay. The salinities in Bay waters were reduced substantially and remained low over the summer, recovering to normal values in September. Several consequences which would influence the fate of pollutants are evident. The reduced salinities can decrease primary productivity. The increased net seaward flow of waters carried pollutants to the ocean. Pollutants at the sediment/water interface in the upper reaches of the Bay were covered over by the river solids.

On a somewhat longer time scale, alterations in primary productivity as a consequence of the El Nino during February and March 1983 were observed around the Galapagos Islands (Feldman *et al.*, 1984). The El Nino is an oceanic response to atmospheric forcing functions. The satellite measured pigment concentrations indicated a decrease in primary productivity in the region. This may explain the observed reproductive failures of seabirds and marine mammals noted in the area. Clearly, such an event also can be involved in the redistribution of pollutants over periods of years.

There are other climate-controlled parameters that can change water temperatures in the photic zone, concentrations of plant nutrients (such as phosphate and nitrate) through alterations in patterns of water movements, and the flow of continental waters into the oceans. Tsunamis, storms, tidal amplitude changes, and upwelling intensity changes are episodic events affecting pollutant behaviour. Climate-controlled changes in surface water temperature of the photic zone can alter primary productivity and hence the fate of pollutants. The impact of temperature on gaseous oxygen solubility at the surface waters can affect the rate of combustion of organic phases, both natural and anthropogenic.

Overall, the persistence of pollutants and the production of interacting chemicals are climate controlled. Our abilities to predict the course of climate will be reflected in our abilities to predict the fates of pollutants.

2.4.8 REFERENCES

Bacon, M. P., and Anderson, R. F. (1982). Distribution of thorium isotopes between dissolved and particulate forms in the sea. *J. Geophys. Res.*, **87**, 2045–2056.

Bruland, K. W., Bertine, K., Koide, M., and Goldberg, E. D. (1974). History of metal pollution in the Southern California Coastal Zone. *Environ. Sci. Technol.*, **8**, 425–432.

Eppley, R. W., Stewart, E., Abbott, M. R., and Heymann, V. (1985). Estimating ocean primary production from satellite chlorophy. Introduction to regional differences and statistics for the Southern California Bight. *J. Plankton Res.*, **7**, 57–70.

Farrington, J. W., Goldberg, E. D., Risebrough, R. W., Martin, J. H., and Bowen, V. T. (1983). U.S. Mussel Watch 1976–1978: an overview of the trace metal, DDE, PCB, hydrocarbon and artificial radionuclide data. *Environ. Sci. Technol.*, **17**, 490–496.

Feldman, G., Clark, D., and Halpern, D. (1984). Satellite color observations of the phytoplankton distribution in the eastern Equatorial Pacific during the 1982–1983 El Nino. *Science*, **226**, 1069–1071.

Florence, T. M., Lumsden, B. G., and Fardy, J. J. (1984). Algae as indicators of copper speciation. In: Kramer, C. J. M., and Duinker, J. C. (eds.), *Complexation of Trace Metals in Natural Waters*, pp. 317–327. Martinus Nijhoff/Dr. W. Junk Publishers, The Hague.

Hedges, J. I., and Mann, D. C. (1979). The characterization of plant tissues by their lignin oxidation products. *Geochim. Cosmochim. Acta*, **43**, 1803–1807.

Hodge, V., Johnson, S. R., and Goldberg, E. D. (1978). Influence of atmospherically transported aerosols on surface ocean water composition. *Geochim. J.* , **12**, 7–20.

Koblentz-Mishke, O. J., Volkovinsky, V. V., and Kabunova, J. G. (1970). Plankton primary production in the world ocean. In: Wooster, W. S. (ed.), *Scientific Exploration of the South Pacific*, pp. 183–193. U.S. National Academy of Sciences, Washington, D.C.

Laegreid, M., Alstad, J., Klaveness, D., and Seip, H. M. (1984). Metal speciation—biological response. In: Kramer, C. J. M., and Duinker, J. C. (eds.), *Complexation of Trace Metals in Natural Waters*, pp. 419–424. Martinus Nijhoff/Dr. W. Junk Publishers, The Hague.

Martin, J., and Meybeck, M. (1979). Elemental mass balance of material carried by major world rivers. *Marine Chem.*, **7**, 173–206.

Meybeck, M. (1982). Carbon, nitrogen and phosphorus transport by world rivers. *Am. J. Sci.*, **282**, 401–450.

Raspor, M., Nurnberg, H. W., Valenta, P., and Branica, M. (1984). Significance of dissolved humic substances for heavy metal speciation in natural waters. In: Kramer, C. J. M., and Duinker, J. C. (eds.). *Complexation of Trace Metals in Natural Waters*, pp. 317–327. Martinus Nijhoff/Dr. W. Junk Publishers, The Hague.

Richey, J. E. (1983). Interactions of C, N, P, and S in river systems: a biogeochemical model. In: Bolin, B., and Cook, R. B. (eds.), *The Major Biogeochemical Cycles and Their Interactions*, SCOPE 21, pp. 363–383. John Wiley & Sons, Chichester.

Ryther, J. H. (1969). Photosynthesis and fish productivity in the sea. *Science*, **166**, 72–76.

Schubel, J. R. (1974). Effects of tropical storm Agnes on the suspended solids of Northern Chesapeake Bay. In: Gibbs, R. J. (ed.), *Suspended Solids in Water*, pp. 113–132. Plenum Press, New York.

Toggweiler, J. R. (1983). A simple model solution for the bomb-plutonium maximum observed in the Pacific Ocean. *EOS*, **64**, 1089.

Wollast, R. (1983). Interactions in estuaries and coastal waters. In: Bolin, B., and Cook, R. B. (eds.), *The Major Biogeochemical Cycles and their Interactions*, pp. 385–407. John Wiley & Sons, Chichester.

Chapter 3

Environmental Fate of Chemicals and Chemical Exposure in Tropical, Arid and Cold Regions

Ecotoxicology and Climate
Edited by P. Bourdeau, J. A. Haines, W. Klein and C. R. Krishna Murti

3.1 Mobility of Environmental Chemicals, Including Abiotic Degradation

W. KLEIN

3.1.1 INTRODUCTION

The mobility of chemicals released into the environment governs their bioavailability to organisms and their partitioning between environmental compartments, and thus represents a direct link between environmental pollution and ecotoxicity. The micro- and mesoscale transport processes involved are related to the properties of the chemicals themselves, but also to environmental compartments. Thus, the criteria of mobility, dispersion, and bioavailability are closely related, having partly a common set of influencing variables. A number of these variables depend on climatic factors, especially temperature and water regimes. A knowledge of the relationships would permit prediction of the relevant descriptors for any climatic conditions without performing the respective experiments.

3.1.2 ENVIRONMENTAL VARIABLES AND CHEMICAL PROPERTIES RELEVANT TO MOBILITY OF CHEMICALS

Environmental compartments are multivariate systems, both in space and time. From the point of view of chemical composition it may be stated that in general soil shows the greatest variation, and the air compartment the lowest. The relevant properties for the compartments may be summarized as follows:

For soil the major dispersion-relevant properties reflecting the influence of the matrix on partitioning, diffusion, sorption, and binding processes of a chemical are:

(a) types and amounts of clay materials (variability and three-dimensional structure);
(b) particle size and porosity;
(c) exchangeable ions (including buffering capacity);
(d) pH, conductivity of soil solution;
(e) redox potential;

(f) temperature;
(g) air and water regime;
(h) composition and amount of organic matter;
(i) biological activity and biomass.

The significance of these properties differs with the geographical region and with the nature of the chemical involved.

For the freshwater aquatic environment, apart from the water itself as transport medium, further properties of interest are:

(a) sediment and its composition (see soil properties above);
(b) suspended solids and their properties with respect to sorption, solution, and partitioning;
(c) dissolved organic matter with respect to changed water solubility and partitioning behaviour of chemicals (including chelating regime);
(d) temperature;
(e) pH;
(f) buffering capacity;
(g) redox potential;
(h) biota.

For the atmospheric environment, apart from transport (discussed in Section 2.3), relevant variables are particulate matter differentiated for large size and inhalable aerosols, and also differentiated for physical and chemical properties (corresponding to suspended solids in aquatic systems).

The major properties and structural characteristics of chemicals involved in the mobility behaviour are:

(a) physical state;
(b) volatility;
(c) molar mass/molecular size;
(d) dissociation constant;
(e) complexation constant;
(f) vapour pressure;
(g) water solubility;
(h) lipophilicity, water/lipids partitioning;
(i) reactivity;
(j) electronic, geometrical, and topological characteristics.

All variables mentioned above interact in complexity and differently for specific compound-compartment-organism systems.

For the ecotoxicological assessment of the occurrence of chemicals in environmental compartments, the composition of the abiotic and biotic sub-compartments is important and should be taken into consideration with regard to receiving masses and properties e.g. of sediment and suspended solids or of biota in running versus stagnant aquatic systems.

In the soil environment the respective subsystems are organic matter, minerals, soil solution, and soil atmosphere, as well as soil biomass.

Apart from the variables mentioned, processes determining the fate and availability of chemicals have to be considered. These are:

(a) processes involved in changing mobile portions of chemicals apart from degradation: binding to persistent organic matter, 'irreversible' binding to clay minerals;

(b) processes involved in the movement into organisms (resorption/transport through membranes): partitioning, diffusion, active transport.

For the retrospective evaluation of the dispersion of chemicals based on monitoring data it should be mentioned that in general the production of analytical data is aimed at maximum recovery from a matrix and does not consider the specific state(s) of the chemical analysed. Therefore, a further analytical characterization would be needed:

(a) total analysis versus solubilized or naturally solubilizable portions;

(b) speciation for both inorganic and organic chemicals to be monitored.

To arrive at the required levels of precision the environmental compartment properties as well as the biological processes may be estimated or measured.

Furthermore, several types of interactions need to be considered. They include physico-chemical processes of complexation and changes in solubility and sorption phenomena, but also chemical reactions between chemicals in the environment, including chemical–nutrient interactions.

Water solubility is one of the key properties of a chemical which influences its mobility. Solubility of all chemicals is temperature dependent; for electrolytes involving the solubility product which depends on water hardness. The solubility of electrolytes also depends on ionic species present and thus on geological conditions and climate.

As a general descriptor for the partitioning behaviour (hydrophilicity) of a chemical, the octanol/water partitioning coefficient is used, which is only slightly temperature dependent (Mackay, 1980). The effect of temperature on K_{ow} is in the range of 0.001–0.01 log K_{ow} units and may be positive or negative (Lyman *et al.*, 1982). It is inversely related to water solubility.

In considering the importance of soil organic matter in immobilization of chemicals in soils it would be interesting to know whether there are specific climatic zone differences in their sorption-binding characteristics.

Various humic acids isolated from cool-temperate acidic soils, subtropical neutral, and temperate neutral soils and one fulvic acid from a Canadian spodosol were analysed by NMR. Neither in aromaticity (varying between 35 and 92%) nor in total acidity (varying between 6.3 and 8.9 meg/g; fulvic acid 12.4) or carboxyl and phenolic strength were there differences which could be related to a climate zone. Only the carbonyl equivalents were higher in the cool-temperate than in the subtropical samples investigated (Hatcher *et al.*, 1981).

Turnover times for soil organic matter, which are important in the kinetics of behaviour of chemicals, vary from days to months in the case of plant residues, and are in the range of hundreds of years for fulvic acids, and of thousands of years for humic acids and humins. Being related to microbial activity turnover they depend significantly on temperature and humidity (Paul and Huang, 1980).

Sorption processes in soils play a major role in the availability of pollutants to flora and soil inhabitants, as well as in their movement as compared to the bulk movement of water. The most important soil (and sediment) property in this respect is the organic carbon content, but clay minerals and amorphous oxides (surface area and cation exchange capacity) may also play a role, especially with increasing polarity of the chemicals and decreasing soil organic carbon content. For nonpolar compounds K_{oc} is a frequently used normalized characteristic, and empirical relationships have been developed between organic carbon and lipophilicity (solubility, octanol/water partition coefficient) and also including the conditions for mineral sorption to become significant (Karickhoff, 1981; 1984). For rather lipophilic chemicals, such as *o*-chlorotoluene at a clay : total organic carbon ratio of 60, mineral contribution to sorption was indicated (Banerjee *et al.*, 1985). In view of the frequently low carbon content of soils in hot climates, e.g. for sandy soils in arid regions, either due to low long-term biomass production or to its rapid turnover, it must be assumed that at least the capacity of these soils for immobilizing less polar chemicals is largely reduced. Thus, the hazard for movement into groundwater is increased. It should be mentioned that sorption of chemicals to colloidal organic macromolecules and clay particles plays an important role in transport in surface waters (*see* Section 3.1.5).

Diffusion of chemicals is of interest with respect to bioavailability and volatilization. The apparent total diffusion coefficient in soil is composed of vapour and non-vapour diffusion (theory from Ehlers *et al.*, 1969a). It is influenced by water content of the soil, bulk density, and temperature. Depending on soil sorption characteristics, water content modifies soil diffusion only to a maximum of about 10%. For lindane and a silt loam it was constant above 5%, going via a maximum to zero at lower water content. Changing bulk density from 1 to 1.5 g/cm^3 reduces the total diffusion coefficient of lindane by about a factor of two. Increasing temperature from 20 to 30°C increases the total diffusion coefficient of lindane in soil by a factor of 2.5 at water saturation, which is explained by a temperature effect on water viscosity and on the interaction between chemical and soil (Ehlers *et al.*, 1969b). Since the total diffusion coefficient increases exponentially with temperature at constant humidity, increased micromobility of chemicals in hot-humid climates, and marked seasonal variations between rainy and dry, winter and summer seasons have to be considered. A more detailed description of partitioning processes in soil has been given by Huang (1980).

The influence of temperature on diffusion in a water-sediment system has been reported for bromacil (Corwin and Farmer, 1984). There is a direct and positive influence of temperature via the viscosity coefficient. Additionally, adsorption is reduced at elevated temperatures at constant humidity.

Chemicals present in soil may be subject to transport into other media by erosion/surface runoff. This depends on topography, and wind and water regimes, but also to a large extent on agricultural practice and soil properties. As an example, surface runoff from ferralite soils in a humid tropical region of the Ivory Coast was found to be 62–70% of the rain from bare soil, 26–39% under pineapple with residues burnt, and 0–2% under pineapple with residues mulched (Roose and Fauck, 1981).

Models have been developed to assess the evaporation rate of chemicals from water surfaces. The rate of evaporation may be calculated as a function of diffusion coefficients in air and water, the Henry constant, water temperature, and water depth, and further as a function of wind speed, and resulting turbulence with qualitative changes depending on water turbulence. Feasible pragmatic models developed by Wolff and van der Heijde (1982) and by Mackay and Yeun (1983) may be used to calculate volatilization from water, especially for temperature and wind regime changes.

Models for environmental compartmentalization, movement and estimation methods, including the relevant mobility parameters, volatility, and partitioning behaviour have been compiled in various handbooks (Baughman and Burns, 1980; Mackay, 1980; Paterson and Mackay, 1985; Lyman *et al.*, 1982).

For temperature-dependent physico-chemical properties the models generally use constant values, most frequently for 25°C. If temperature dependencies are known it is easy to make adjustments for non-temperate conditions. According to presently known relationships it would be difficult to estimate the influence of non-temperate conditions on bioaccumulation, which is an important descriptor in environmental compartmentalization models. If measured values are not available, bioconcentration factors are frequently estimated from log K_{ow}, which has little temperature dependence. However, there is a greater effect of temperature on bioaccumulation via effects on passing biological barriers and on metabolism, so that modified constants should be used in the empirical correlation equations. This influence, however, has to be weighed against species differences and lipid content variations to judge the significance. Examples are known of slight temperature differences measurably changing toxicity. The few data available show a high influence of temperature on bioaccumulation in fish, but with no general direction (Bude, 1981).

3.1.3 MOBILITY IN SOILS: EXAMPLES OF INTERACTIONS WITH CLIMATE-RELATED FACTORS

The mobility of chemicals in soils is of interest with respect to leaching, volatilization, and bioavailability to flora and soil inhabitants.

There is a marked spatial variation in soil properties relevant to the dispersion of chemicals even for small areas, e.g. 1 ha. The water-soluble carbon was determined in a 200 point grid on a 1.4 ha cropped field. The concentration ranged from 24 to 274 mg/kg, with over 90% of the values around the mean of 40 ppm. Spatial dependence could not be demonstrated in a transect with sampling points every 1.37 m (Liss and Rolston, 1983).

There is additionally a seasonal variation, with maxima of the mean in spring and fall (up to 25% higher than in summer).

Mobility of chemicals in soils is largely dominated by sorption processes following various mechanisms. These processes depend on the polarity (lipophilicity, ionic/non-ionic) of the chemical, soil mineralogy and organic matter content, and especially soil humidity between zero and monolayer of water. Adsorption being generally an exothermic process (decrease at higher temperatures), competitive adsorption, distribution, and ion-exchange processes in the multivariate soil systems result in complex relationships. There are some general tendencies, which, however, need to be confirmed for specific cases:

(a) decreasing water content increases adsorption and decreases mobility;
(b) increasing temperature decreases adsorption and increases mobility;
(c) increasing clay mineral content and organic matter increase adsorption and decrease mobility;
(d) plant cover increases metabolism, and decreases leaching.

Water regime and temperature change the mobility of nutrient and other ions directly, and also that of non-ionic organic compounds via varying soil properties, such as carbon content and microbial biomass. The availability of Zn in different soils increased by 46% when elevating the temperature at field capacity moisture from 10°C to 45°C, while submerging increased the available portion by 20% (Budhewar and Omanwar, 1980).

Changes of climatic factors in one direction may have opposite effects on the movement of different chemicals, e.g. the leaching from soil of the highly mobile cations Na^+ and K^+ may be augmented for both with increasing rainfall, but increasing temperature (4°C to 15°C) may augment the leaching of K^+ and decrease that of Na^+. Low temperatures reduce nitrate losses (Buldgen, 1982).

As an example, for ionic organic compounds, much work has been performed on paraquat and diquat. High humidity and high temperature and low light intensity increase foliar uptake. In soil, diquat and paraquat cations are tightly adsorbed to clay minerals and organic matter, and thus are immobile at levels below the strong adsorption capacity of soil. The ion-exchange process is independent of pH, temperature, and exposure time (Ackhavein and Linscott, 1968).

For dry desert soils with prolonged periods of high temperature and intensive sunlight, although volatilization is lower than in humid conditions, it may be assumed that disappearance of chemicals is nevertheless effective. The loss of DDT and methyl parathion was studied in three dry soils at 30°C and 50°C with low-energy UV irradiation (300–400 nm) and in the dark. About 50% of initial 5 ppm methyl parathion was lost within 50 days at 30°C, 65% at 50°C in the dark, and irradiation added 10 to 17% to the loss. Whereas the same change in temperature doubled the loss of DDT from 25% to about 50%, the effect of light was only about 8% (Baker and Applegate, 1970).

In three Israeli soils with highly differing clay (6–56%), cation exchange capacity (CEC 4–63 m/100 g), surface area (39–410 m^2/g), and organic matter (0.3–1.9%), the influence of water and temperature on parathion adsorption was measured (Yaron and Saltzmann, 1972). In agreement with other findings (Mills and Biggar, 1969) it may be concluded from the results that for slightly polar compounds the effect of temperature on adsorption isotherms under saturated conditions increases with a lowering of temperature and also with soil quality. Using a three-phase solid-hexane-parathion system, varying humidity between 0 and 98% RH, and temperatures of 10, 30, and 50°C, an increase of adsorption with increasing humidity was found, dominated by water desorption. The initial water content had much less influence at the higher than at the lower temperatures. For the clay, the effect of soil humidity at the highest temperature is several times greater than for the soils poorer in nutrients.

Foliar absorption of chemicals is of interest for several reasons: elimination from the atmosphere, uptake into crops, and phytotoxicity of airborne substances. Increasing temperatures increase uptake of foliar deposits, but higher volatility may lead to an opposite overall temperature effect. High humidity in general facilitates leaf uptake into plants. With significant quantities of rainfall and depending on its components both leaching and runoff may occur, influenced by preceding leaf water loss.

Although light intensity does influence foliar uptake, no general tendency can be given (Hull, 1970). At higher light intensities, which result in increased photosynthetic activity, carbohydrate transport from the leaves is increased, moving away other chemicals in the assimilate stream (Morrison and Cohen, 1980).

A study was made of soil-applied phorate uptake by plants in relation to light intensity and temperature. By increasing light intensity (500 or 3000 foot candles) and/or temperature (13°C and 28°C) separately and in combination, translocation into plants was increased (Anderegg and Lichtenstein, 1984).

Complex interactions of physical, chemical, and biological environmental variables may eliminate the influence of temperature on the fate of the chemicals. The large number of investigations with 'classical' pesticides like aldrin/dieldrin may be used for demonstration (Weisgerber *et al.*, 1974). In temperate climates aldrin residues, including dieldrin, were found to persist for periods of 6 months to more than 15 years (using the detection systems available at the time) in different soils and under different environmental conditions. Under the agroclimatic conditions of the soils of northern India persistence is normally about 6 months. By changing the water regime, under flooded sandy loam soil, it is increased to 9 months. This has been explained by higher movement into soil and increased sorption, less volatilization, decrease of soil temperature due to stagnant water, and reduced biodegradation (*see* Section 3.2; Lichtenstein *et al.*, 1971; Singh *et al.*, 1985).

In a comparative long-term outdoor lysimeter experiment using local soils, which was performed concurrently in the south of England, south of Spain, California, and West Germany, the disappearance of residues from soil, including transformation products, was greatest in Spain, a little less in California, and lowest in England. Residues in the dosed 0–10 cm soil layer were least in Germany and highest in England. Leaching into deeper soil layers was highest in Germany, low in Spain, and insignificant in England and California (Klein, 1975; Scheunert *et al.*, 1977).

In order to arrive at a better understanding of these interactions and for improvement or validation of environmental fate models, a detailed description of all relevant variables in field and environmental simulation experiments would be needed.

In an early review of insecticide residues in soil (Edwards, 1966), empirical conclusions were drawn on the importance of chemical and soil/environmental characteristics, which—as far as movement of the classes of chemicals under consideration is concerned—are still valid, despite relatively poor analytical techniques compared with current ones. The direct influence of climatic factors on mobility is in general the acceleration of volatilization and desorption with increasing temperature. Volatilization is markedly retarded when soil becomes dry. This is less pronounced for poor sandy soils (low content of organic carbon and clay minerals).

Increasing soil temperature from 10°C to 30°C increased aldrin volatilization by a factor of about 4 at constant humidity, and dieldrin volatilization by a factor of 2 to 4, depending additionally on initial concentrations (Harris and Lichtenstein, 1961; Farmer *et al.*, 1972). Increasing the air flow-rate over soil by a factor of 4 roughly doubled dieldrin volatilization, which, however, approached zero for the soil investigated when its water content was below 3% (Igue *et al.*, 1972). Wind speed increases volatilization depending on upward movement of soil solution, codistillation, and diffusion. Volatility as influenced by vapour pressure and water regime also determines whether and to what extent movement in soil occurs within the soil solution or soil atmosphere. Apart from surface runoff, rain may cause leaching into deeper soil layers of the chemicals dissolved in water, or it may move particle-bound material through cavities. In an experiment with Indian loam soils aldrin volatilization was highest at 25°C as compared to 35°C and 45°C due to drying of soils during incubation (Kushwara *et al.*, 1978). Volatilization is reduced by cover crops due to reducing air movement on the soil surface, although increased release into the atmosphere by covering plants has also been reported (Klein *et al.*, 1968; 1973).

3.1.4 ABIOTIC DEGRADATION

The significance of abiotic degradation with respect to the exposure assessment component of ecotoxicology as well as experimental methods and their predictability have been discussed in SCOPE 12 and 25; therefore only those aspects which are related to climate factors are summarized here.

Photochemical processes relevant to the degradation of xenobiotics have been reviewed in SCOPE 25 (1985) and in the *Handbook of Environmental Chemistry*, Vol. 2 (Hutzinger, 1982).

For the present purpose, it should be mentioned that light energy and intensity and sunshine (direct or indirect) duration enhance the rates of degradation, either directly or indirectly via the formation of active species. Radiation and energy transport and their global distribution have been reviewed by Bolle (1982). In the northern temperate zones with higher concentrations of pollutants providing reactive species upon photochemical reactions, these rates may be enhanced. Although chemistry in the lower atmosphere differs over the various types of world ecosystems (Graedel, 1980) no conclusions on degradation of xenobiotics can be drawn from this.

For oxidations in water mainly reactions with RO_2 and singlet oxygen are of importance resulting from photolysis of e.g. humic materials. Thus, apart from the effect of temperature on reaction kinetics it is mainly the climatic influences on the occurrence of donors for the reactive species which are important. In specific cases the respective rate constants will show whether the process is significant ($k > 10^{-3}$ Mol sec^{-1}) (Mill, 1984).

The hydrolysis rate of chemicals usually following first-order kinetics depends directly on temperature and is of importance both in water and also in soils with interstitial soil solution. Apart from temperature, the pH of the medium plays a major role, reflecting acid- and base-catalysed and non-catalysed reactions. The finding that the non-persistent pesticide aldicarb may penetrate groundwater at low pH and sandy soil conditions (Bull *et al.*, 1970) has drawn attention to the importance of ambient pH changes resulting in half-lives from days to months for several pesticides. Many measurements exist on the influence of these variables, including structure-reactivity correlations, the latter being generally validated for temperatures between 5°C and 30°C and a pH range of 5–9 (Mill, 1980; Mabey and Mill, 1978).

For quantitatively relating hydrolysis rate with temperature the Arrhenius approximation ($k = A_e(-E_A/RT)$, where $E_A =$ activation energy and other equations are in use which all fit rather well with experimental data, although they neglect second-order temperature dependencies. As a rough approximation of the influence of temperature on hydrolysis rate in the range 0–50°C, a difference of 10 °C results in a change in k by a factor of 2.5, a 25 °C difference by a factor of 10 (Lyman *et al.*, 1982). This shows that the significance of hydrolytic transformation of chemicals, leading to derivatives with unpredicted environmentally relevant physico-chemical and toxicological properties, drastically differs between cold and hot climates. In hot climates hydrolysis thus may become a significant parameter for assessing the fate of chemicals, whereas it is less significant compared to other loss processes in temperate zones, and negligible in cold zones. For persistent chemicals, e.g. bis-chloro ethers and chlorinated aliphatics, hydrolysis seems to be the major loss process in hot (humid) climates (Suffet and Harris, 1984), although it is very slow in temperate climates.

The influence of other environmental variables catalysing or retarding hydrolysis is poorly understood. There are reports that humic materials catalyse the hydrolysis of pesticides, although retarding of the rate has also been reported. Ion strength, heavy metals content, and suspended solids also may change the rate of hydrolysis (Wolfe, 1980). For chloropyrifos a change of the hydrolysis rate by a factor of 15 has been reported. Apart from retardation by sorption and binding to humic materials and solids, the mechanisms are unknown (Suffet and Harris, 1984; Lyman *et al.*, 1982). Given the wide variation of these environmental factors in the world's ecosystems the range of their effects should be investigated and integrated into environment fate models.

Although hydrolysis is in general a transformation process only, for a number of pesticides it presents a potent detoxification mechanism and reduces their bioaccumulation potential. This has been demonstrated, e.g. for fenitrothion in an aquatic microcosm (Fisher, 1985).

Since pH values vary locally or regionally within the range of concern, pH does not appear to be an important variable within the scope of this report.

3.1.5 EXPERIMENTAL METHODS

The general methodology for studying the fate of chemicals in the environment includes investigations on their bioavailability. This methodology also includes experiments giving fragmentary information, such as transfer rates, total uptake by a species, resorption, binding to macromolecules, leaching in soil and partitioning between the environmental media.

The most comprehensive methodology has been developed for studying the environmental behaviour of radionuclides, pesticides, and heavy metals, and has been adopted for a few further persistent or hazardous groups of chemicals. Most experiments with organic chemicals are performed with labelled test chemicals to facilitate quantification and to allow for assessment of conversion products (Korte, 1980). This methodology covers a wide range of sophistication according to the specificity of problems investigated: single factor testing, multiple factor testing, and experimental models; outdoor investigations, field studies, and environmental monitoring. The respective methods have been broadly reviewed and appraised elsewhere (SCOPE 25, 1985; SCOPE 12, 1978; Nürnberg, 1985).

Given the complexity of interacting chemical and environmental factors, experiments with appropriate model ecosystems would be most helpful. A great number of terrestrial, aquatic, and mixed systems—both of laboratory and outdoor type— have been developed for studying mobility, transformation, and degradation, all three being important criteria within the area of study. A critical evaluation of feasibility and predictability of model ecosystems revealed—despite the use of labelled chemicals—common limitations with respect to number of species and environmental factors, lack of representativity of food chain organisms, and lack of equilibrium of trophic levels (lower trophic organisms are frequently consumed in too short a period, physiological conditions for component organisms are not optimal). Larger open and outdoor models may overcome these limitations but their costs are very high, and they do not provide reproducible results (SCOPE 25, 1985; IUPAC, 1985).

Soil- and plant-bound residues—a phenomenon which is not restricted to pesticides (Führ, 1984)—are of great importance and due to respective variations of the relevant environmental properties are to some extent climate-zone dependent. Published data on pesticide-bound residues reveal that their bioavailability cannot be excluded a priori (Review IUPAC, 1984), and a sequential approach for studying these residues based on their bioavailability has been proposed (Huber and Otto, 1983).

3.1.6 CONCLUSION

Standardized and internationally harmonized test methods for investigating the properties of chemicals relevant to their mobility mostly use temperate climate conditions as the standard. Partial exceptions are hydrolysis for which different temperatures are considered, and direct and indirect photolysis, which is investigated at various light intensities and duration (OECD, 1981).

To permit the extrapolation of test data to real field conditions and to generalize specific data from simulation and outdoor experiments, the real influence of the discussed environmental variables on the fate of chemicals should be known on the basis of systematic or comparative experiments. This would involve the understanding of the respective prevailing processes and thus provide the basis to improve considerably the environmental fate models and consequently the prediction of exposure to chemicals under the environmental conditions under study. A tremendous research effort for the advancement of the understanding of the behaviour of chemicals in the environment is lost worldwide by repetition of experiments, finding differences without elaborating the causes. It should be, therefore, a task for research planners to emphasize and support:

(a) systematic investigations on the influence of single climate or climate-related factors;
(b) systematic investigations on the interactions of these factors as they modify their influence;

(c) improved predictions by data interpretation and modelling, on the basis of reliable and quantified inclusion of the variations of the relevant processes.

3.1.7 REFERENCES

Akhavein, A. A., and Linscott, D. L. (1968). The dipridylium herbicides, paraquat and diquat. *Residue Revs.*, **23**, 97–145.

Anderegg, B. N., and Lichtenstein, E. P. (1984). Effects of light intensity and temperature on the uptake and metabolism of soil-applied (^{14}C)phorate by plants. *J. agric. Fd. Chem.*, **32**, 610–614.

Baker, R. D., and Applegate, H. G. (1970). Effect of temperature and ultraviolet radiation on the persistence of methyl parathion and DDT in soils. *Agric. Journal*, **62**, 509–512.

Banerjee, P., Piwoni, M. D., and Ebeid, K. (1985). Sorption of organic contaminants to a low carbon subsurface core. *Chemosphere*, **14** (8), 1057–1067.

Baughman, G. L., and Burns, L. A. (1980). Transport and transformation of chemicals: a perspective. In: Hutzinger, O. (ed.), *The Handbook of Environmental Chemistry*, Vol. 2, part A, Reactions and Processes, pp. 1–17. Springer-Verlag, Berlin, Heidelberg, New York.

Bolle, H. J. (1982). Radiation and energy transport in the earth atmosphere system. In: Hutzinger, O. (ed.), *The Handbook of Environmental Chemistry*, Vol. 1, part B, The Natural Environment and the Biogeochemical Cycles, pp. 131–303. Springer-Verlag, Berlin, Heidelberg, New York.

Bude, A. (1981). Bestimmung des quantitativen Einflusses von biologischen, physikalischen und chemischen Faktoren auf die Bioakkumulation in Fischen am Beispiel von 2,5,4′-Trichlorbiphenyl und Pentachlorphenol. Dissertation, TU Munich.

Budhewar, L. M., and Omanwar, P. K. (1980). Effect of temperature and moisture regimes on DIPA extractable zinc in some Tarai soils. *J. Indian Soc. Soil Sci.*, **28** (1), 125–128.

Buldgen, P. (1982). Features of nutrient leaching from organic soil layer microcosms of beech and spruce forests: effects of temperature and rainfall. *Oikos*, **38** (1), 99–107.

Bull, D. L., Stokes, R. A., Coppedge, J. R., and Ridgeway, R. L. (1970). Further studies on the fate of aldicarb insoil. *J. econ. Ent.*, **63**, 1283–1289.

Corwin, D. L., and Farmer, W. J. (1984). An assessment of the significant physico-chemical interactions involved in the pesticide diffusion within a pesticide-sediment-water system. *Chemosphere*, **13** (12), 1295–1317.

Edwards, C. A. (1966). Insecticide residues in soils. In: Gunther, F. A. (ed.), *Residue Reviews, Residues of Pesticides and other Foreign Chemicals in Foods and Feeds*, **13**, 83–132.

Ehlers, W., Farmer, W. J., Spencer, W. F., and Letey, J. (1969a). Lindane diffusion in soils: I. Theoretical considerations and mechanism of movement. *Soil Sci. Soc. Amer. Proc.*, **33**, 501–504.

Ehlers, W., Farmer, W. J., Spencer, W. F., and Letey, J. (1969b). Lindane diffusion in soils: II. Water content, bulk density, and temperature effects. *Soil Sci. Soc. Amer. Proc.*, **33**, 505–508.

Farmer, W. J., Igue, K., Spencer, W. F., and Martin, J. P. (1972). Volatility of organochlorine insecticides from soil: I. Effect of concentration, temperature, air flow rate, and vapour pressure. *Soil Sci. Soc. Amer. Proc.*, **36**, 443–447.

Fisher, W. S. (1985). Effects of pH upon the environmental fate of (^{14}C)fenitrothion in an aquatic microcosm. *Ecotoxic. and Environment. Safety*, **10**, 53–62.

Führ, F. (1984). Agricultural pesticide residues. In: Annunziata, M. L., and Legg, J. O. (eds), *Isotopes and Radiation in Agricultural Science*, II pp. 239–269. Academic Press, London.

Graedel, T. E. (1980). Atmospheric photochemistry. In: Hutzinger, O. (ed.), *The Handbook of Environmental Chemistry*, Vol. 2, part A, Reactions and Processes, pp. 107–143. Springer-Verlag, Berlin, Heidelberg, New York.

Harris, C. R., and Lichtenstein, E. P. (1961). Factors affecting the volatilization of insecticidal residues from soil. *J. econ. Ent.*, **54**, 1038–1045.

Hatcher, P. G., Schnitzer, M., Dennis, L. W., and Maciel, G. E. (1981). Aromaticity of humic substances in soils. *Soil Sci. Soc. Am. J.*, **45** (6), 1089–1094.

Huang, P. M. (1980). Adsorption processes in soil. In: Hutzinger, O. (ed.), *The Handbook of Environmental Chemistry*, Vol. 2, part A, Reactions and Processes, pp. 47–59. Springer-Verlag, Berlin, Heidelberg, New York.

Huber, R., and Otto, S. (1983). In: Miyamoto, J., and Kearney, P. C. (eds), *Pestic. Chem.* Vol. 3, pp. 357–362. Pergamon Press.

Hull, H. M. (1970). Leaf structure as related to absorption of pesticides and other compounds. In: Gunther, F. A. (ed.), *Residue Reviews, Residues of Pesticides and Other Foreign Chemicals in Food and Feeds*, **31**, 76–80. Springer-Verlag, Berlin, Heidelberg, New York.

Hutzinger, O. (ed.) *Handbook of Environmental Chemistry*, Vol. 2, part A (1980); Vol. 2, part B (1982). Reactions and Processes. Springer-Verlag, Berlin, Heidelberg, New York.

Igue, K., Farmer, W. J., Spencer, W. F., and Martin, J. P. (1972). Volatility of organochlorine insecticides from soil: II. Effect of relative humidity and soil water content on dieldrin volatility. *Soil Sci. Soc. Amer. Proc.,* **36**, 447–450.

IUPAC Reports on Pesticides (17) (1984), Non-extractable pesticide residues in soils and plants, *Pure & Appl. Chem.*, **56** (7), 945–956.

IUPAC Reports on Pesticides (20), Critical evaluation of model ecosystems, *Pure & Appl. Chem.*, **57** (10), 1523–1536.

Karickhoff, S. W. (1981). Semi-empirical estimation of sorption of hydrophobic pollutants on natural sediments and soils. *Chemosphere*, **10**, 833–849.

Karickhoff, S. W. (1984). Organic sorption in aquatic systems. *J. Hydraul. Enginrng.*, **110** (6), 707–735.

Klein, W. (1975). Habilitationsschrift, TU Munich.

Klein, W., Kohli, J., Weisgerber, I., and Korte, F. (1973). Fate of aldrin-^{14}C in potatoes and soil under outdoor conditions, *J. agric. Fd. Chem.*, **21**, 152.

Klein, W., Korte, F., Weisgerber, I., Kaul, R., Müller, W., and Djirsarai, A. (1968). Über den Metabolismus von Endrin, Heptachlor und Telodrin. *Qualitas Pl. Mater. veg.*, **15**, 225.

Korte, F. (ed.) (1980). *Ökologische Chemie: Grundlagen und Konzepte für die ökologische Beurteilung von Chemikalien* Thieme Verlag, Stuttgart, New York.

Kushwara, K. S., Gupta, H. C. L., and Kavadia, K. S. (1978). Effect of temperature on the degradation of aldrin residues in sandy loam soil. *Ann. Arid Zone*, **17** (2), 200–206.

Lichtenstein, E. P., Fuhremann, T. W., and Schulz, K. R. (1971). Persistence and vertical distribution of DDT, lindane and aldrin residues 10 and 15 years after a single soil application. *J. agric. Fd. Chem.*, **19**, 718–721.

Liss, H. J., and Rolston, D. E. (1983). Spatial and temporal variability of water soluble carbon for a cropped field. In: *Agrochemical-Biota Interactions in Soil and Water Using Nuclear Techniques* IAEA-TECDOC-283.

Lyman, W. J., Reehl, W. F., and Rosenblatt, D. H. (eds) (1982). *Handbook of Chemical Property Estimation Methods, Environmental Behaviour of Organic Compounds* McGraw-Hill, New York.

Mabey, W., and Mill, T. (1978). Critical review of hydrolysis of organic compounds in water under environmental conditions. *J. Phys. Chem. Ref. Data*, **7**, 383.

Mackay, D. (1980). Solubility, partition coefficients, volatility, and evaporation rates. In: Hutzinger, O. (ed.), *The Handbook of Environmental Chemistry*, Vol. 2, part

A, Reactions and Processes, pp. 31–45. Springer-Verlag, Berlin, Heidelberg, New York.

Mackay, D., and Yeun, A. T. K. (1983). Mass transfer coefficient correlations for volatilization of organic solutes from water. *Environ. Sci. Technol.*, **17**, 211–217.

Mill, T. (1980). Data needed to predict the environmental fate of organic chemicals. In: Hawue, R. (ed.), *Dynamics, Exposure and Hazard Assessment of Toxic Chemicals*. Ann Arbor Science.

Mill, T. (1984). Oxidation in water. In: *Summary Report of the Workshop on Structure Activity Concepts in Environmental Sciences*. U.S. Dept. of Commerce, Nat. Tech. Inform. Service, PB85-137239.

Mills, A. C., and, Biggar, J. W. (1969). Solubility-temperature effect on the absorption of gamma- and beta-BHC from aqueous and hexane solutions by soil materials. *Soil Sci. Soc. Amer. Proc.*, **33**, 210–216.

Morrison, I. N., and Cohen, A. S. (1980). Plant uptake, transport and metabolism. In: Hutzinger, O. (ed.), *The Handbook of Environmental Chemistry*, Vol. 2, part A: Reactions and Processes, pp. 193–219. Springer-Verlag, Berlin, Heidelberg, New York.

Nürnberg, H. W. (ed.) (1985). *Pollutants and their Ecotoxicological Significance*. John Wiley & Sons, Chichester, New York, Brisbane, Toronto, Singapore.

OECD (1981). *Guidelines for Testing of Chemicals*. OECD, Paris.

Paterson, S., and Mackay, D. (1985). The fugacity concept in environmental modelling. In: Hutzinger, O. (ed.), *The Handbook of Environmental Chemistry*, Vol. 2, part C, pp. 121–140. Springer-Verlag, Berlin, Heidelberg, New York, Toronto.

Paul, E. A. and Huang, P. M. (1980). Chemical aspects of soil. In: Hutzinger, O. (ed.), *The Handbook of Environmental Chemistry*, Vol. 1, part A: The Natural Environment and the Biogeochemical Cycles, pp. 69–86. Springer-Verlag, Berlin, Heidelberg, New York.

Roose, E. J., and Fauck, R. F. (1981). Climatic conditions set limits to the cultivation of ferrallitic soils in the humid tropical regions of Ivory Coast. *Cah. Orstom, Ser. Pedol.*, **18** (2), 153–157.

Scheunert, I., Kohli, J., Kaul, R., and Klein, W. (1977). Fate of aldrin-^{14}C in crop rotation under outdoor conditions. *Ecotoxicology and Environmental Safety*, **1**, 365–385.

SCOPE 12 (1978). *Principles of Ecotoxicology*, Butler, G. C. (ed.). John Wiley & Sons, Chichester, New York, Brisbane, Toronto, Singapore.

SCOPE 25 (1985). *Appraisal of Tests to Predict the Environmental Behaviour of Chemicals*, Sheehan, P., Korte, F., Klein, W., and Bourdeau, P. (eds). John Wiley & Sons, Chichester, New York, Brisbane, Toronto, Singapore.

Singh, G., Kathpal, T. S., Kushwaha, K. S., and Yadav, G. S. (1985). Persistence of aldrin in flooded soil under the cover of rice. *Ecotoxicol. and Environment. Safety*, **9**, 294–299.

Suffet, M., and Harris, J. (1984). Hydrolysis in aquatic systems. In: *Summary Report of the Workshop on Structure Activity Concepts in Environmental Sciences*, U.S. Dept. of Commerce, Nat. Tech. Inform. Service, PB85-137239.

Weisgerber, I., Kohli, J., Kaul, R., Klein, W., and Korte, F. (1974). Fate of aldrin-^{14}C in maize, wheat, and soils under outdoor conditions. *J. agr. Fd. Chem.*, **22**, 609.

Wolfe, N. L. (1980). Determining the role of hydrolysis in the fate of organics in natural waters. In: Haque, R. (ed.), *Dynamics, Exposure and Hazard Assessment of Toxic Chemicals*. Ann Arbor Science, Collingwood, Michigan.

Wolff, C. J. M., and van der Heijde, H. B. (1982). A model to assess the rate of evaporation of chemical compounds from surface waters. *Chemosphere*, **11** (2), 103–117.

Yaron, B., and Saltzmann, S. (1972). Influence of water and temperature on adsorption of parathion by soils. *Soil Sci. Soc. Amer. Proc.*, **36**, 583–586.

Ecotoxicology and Climate
Edited by P. Bourdeau, J. A. Haines, W. Klein and C. R. Krishna Murti

3.2 Biotic Degradation of Pollutants

F. MATSUMURA

Pollutants in the environment are exposed to various degradative forces. Among them biotic degradation, or metabolic processes, are known to play a vital role in deciding overall fates of organic pollutants. They not only contribute to the disappearance of the original form of pollutants, but also change their physico-chemical properties, and thus affect their transport and distribution behaviour among various compartments in the environments.

Most forms of living organisms are capable of directly interacting with pollutants, and some of them are capable of metabolizing even very recalcitrant pollutants. Metabolic processes found in microorganisms, plants, and animals are, however, qualitatively different.

Microorganisms are known to play major roles in metabolizing chemicals in the environment (Hill and Wright 1978; Matsumura and Krishna Murti, 1982). Their contributions to the metabolic alteration of pollutants in the environment are aided by the phenomenon that the bulk of pollutants are found in soil and aquatic sediment (excluding open ocean floors) loaded with microorganisms. Therefore, in this section, a great emphasis will be given to microbial degradation of pollutants. This does not imply that other types of living organisms do not contribute to overall metabolic activities. They indeed do play major roles. However, the emphasis of this section is on the metabolic events occurring on pollutants before they enter food chains and the effects of climate on such processes. Since most plant and animal species may be considered members of food chains, their exclusion from this section may be justified. The reader who is interested in animal and plant metabolism of pollutants is referred to more specialized textbooks (e.g., Matsumura and Krishna Murti 1982).

3.2.1 CHARACTERISTICS OF MICROBIAL METABOLISM

It was originally assumed by many pesticide scientists that the patterns of microbial metabolism were in general very similar to the ones already found in animals, particularly the mammalian species, as studies on microbial metabolism of pesticides were lagging far behind comparable studies in mammalian species. However, as knowledge on microbial degradation has advanced, it has become apparent that in many cases the patterns of degradation in these two different groups of organisms are often very different.

First of all, the purpose of all metabolic reactions on xenobiotics in higher animals is to eventually convert them into polar and, therefore, excretable forms. Second, in higher animals the processes of primary metabolism of xenobiotics are centralized in a few specialized organs. In the case of the liver, its metabolic pattern is largely determined by the activity of an oxidative detoxification system, generally termed mixed-function oxidase.

On the contrary, the predominant metabolic activities in the microbial world are meant for production of energy. In this respect, it is not even possible to define xenobiotics here, since most organic materials can serve as the source of energy to at least some microorganisms. Hence only a few groups of chemicals may be regarded as foreign to microorganisms. Among insecticidal compounds, halogen-containing chemicals, particularly halogenated aromatics, must be regarded as generally foreign (or unusable) material to microorganisms.

Another characteristic of microbial metabolism is the adaptability of microorganisms to changing environments through mutation and induction, particularly toward chemicals that are initially toxic to them. The case of penicillin resistance in bacteria through induction of penicillinase is well known.

The metabolic activities of microorganisms encompass many different types of biological processes not found in any other organism. They include fermentation, some anaerobic metabolism, chemolithotrophic metabolism, and metabolism through exoenzymes.

In general, microbial contributions to metabolic alteration of insecticides may be classified in several categories, as shown in Table 3.2.1. It must be stressed that the main purpose of such a classification effort is to present clearly the types of microbial degradation according to their final manifestations. They are not classified according to the intrinsic mechanisms by which they degrade pesticides. Various reactions which involve different enzymatic mechanisms and yet are known to behave in similar patterns have been grouped together.

3.2.2 WIDE-SPECTRUM METABOLISM: INCIDENTAL METABOLISM

The key characteristic of incidental metabolism is that the pesticides themselves, including any part of the pesticide molecules, do not serve as the energy and carbon source for the microorganisms. Therefore, addition of pesticides does not affect their growth, which is always controlled by other nutrients. Two completely different subgroups of metabolic activities are present in this type. In the first subgroup, pesticides are degraded by enzymes which are not specifically related to pesticidal molecules (Table 3.2.1, categories a and b).

Table 3.2.1 General classification of microbial metabolism of pesticides. Modified from Matsumura and Krishna Murti (1983)

ENZYMATIC
(A) Incidental metabolism: pesticides themselves cannot serve as energy sources
  (i) Wide-spectrum metabolism: metabolism by generally available enzymes
    (a) Metabolism due to generally present broad-spectrum enzymes (hydrolases, oxidases, etc.)
    (b) Metabolism due to specific enzymes present in many microbe species
  (ii) Analogue metabolism (co-metabolism)
    (c) Metabolism by enzymes utilizing substrates structurally similar to pesticides
(B) Catabolism: pesticides serve as energy sources
    (d) Pesticides or a part of the molecule are the readily available source of energy for microbes
    (e) Pesticides are not readily utilized. Some specific enzymes must be induced.
(C) Detoxification metabolism
    (f) Metabolism by resistant microbes

NON-ENZYMATIC
(A) Participation in photochemical reactions
(B) Contribution through pH and other physico-chemical changes of microenvironment
(C) Through production of organic and inorganic reactants
(D) Through production of cofactors

In these cases pesticides are degraded by either broad-spectrum enzymes, such as hydrolases, reductases, and oxidases, or by specific enzymes commonly present in high percentages of microorganisms. In either case the pesticidal substrates are metabolized as a result of general microbial activities. In the environment, therefore, those types of reactions are observed in which microbial activities are stimulated by the availability of nutrients and moisture, and by the right temperature and pH.

3.2.3 ANALOGUE METABOLISM: INCIDENTAL METABOLISM

The term co-metabolism has been used in the past to include all types of metabolism in categories a, b, and c (Table 3.2.1). However, in this section I have decided to apply the term in a more restricted manner to include only the cases where the microorganisms are induced by chemicals which structurally resemble the pesticide molecules. Thus, to avoid confusion, I shall use the term analogue-induced metabolism here. In such a case the microorganisms can grow on the given chemical (analogue) but not on the pesticide, despite their capability to at least partially metabolize the latter. Good examples are those produced by Focht and Alexander (1970a, 1970b), who obtained DDT-degrading *Hydrogenomonas* sp. (*Pseudomonas* sp.) by using diphenylmethane as a carbon source, and those produced by Ahmed and Focht (1972) and Furukawa and Matsumura (1975), who selected PCB-metabolizing microorganisms by using biphenyl as a carbon source. The important feature distinguishing this type from other incidental metabolisms is that these microorganisms are purposely selected by non-pesticidal analogues. The enzyme systems induced do not initially recognize the difference in the substrate molecules and, therefore, partially degrade the pesticide molecules. However, in such cases the microorganisms are incapable of completing the metabolism necessary to receive energy for growth.

3.2.4 CATABOLISM

In catabolism microorganisms are capable of deriving energy from the pesticide molecules and, therefore, can grow on them. I include here cases in which only a part of the pesticide molecule is utilized for growth from a practical standpoint. Thus a complete mineralization of the pesticide is not the absolute requirement so long as growth is observed by using the pesticide as a sole carbon source.

3.2.5 NON-ENZYMATIC PROCESSES

The processes by which microbial activities contribute to the overall alteration of insecticidal molecules by non-enzymatic mechanisms are less well studied than those involving enzymatic reactions. It is known that some pesticidal chemicals can be photochemically altered in the environment, and microbial products can promote photochemical reactions in two ways. First, microbial products may act as photosensitizers by absorbing the energy from light and transmitting it to the insecticidal molecule. We have been able to show, for instance, that an aqueous extract from heat-sterilized blue-green algal cultures promoted photochemical degradation of DDT (Esaac and Matsumura, 1980). Another way that microbial products can facilitate such photochemical reactions is to serve as donors or acceptors of electrons and/or reacting groups of chemicals, for example, hydrogen and OH-, which are often needed for photochemical reactions.

Recently, Esaac and Matsumura (1980) demonstrated that ferridoxin and flavoproteins isolated from algae are powerful photosensitizers. These are known to play important roles in electron transfer systems in algae. Since they are quite stable molecules, it would not be surprising if they persist long enough in the environment after the death and lysing of algae cells to become a factor in pesticide degradation.

The effect of pH is often neglected in the field of pesticide metabolism despite numerous reports on the pH-dependent reactions of relatively labile molecules, both in soil and *in vitro*. Large pH changes are often associated with microbial activities, together with changes in nutritional sources, particularly in aqueous media. Initially degradation of proteins causes alkaline pHs, and with carbohydrate metabolism the pH becomes acidic. While the actual occurrence of microbial pH effects in nature might be difficult to document, it is certainly easy to demonstrate the phenomenon *in vitro*, where during the decay period the pH of spent culture media often becomes very low. Incubation of labile insecticides, such as tetraethyl pyrophospate (TEPP), with such spent medium under sterile conditions would certainly cause breakdown of the insecticides.

Little attention has been paid so far to the importance of the microbial formation of organic products capable of reacting with pesticides. Such reactants of microbial origin can be postulated to include amino acids, peptides, alkylating agents such as acetyl CoA, methylcobalamine, adeosylmethionine, and organic acids.

3.2.6 GENERAL EFFECTS OF ENVIRONMENTAL CONDITIONS ON SOIL MICROBIAL ACTIVITIES

Many environmental factors are known to affect the rate of metabolic conversion of pollutants. In the case of soil microbial activities, water content of soil greatly affects and in extreme cases limits the type of microbial activities. Waterlogged conditions in combination with high nutrients could promote growth of anaerobic microbial species. Extremely dry conditions limit microbial activities. The tendency is greatest with sandy soil, with least water-holding capacity among various soil types. In very dry conditions *Aspergillus* and *Penicillium* species show abilities to persist, while streptomycetes show moderate degrees of tolerance to dryness. Bacteria in general require high moisture, though some (e.g. *Bacillus* spp.) are capable of surviving dry periods in the form of spores.

Acidity also affects soil fauna. In general acidity depresses the growth of bacteria more than that of fungi. For instance, below pH 5 bacterial activities are usually very low.

Soil is generally considered to be nutrient deficient as a normal state (with some exceptions, such as forest soil). Therefore, addition of exogenous nutrients causes a sudden burst of microbial activities. Yet, increase in total microbial activities alone may not result immediately in increased metabolism of pollutants as will be explained later.

Temperature is the major contributing factor in deciding the characteristics of soil biota. Each microbial species has a definite range of temperature preference for significant activity. Thus composition of active microbial species varies in a given soil as temperature changes. For instance, Okafor (1966) has found that chitin metabolizing organisms in tropical soil at 28–30°C were predominantly actinomycetes, nematodes, and protozoans, while at 2–15°C, fungi and bacteria were more important. It must be emphasized, however, that higher temperature does not automatically guarantee higher microbial populations and activities. Certain microorganisms are well adapted to colder climates and as such would flourish even at seemingly low temperatures. In temperate zones with moderate to severe seasonal variations, for instance, it is common to observe maximum microbial activities in spring and fall rather than during the summer, even where adequate moisture is maintained throughout the year.

Other determinants, such as redox potential, nutrients, and physico-chemical characteristics of soil, also play very important roles. However, they play similar roles both in the tropics and temperate zones. Among these factors there are certain known joint actions. For example, the availability of nutrients with water often creates low redox potential, and at the same time, in those environments where nutrients and water are constantly available, the soil characteristics become organic.

In arid environments microbial activities are naturally low, even with irrigation, since soil lacks organic constituents. Under such conditions organic pollutants tend to vertically migrate to lower layers because of low holding capacity of sandy soil, where microbial activities are low, causing groundwater contamination where there is little microbial action.

3.2.7 COMPARISON OF PESTICIDE DEGRADATION PATTERNS BETWEEN TEMPERATE AND TROPICAL REGIONS

Despite the wealth of information on metabolic degradation of organic chemicals, mostly in temperate zones and to a lesser extent in tropical and subtropical regions, there is a striking absence of actual scientific data to indicate comparative differences in patterns of microbial degradation of organic chemicals. There are a number of inferences, suggestions, and generalizations regarding higher levels of microbial activities in tropical environments as compared to temperate areas and hence faster disappearance of organic chemicals in the former environments. Yet, there is little evidence to prove such a point.

In Table 3.2.2 I have summarized some of the pesticide residue data gathered for registration purposes in the USA under the IR-4 Program (Interregional Project No. 4). These are the cases where a given pesticide is applied to an identical crop under an identical application protocol in two or three regions with different climatic conditions within the USA. The resulting residues in the crop should, therefore, indicate the relative levels of pesticides remaining in the soil.

In the case of iprodione application on broccoli there was a striking difference between north and south in its residue levels. In Michigan, which is a northern state, the residue levels in broccoli were much higher than those found in Georgia despite the fact that iprodione was applied for fewer days. Yet, the tendency is not consistent. In the case of permethrin application on cantaloupe, the highest residue level was found in those samples from Texas at 0.4 pounds per acre (450 gm/ha). Under identical test conditions cantaloupe from Indiana, which is a northern state, showed a lower residue level. The fact that the same tendency is observed at 0.2 pounds per acre shows that this tendency is not a coincidence.

Table 3.2.2 Comparative residue data from some pesticide-crop combinations among various regions of the United States

Project Number	Pesticide	Commodity	State	PHI day	RATE lb/A	RESIDUE ppm	AVE. ppm
1545	Chlorpyrifos	Cantaloupe	IN	31	2.0	ND	<0.01
			TX	35	4.0	ND	<0.01
			CA	34	2.0	ND.-0.11	0.08
1545	Chlorpyrifos	Summer squash	IN	31	2.0	ND	<0.05
			CA	31	2.0	ND	<0.05
2194	Iprodione	Broccoli	MI	11	1.0	1.77–6.15	4.30
				11	2.0	3.41–11.94	6.81
			GA	17	1.0	0.03–0.06	0.04
				17	2.0	0.05–0.08	0.06
			CA	30	1.0	0.03–0.06	0.04
				30	2.0	0.03–0.05	0.04
1730	Permethrin	Cantaloupe	IN	1	0.2	0.26–0.57	0.39
				3	0.2	0.22–0.36	0.31
				5	0.2	0.17–0.49	0.29
				1	0.4	0.20–0.77	0.50
				3	0.4	0.57–1.42	1.00
				5	0.4	0.28–0.66	0.47
			TX	1	0.2	0.70–0.95	0.80
				3	0.2	0.30–1.31	0.91
				5	0.2	0.48–0.73	0.63
				1	0.4	1.51–1.92	1.73
				3	0.4	0.77–2.55	1.48
				5	0.4	1.26–2.67	1.75
			CA	1	0.2	0.13–0.41	0.22
				4	0.2	ND–0.35	0.17
				6	0.2	0.10–0.14	0.11
			FL	1	0.2	0.06–0.26	0.13
				3	0.2	0.11–0.24	0.16
				5	0.2	0.60–0.15	0.12
				1	0.4	0.39–0.61	0.48
				3	0.4	0.23–0.32	0.27
				5	0.4	0.17–0.41	0.27
1886	Monitor	Endive	NY	28	1.0	0.08–0.60	0.24
			FL	28	1.0	0.56–1.30	0.99

IR-4 pre-registration data (1984) for EPA. Compiled by Dr. R.A. Leavitt, Pesticide Research Center, Michigan State University, East Lansing, Michigan.

Since in these studies protocols in the field experiments as well as for the analytical procedures are rigidly defined and the quality control is achieved via sample exchanges, a direct comparison of the residue levels, and hence the probable rates of pesticide disappearance is possible. Upon examination of the entire data presented in Table 3.2.2, one is forced to conclude that there is no consistent trend to indicate that the disappearance of pesticides in one region is faster than that in other regions.

The foregoing discussion illustrates the difficulties and the problems associated with generalizing certain trends in microbial degradation of pollutants. The rates by which organic substances are degraded in soil and aquatic environments are governed by many complex environmental factors. Temperature, exclusive of extremes, is only one of these rate-determining factors.

Having failed to demonstrate simple trends among microbial degradation patterns in different climates, a question may be raised whether there could be other ways to approach the subject area. In the absence of truly comparable data among various climatic zones, one is forced to examine the data addressing seasonal variations in disappearance of organic substances (e.g. pesticides) in the same locality or experimental plot. This approach has merit in that the basic soil characteristics during the experimental period may be assumed to be relatively constant. The shortcomings of this approach are: (a) often temperature changes come with other climatic changes, such as humidity (e.g. rainy season), day length, and nutritional inputs; and (b) the findings from such studies may not be applicable where seasonal variation in climatic conditions is minimal.

Telekar *et al.* (1977; 1983a; 1983b) have studied persistence of several insecticides in southern Taiwan for several seasons. In the first study, DDT, dieldrin, fonofos, phorate, and carbofuran were applied at 5 kg/ha or at 10 kg/ha on 10 m × 10 m plots four times (on 9 December, 1974, 15 April, 1975, 30 October, and 26 April, 1976) and the residue levels in soil were monitored. They found that for all compounds the residue levels declined rapidly during May through August (summer months) for two successive years, while they increased during winter months. The summer months in Taiwan also represent the wet season, and, therefore, the seasonal variation may be a result of the combined effects of temperature and humidity. Also, it must be noted that in the case of carbofuran, phorate, and fonofos their degradation may have been aided by the alkalinity of the soil (pH 8.5–8.7). On the other hand, for these chemicals there was a trend of soil acclimatization leading to faster rate of disappearance of the parent compounds in the second season than at the initial time of application, indicating microbial participation in the process of pesticide degradation. An interesting observation was that under the application rates neither DDT nor dieldrin showed a tendency to accumulate over the test period. These authors have calculated that the time required for 50% disappearance of DDT and dieldrin during the hot and rainy season was 3.5 and 3 months respectively, which are considerably shorter times than those reported from temperate zones.

A similar pattern was observed for fenvalerate, a pyrethroid insecticide (Telekar *et al.*, 1983a) applied at 1 kg/ha five times in two years. After 2.5 years only 2% of the applied fenvalerate could be recovered. The pesticide residues declined rapidly during the hot and rainy months.

In the second study Telekar *et al.* (1983b) applied heptachlor, triazophos, and fenvalerate biweekly for a two-year period to small plots (5 × 4 m) with three different soil types. Residue patterns were monitored for three seasons. Again, the pattern of seasonal variations in residue levels was identical to the one observed in the first two studies. Comparing these data to those obtained in temperate regions these authors concluded that pesticides in general disappear faster in subtropical and tropical environments.

The above studies have clearly demonstrated that under their experimental conditions all pesticides tested degraded faster at high temperature and humidity. Care must be taken, however, not to generalize that pesticide degradation is always high at higher temperatures. In many parts of the world, microbial activities are known to decline during hot summer months and increase during cool spring and fall months. The tendency is particularly pronounced in areas where summer is the dry period of the year.

Microbial degradation of pollutants and pesticides in soil is influenced by many environmental factors. Thus there is a danger in simply assuming that a given compound would degrade faster under hot tropical conditions than it would under a cooler temperate environment. On the other hand, there are a few practical considerations which could help the reader to judge the conditions which favour or disfavour microbial degradation of organics in non-temperate climates.

First, microbial degradation of organics decreases dramatically below 10°C and practically ceases to operate at temperatures below 5°C in all environments. Second, there are much longer growing seasons in warmer climates than in colder areas. And third, high microbial activities are always associated with high moisture. Thus, microbial contributions to degradation of chemicals in the environment would be negligible in periods or areas where temperature stays below 5°C, or in places receiving very little water through rain, flooding, or irrigation. Therefore, at least in those extreme environments, microbial degradation of pollutants would be minimal.

In tropical zones the intensity of sunlight is higher, in addition to higher average temperatures throughout the year. Knowing that many pollutants are degraded by combinations of biotic and abiotic factors (*see* Section 3.1), these two factors would a priori promote overall reaction processes of pollutant degradation.

There are a number of examples indicating that the overall carbon utilization rates per year by soil microorganisms are higher in tropical zones compared to temperate zones. A good example is the case of coniferous forest litter, which tends to accumulate up to a few years in the extreme north, and to disappear in a few months in the south. As I indicated at the beginning of this chapter, some types of pollutant degradation are closely associated with overall microbial activities and/or biomass (e.g. wide-spectrum metabolism). Therefore, those types of metabolic activities are expected to be high in the south, when the rate of disappearance is considered on an annual basis (i.e. instead of comparing a given time period within a growing period of the year between north and south).

Another way to assess the overall differences between north and south in the rate of disappearance of pesticides is to examine dosages or frequency of pesticide application for comparable crops and pests. Certainly this approach is not foolproof as the variety of crops, even if the species is the same, as well as pest species or strains may be quite different in the south from those in the north. Also, such an approach does not take into account the role of abiotic factors as explained before. Nevertheless, taken together with other observations such an approach could be used as secondary evidence to indicate the north–south difference in this regard.

3.2.8 CONCLUSION

From the above discussion on relative microbial contribution on pesticide degradation between tropical and temperate zones, it has become evident that there is no simple straightforward trend that could be used to assess regional differences due to differences in temperature and associated climatic factors.

Only under extreme conditions, such as temperatures below 5 °C, low humidity, lack of organic matter, etc., may a low rate of pesticide degradation be predicted. While there are some supporting factors (such as longer growing season, etc.) to the generally held belief that pesticides degrade faster in tropical zones than in temperate regions, one must be cautious in extending such circumstantial evidence to predicting pesticide degradation rates in any region. After all, high DDT residues are found in food chains and in man in India, where a short half-life of DDT had been predicted because of the country's predominantly tropical climate. Such an observation points to the need for cautious approaches in assessing pollutants' fates in the environment in non-temperate zones. Thus the only sensible approach is to assess residue characteristics and fate for each pollutant in each locality where pollution is suspected, until a reliable data base has been accumulated. One helpful aspect in this regard has been that microbial degradation pathways and metabolic products formed in tropical regions appear to be qualitatively identical or very similar to those found in temperate zones (Sethunathan *et al.*, 1982), suggesting involvement of similar metabolic activities among different zones and regions in this regard. Thus there is a good possibility of developing logical approaches to assessing basic mechanisms of pollutant degradation by microorganisms in each locality by applying basic knowledge accumulated in this field in the last few decades.

3.2.9 REFERENCES

Ahmed, M. K., and Focht, D. D. (1972). Degradation of polychlorinated biphenyls by two species of *Achromobacter*. *Can. J. Microbiol*, **19**; 47.

Esaac, E. G., and Matsumura, F. (1980). Metabolism of insecticides by reductive systems. *Pharmacol. Ther.* **9**, 1–35.

Focht, D. D., and Alexander, M. (1970a). DDT metabolites and analogues by *Hydrogenomonas* sp. *J. Agric. Fd. Chem.*, **19**, 20.

Furukawa, K., and Matsumura, F. (1975). Microbial metabolism of PCBs. *J. Agric. Fd. Chem.*, **42**, 543.

Hill, I. R., and Wright, S. J. L. (eds) (1978). *Pesticide Microbiology*. Academic Press, New York.

Matsumura, F., and Krishna Murti, C. R. (eds) (1982). *Biodegradation of Pesticides*. Plenum Press, New York.

Okafor, N. (1966). Ecology of microorganisms on chitin buried in soil. *J. Gen. Microbiol*, **44**, 311–327.

Sethunathan, N., Adhya, T. K., and Raghu, K. (1982). Microbial degradation of pesticides in tropical soils. In: Matsumura, F., and Krishna Murti, C. R., (eds), *Bidegradation of Pesticides*, pp. 91–116. Plenum Press, New York.

Telekar, N. S., Sun, L. T., Lee, E. M., and Chen, J. S. (1977). Persistence of some insecticides in subtropical soil. *J. Agr. Fd. Chem.*, **25**, 348–352.

Telekar, N. S., Chen, J. S., and Kao, H. T. (1983a). Persistence of fenvalerate in subtropical soil. *J. Econ. Entomol.*, **76**, 223–226.

Telekar, N. S., Kao, H. T., and Chen, J. S. (1983b). Persistence of selected insecticides in subtropical soil after repeated biweekly applications over two years. *J. Econ. Entomol.*, **77**, 711–716.

Ecotoxicology and Climate
Edited by P. Bourdeau, J. A. Haines, W. Klein and C. R. Krishna Murti

3.3 Chemicals in Cold Environments

D. R. MILLER

3.3.1. INTRODUCTION

This section has been prepared using primarily Canadian materials and references. Thus, although the title refers to 'cold environments', most of our attention will be devoted to northern circumpolar regions, i.e. to the Arctic, specifically the Canadian Arctic. Naturally, it is expected that much of the material will be relevant to other circumpolar countries and to alpine regions. Comparisons to such other areas will be drawn from time to time in the text.

3.3.2 BACKGROUND: GEOLOGICAL HISTORY

On a geological time-scale, the arctic areas have not been stable, but have been subjected to migration of continental land masses and also to vast fluctuations in their ambient climate. The same has been true, indeed, for most regions of the planet; however, it is essential to take note of some of these changes for arctic regions in particular since the ecosystems involved take a great deal of time to recover their 'stable' state after major perturbations. It has been suggested that only rarely in history have they been able to enjoy long periods of stability, free from the profound effects of major climatic variations and continental drift.

The lands now surrounding the northern Polar Sea did not, of course, originate there, but formed part of a large land mass, now referred to as Pangea, lying largely in temperate and subtropical regions. It was scarcely 200 million years ago that the mid-ocean ridges opened and the continents moved toward higher latitudes. The North Atlantic Ocean opened, separating North America from Europe only at the end of the Cretaceous period, some 80 million years ago, and the Labrador Sea opened about 50 million years ago. The Greenland Sea appears to have opened a mere 1.5 milion years ago.

Ice ages may not seem relevant to the high Arctic, but the surrounding regions, sources of inward migration of species of all kinds, have been profoundly affected many times in the past 200 000 years. The most recent, a period of extreme glaciation between 20 000 and 10 000 years ago, established the present physiognomy of the Arctic. Its lands, soils and ecosystems are actually quite young, in spite of the fact that in geological terms the underlying rock, particularly in the Canadian Shield, is some of the oldest exposed rock on the planet.

3.3.3 GEOGRAPHICAL AND PHYSICAL CHARACTERISTICS

The North Pole is not, of course, located on land. It lies in a large central ocean of more than 9.5 million km^2, including several basins of depths of up to 5000 m, surrounded by large, relatively shallow epicontinental coastal seas. These seas cover almost 36% of the area, but hold only about a fiftieth of the water. These seas are, furthermore, in relatively poor contact with the worldwide ocean system. The channels connecting the Arctic Sea with more southerly waters are narrow or shallow, with the exception of the Fram Strait, between Greenland and Spitzbergen, which accounts for three-quarters of the flow both in and out of the Arctic Sea.

The main flow is outward; although an influx of (largely surface) waters does take place, the inflow of water from the sizeable northern rivers makes the water balance significantly positive (the northern rivers input about 90 000 m^3/s, which is about half the flow of the Amazon).

This combination of a deep central sea and positive water balance leads to a most important consideration for the Arctic, namely that water circulates in the Polar Sea for a significant time before leaving. This means that an input of heat from the warmer oceans hardly exists at all; it also means that any chemical contaminant has ample opportunity to circulate to all countries bordering on the Arctic Ocean before it can be expected to leave the region. It is very much a situation where the several countries involved have to live with each others' pollution, and for a long period of time at that.

On the lands surrounding the sea, particular conditions must also be considered. The dominant consideration is probably the permafrost, by which we describe the phenomenon of frozen layers of ground, ranging sometimes many metres in depth. This phenomenon makes our conventional ideas of groundwater transport and drainage totally inappropriate. In large regions of the North, there simply is no subsurface water flow, and this in turn must remove from our consideration any thoughts about the earth itself as removing toxic chemicals from areas of immediate concern, as happens to a very large extent in warmer regions.

The other consideration is that rivers flowing from south to north tend to melt at their headwaters earlier in the spring than at their downstream portions. The result is that extensive flooding routinely takes place along the length of these rivers, further complicating the question of environmental behaviour of chemicals and, in fact, making all sorts of activities, including scientific research, immensely difficult.

Finally, we should note that, perhaps surprisingly to many, most northern regions have quite a low level of precipitation. Rey (1983) has pointed out that Phoenix, Arizona has more than three times the average annual precipitation of the interior of Greenland. Although the flow from the rivers draining northward is substantial, most of the water comes from much further South. The amount of water available for a general irrigation of the vast expanses of the North is very small. This also has an effect on the behaviour of pollutants, by making a 'flushing' or 'leaching' phenomenon relatively less important in the process of pollutant or chemical removal.

3.3.4 GEOGRAPHICAL EXTENT

The phrase 'cold environment' is hard to give a definite meaning to, and each of the definitions that might be considered has its shortcomings. There seems little point in using an arbitrary geographical definition, such as all area north of the Arctic Circle or any other latitude, since there is so much variation in climate at the same latitude around the world. In the North Atlantic year-round navigation is possible up to about 75°N, whereas the second coldest capital city in the world is reputed to be Ottawa, Canada (the coldest is Ulan Bator), which lies at approximately 45°N.

Other definitions may be suggested, such as one depending on average annual temperature, maximum yearly temperature, or a temperature-related characteristic, such as the existence of permafrost or, less severely, the guaranteed existence of a month or more of freezing weather every year. Alternatively, an ecological feature could be used, such as defining cold regions as those beyond the tree line. The latter has the advantage of being applicable also to alpine regions.

The author believes that an appropriate definition should be made in the context of whatever feature is being discussed. In what follows, we shall try to make clear what we mean by 'how cold is cold' in the contexts of the various sections, rather than ask the reader to adopt a rigid definition for the entire work. Suffice it to say we regard a region as cold if it is influenced by one or more of the factors we have mentioned.

3.3.5 POLLUTANT TRANSPORT

The entry and initial distribution of chemical pollutants in northern systems also has some features which, although not totally unique, are sufficiently different from temperate and tropical systems to be worthy of examination.

To begin with, northern climates often have fixed or prevailing weather patterns. Meteorologists are quite familiar with the Aleutian Low, the Icelandic Low, and the Canadian and Asiatic Highs. These patterns persist for lengthy periods, providing more or less fixed pathways for airborne contaminants into the northern regions. The result, of course, is that diffusion over large areas is a more limited phenomenon than in temperate regions, and local concentrations of any given chemical may, therefore, be noticeably higher than if it were evenly distributed over the area into which it was introduced.

Another feature of these weather patterns and the air flows to which they give rise is that, as already noted, there is very little precipitation involved. Since many pollutants are removed from air masses primarily by precipitation events, this means that chemicals may be transported for surprisingly long distances before being deposited, and furthermore the low temperatures involved help the chemical to resist degradation to perhaps more innocuous forms.

These general observations lead to predictions that have been clearly verified. Although there is still some dispute about the effectiveness of acid rain at destroying ecosystems, there is no question that it is there. Large quantities of acidic material have been observed in northern regions farther from civilization than would be expected. Many lakes in Canada, Sweden and Norway have been seriously affected by acidic material that is known to originate in regions further south, in the industrial areas of southern Canada and the United States, and in the power generating stations and steel mills of the Ruhr valley and Eastern Europe.

It is interesting to note that the industrial areas of eastern North America, and Japan and Korea, do not seem to contribute to an international problem. This appears to be because they produce pollutants which quickly move over oceans, and the relatively high precipitation rates remove most of the noxious materials. An interesting exception appears to be mercury, for which there is increasing evidence that it travels through the atmosphere in the form of elemental mercury vapour rather than associated with particulate matter, and is hence less easily removed by wash-out. The finding of significant quantities of mercury in the Greenland ice cap lends support to this concept of its transport.

The transport over long distances in the northern regions is not simply due to the general northeasterly drift of air masses in the northern temperate and sub-arctic zones. The stationary weather patterns in the North are now known to provide an actual pathway from European and western USSR sources northward near Novaya Zemlya, over the Arctic Ocean and then westerly to northern Canada and the northern parts of Scandinavia and Greenland. Rey (1983) points out that these aerosols have been carefully examined and are known to be of continental origin because of their ^{210}Pb content; they are known to be anthropogenic because of the differences in the vanadium : aluminium ratio to that in the earth's crust (this ratio is widely used in tracing the sources of airborne contaminants). Furthermore, examination of the ageing of these aerosols by measuring their $SO_4 : V$ ratio shows that they have been airborne for periods of more than 2 weeks (generally, a precipitation event will remove an atmospheric aerosol within 4 or 5 days).

An example of a location affected by such a pattern is Barrow, Alaska: here winds from the Pacific are essentially unpolluted, but northeasterly winds can carry SO_4 at a level of 2 $\mu g/m^3$, a level which would be regarded as worthy of concern in a highly industrialized southern city. We might add that the phenomenon is undoubtedly not limited to Barrow; if sufficient analytical facilities were available, the same situation would surely be found over a wide area of the northern land masses.

Not all movement of pollutants through the atmosphere takes place in a single step, since not all deposition is the end event of a geographical movement process. When relatively volatile substances are deposited, e.g. substances like mercury or chlorinated hydrocarbons, the deposition is often quickly followed by a volatilization and re-suspension. Such substances may make many such short trips before they are removed from the system, either by being deposited into the ocean or by being firmly bound to the surface of the earth—e.g. by being taken up by some kind of biota. As this multiple-step process moves a substance toward colder regions, of course, the re-suspension through volatilization becomes less effective simply because of the lower temperatures encountered. The conclusion is clearly that the northern, colder regions will by diffusion alone become a sink for many pollutants, and again this has been verified by the discovery of relatively large concentrations of such chemicals as DDT in Arctic and Antarctic regions.

3.3.6 DEPOSITION AND SUBSEQUENT MOVEMENT

As pointed out previously, one of the characteristics of low-temperature regions is a shortage of rainfall, and this means that compared to temperate climates a larger fraction of the deposition will be so-called dry deposition, a general settling of dust and small aerosol particles rather than on droplets of rain. The consequence of this is that there is less tendency for pollutants to soak into the soil and be chelated or leached away; direct exposure to plants in particular, through leaves and green surfaces generally, is thereby increased and with it the expectation of direct damage.

That part of the pollutant load which comes directly to the soil will be characterized by a long residence time, both because of little subsequent leaching by rainfall, and because of chemical stability at low temperatures. Since the chemical will also remain close to the surface of the ground, it also follows that it will be more available for uptake by burrowing or browsing animals of all kinds, including insects and birds; the pollutant that is so deposited will, in other words, be more immediately incorporated into the food chain for subsequent bioaccumulation.

Movement in the environment will, of course, take place, but again in a special way which implies that cold environments are in a sense more likely to be damaged by a pollutant burden. This is because movement is most likely to take place suddenly, at a particular time of year, namely at the time of spring thawing and runoff. At this time, there may be extensive flooding because of blocked drainage of the rivers, so the land which has been dry most of the year and on which there has been an almost continuous input of pollutants will suddenly be flushed of its burden. The water which carries out this process will necessarily end up with a large chemical concentration.

In addition, another phenomenon is at work. Such snow and ice as exist on the land surface, and on the surfaces of rivers, will also have a pollutant burden stored from the previous months, and the sudden springtime melting will release this into the waterways and lakes as a sudden large input rather than as a uniform amount spread over the year. Short-term concentrations of pollutants in the spring will, therefore, be noticeably higher than in similar environments without the spring-melt phenomenon.

As if further aggravation of the situation were needed, we have to say that there is yet another consideration. This is the observation that the elevated pollutant concentrations of the spring runoff will typically occur when aquatic organisms are just hatching out or otherwise at their period of maximum growth rate. This in general will mean that they are more sensitive to effects of pollutants than they will be at a larger, older stage. Cases exist where it has been shown quite clearly that fish in a polluted lake appear to be quite healthy, but the population can still be seen to vanish; the explanation has been given that a young generation is not surviving to replace the older adults.

3.3.7 REFERENCES

Rey, L. (ed.) (1983). Arctic Energy Resources: *Proc. of the Comité arctique internat. Conf. on Arctic Energy Resources*, Oslo, Norway, Sept. 22–24, 1982. Elsevier, New York.

Ecotoxicology and Climate
Edited by P. Bourdeau, J. A. Haines, W. Klein and C. R. Krishna Murti

3.4 Chemicals in Tropical and Arid Regions

C. R. KRISHNA MURTI

3.4.1 INTRODUCTION

The aim of this work on ecotoxicology and climate is to review the relevance of the principles of ecotoxicology enunciated in the Seventies, based on experience gained largely in temperate regions, to the situation posed by environmental chemicals in regions which have tropical, semi-tropical, arid, and arctic climates. The conceptual approach to the present study leans heavily on the following premises:

(i) temperature and humidity are two of the well-known climatic factors that significantly affect the nature and course of chemical reactions and, in particular, the interactions between the biotic and abiotic components of the environment;

(ii) the ecosystems associated with the tropical, arid, and subpolar climates may present facets of sensitivity and response to the impact of chemicals that are quite different or distinct from those encountered in temperate climates.

While the basis for (i) is embodied in well-established laws of physical chemistry, the validity of the underlying assumption made in statement (ii) requires more quantifiable information on ecotoxic effects as well as pathways of chemicals in the diverse ecosystems sustained or surviving under tropical, arid, or subpolar conditions. Production and consumption data on chemicals, trends of their use pattern, and existing levels of chemical pollution in these regions are urgently needed to make meaningful estimates of exposures. Compilation of the relevant data is by no means an easy task. First of all, one has to face the paucity of reliable information pertaining to the use of chemicals in the traditional or even the newly emerging community of nations in the tropical and arid zones. The information presented in this section is mostly derived from sources in India and some of the tropical and semi-tropical countries of Asia and Latin America. It is hoped that this overview, made within the above constraints, will at least help in identifying the gaps for further studies.

The data from the Indian subcontinent are also useful for constructing models for comparative effects, given the fact that there are regions in India with widely varying temperate, tropical, and arid types of climates.

3.4.2 SOME ECOTOXIC PROBLEMS OF GLOBAL DIMENSION AND CLIMATIC CHANGES

Many ecotoxic problems transcend geographical and climatic barriers. Indeed the transfrontier migration of some chemicals has been implicated in a few ecotoxic problems of a global dimension. That chemicals do not recognize geographic frontiers is evident from the adverse effects of acid rain in the subpolar regions of North America and subarctic Europe as a result of the export of the oxides of sulphur and nitrogen from their origin elsewhere in a temperate region. Two other problems of atmospheric chemistry of current concern are the depletion of ozone and the accumulation of carbon dioxide, both connected with the fast rate of deforestation in the last three decades. The complexity of the underlying chemical reactions is yet to be resolved. Human intervention unquestionably accounts largely for the extensive denudation of tropical forests (Lanly and Clement, 1979). There are no reliable estimates of the consequent emission of particulate matter including carbon and hydrocarbons of the isoprene and terpene class. The adverse effects of these pollutants on the associated ecosystems and on the heat balance of the atmospheric boundary layer are not adequately understood (Bolin *et al.*, 1983).

The 1960s witnessed the rapid emergence of a group of oil-producing countries in the arid zone of the Middle East, which have exerted a powerful influence on the world economy. The large tankers carrying crude oil from the loading harbours of the Persian Gulf or the Red Sea to the Near and Far East, Europe, and America leave their trails of oil pollution causing potential damage to the coastal ecosystems of Asia, Africa, and Australasia. From a survey of the tar pollution of beaches in the Indian Ocean, the South China Sea, and South Pacific Ocean, Oostdam (1984) calculates the standing crop of beach tar in the Indian Ocean to be in the region of $1.5–0.5 \times 10^6$kg. Buried tar in the coastal tracts of this region is relatively uncommon. Quantitatively, this amount of tar may not stand comparison with the high levels of tar pollution noticed in the coastal zones of the Atlantic, the North Sea, and the Mediterranean.

3.4.3 ECOTOXICOLOGICAL IMPACT OF DEVELOPMENT

Nearly two-thirds of the human race, belonging to what is loosely called the Less Developed or Developing countries, inhabit the tropical and semi-tropical regions of Latin America, Africa, and Asia. The urban clusters in some of these countries are predictably growing at an unprecedented rate, and urban pollution due to the sheer size of the population is likely to touch new heights in the very near future. The developmental strategies adopted by these countries involve the extensive use of chemicals in agriculture, production and storage of food and consumer goods and in meeting the increasing demands from industry, transport, and defence sectors. Very often the environmental impact of such activities has not been assessed prior to their initiation. As a result, acute and chronic problems of environmental pollution have begun to appear in many urban pockets in these countries.

One may refer in this context to the data published by WHO as part of the Global Environment Monitoring System (GEMS). This programme, jointly undertaken with the United Nations Environment Programme, assesses the quality of air, water, and food by measuring the levels of certain chemical pollutants and monitoring the levels of cadmium in human blood and the kidney cortex, or of pesticides in human breast milk. The values of sulphur dioxide and suspended particulate matter in selected samples of the ambient air of Calcutta, Bombay, Sao Paulo, Rio de Janeiro, Seoul, Bogota, and Santiago for the period 1973–1978 (not all cities covered for all the years) indicate levels of air pollution not less than corresponding values in London, Tokyo, or Los Angeles before they were reduced by control measures (WHO, 1976; 1978; 1980; 1983).

3.4.3.1 Modernization of Agriculture

The four major inputs needed for increasing farm productivity in the tropical countries are better seed, water, chemical fertilizers, and pest control chemicals, and these factors do affect the environment adversely (FAO, 1981). Another important feature is that, besides supporting massive human populations, most of the regions where there is a food deficit are also those where vector-borne plant and animal diseases are endemic. This has led to an increasing dependence on chemical weapons to combat pests. Programmes for control of vector-borne diseases in the public health sector are not usually linked to the measures mounted for controlling pests in the agricultural sector. Some of the ecotoxicological problems witnessed in these countries could as well be part of the sequelae of lack of co-ordination and possibly the indiscriminate use of pesticides (Bull, 1982). Although environmental control of vectors of malaria and filariasis has shown promising results (Rajagopalan and Das, 1984), integrated pest management with less reliance solely on chemical control is yet to be accepted as part of the strategy for pest control (Krishna Murti, 1982a).

3.4.3.2 Way of Living and Exposure Levels

Dramatic changes ensuing from development can be seen in the way of living, customs, and food habits. With better means of communication, ways of life associated with urbanization are rapidly introduced into the remotest rural or tribal settlements. The use by rural communities in India of an ever increasing number of consumer products made by modern industrial production techniques is testimony of the influence of the urban way of life on rural people. Exposure to chemicals by human and non-human receptors as a result of shift in life-style in the tropical rural sector remains to be evaluated.

3.4.4 TRENDS AND PATTERNS OF INDUSTRIALIZATION

In the five-year period 1970–75, the rate of growth of industrial activity in some of the developing countries was substantial, although their combined contribution to global output of manufactured goods was only 7%. This figure is expected to rise to between 17.5 and 25% (UNIDO, 1979). Of the total industrial output of the 1970s in the developing countries, Latin America accounted for 50%, Asia and the Middle East together for 33%, and Africa for 13%. The forecast for 2000 AD, however, indicates a shift in favour of Asia. In general, trends of industrialization are evident in 38 countries for which statistics are available, and ten at least out of 38 are in the tropical belt and can be ranked as industrialized on the criterion that manufacturing industries contribute more than 30% of the Gross Domestic Product of the country (UNIDO, 1974).

Certain features characterize the pattern of industrialization in the developing countries:

(a) Industrial activity seems to become concentrated in existing urban settlements and significantly adds to overcrowding and shortage of water, housing, and facilities for sanitary disposal of wastes.
(b) Industrial activity has stimulated the expansion of the unorganized sector, i.e. the sector which relies more on labour-intensive techniques. Thus 40–70% of the urban labour force is employed by this sector.
(c) Industrial growth has attracted hazard-prone technology, which in many instances is banned in the countries from where the technology is imported.
(d) Industrial activity has permeated into regions which have hitherto been free from environmental pollution.

3.4.5 TYPES OF CHEMICALS

Whether they concern the agricultural, the industrial, or the public health sector, development programmes involve the use of a variety of chemicals of which the degree of exposure in the ecosystems differs considerably. Developing countries have adopted diverse measures for generating a continuous supply of industrial chemicals and agrochemicals; these include both indigenous as well as imported production. The control of the use of these chemicals is more often dictated by economic considerations than by a rational consideration of their effect on human health or the ecosystem. In the oil-producing countries of the arid zone of the Middle East, there has been a very rapid expansion of the industries related to refining of crude oil and production of petrochemicals. Most of these industrial units are located in the proximity of harbours of the Persian Gulf. Salt-free water is recovered from seawater, and the waste water of the wells and refineries is discharged back into the Gulf. Marine pollution has begun to surface in this region and the National Institute of Oceanography, Goa, India, has given high priority to related studies in the Arabian Sea and the Indian Ocean.

In the following sections, brief accounts are given of the diversity of chemicals that are diffused into the environment.

3.4.5.1 Chemical Pollution in Energy Production and Domestic Fuel Usage

Coal continues to be the main source of solid fuel for energy and also the raw material for the production of coke used in steel mills. Crude petroleum oil forms the source of liquid fuel. Natural gas is also being used widely in some of the developing countries. Extraction and processing of fossil fuels lead to the release of many by-products, such as phenols, thiocyanates, cyanides, sulphides, ammonia, and others. Approximately 2 tonnes of coking coal are used for the production of 1 tonne ingot steel and with rapid expansion of the steel industry there has been concurrent output of coal and coke.

Carbonization is carried out in coke oven batteries, and the volatile matter containing several fractions of aromatic substances is scrubbed and recovered in by-product recovery systems. The coke oven by-product recovery plant may require 375 litres of water per tonne of steel produced (Parhad *et al.*, 1982). Amounts of waste water to be disposed of are, in low-temperature carbonization, $0.15\,m^3$/tonne coal processed; in high-temperature carbonization 0.3–$0.4\,m^3$/tonne; and in gasification $1\,m^3$/tonne. Waste water effluents emerging from the above processes contains inorganics including free ammonia and fixed ammonia, and organics including phenols and pyridines.

The low-temperature carbonization process is increasingly adopted to produce smokeless domestic coke for use in urban and rural areas using low-grade coal, including lignite; this produces pollutants in the form of waste ammoniacal liquor. Approximately 1 kg of phenol is released into the environment for every tonne of fossil fuel processed.

The producer gas process completely gasifies solid fuels. The fuel is heated by the self-generated gases resulting in significant loss of its volatile matter. The coke residue is converted to products of incomplete combustion, principally carbon monoxide. Using $1\,m^3$ of water for scrubbing the oil, tar, and other aromatic compounds from $10\,m^3$ of gas, waste water discharged from a modestly sized producer gas plant can be around 100–$110\,m^3$ per day.

Coke oven plant effluents can be divided into three categories:
(i) coke quenching water
(ii) benzol plant water
(iii) ammoniacal liquor

A typical effluent from a coal carbonization plant may contain anywhere up to 130 compounds at concentrations more than 1 mg/litre (Pearse and Punt, 1973). The concentration of pollutants in coke oven wastes depends upon the coking coal used for the process. Increasing coking temperatures as a general rule results in a lower phenol and higher cyanide concentration (Kostenbader and Flecksteiner, 1969; Vaidyeswaran, 1977). The general composition of waste water from a metallurgical coke plant is given in Table 3.4.1, and that of a producer gas plant and domestic coking plant (LTC) is given in Table 3.4.2

Table 3.4.1 Characteristics of waste waters from by-product recovery plant in a steel plant. From Parhad *et al.* (1982)

	Cooling, scrubbing, and refining process waste	Ammonia still waste	Phenols tar distillation waste	Combined waste from by-product recovery
	(All values except pH are in mg/l)			
pH	9–9.5	8.6–9.1	7.0–8.9	8.6–9.0
Dissolved solids	165–400	200–700	60–1000	125–800
Suspended solids	20–730	25–200	20–330	50–500
Total alkalinity $CaCo_3$	1200–3000	200–1000	250–1000	450–2000
Phenols	1000–1600	40–170	10–200	100–1000
NH_3-N	100–2700	40–380	100–370	100–1500
Cyanides (CN-)	10–50	1–20	10–20	10–60
Chemical oxygen demand (COD)	800–2000	600–1000	480–1350	800–1800
Biological oxygen demand (BOD)	300–750	350–1300	170–440	200–1100

Table 3.4.2 Characteristics of waste water from producer gas and LTC plants. From Parhad *et al.* (1982)

	Producer Gas	LTC
	(All values except pH are in mg/litre)	
pH	8.2–8.7	9–10.0
Suspended solids	320–680	800
Phenols	870–2020	4692–7360
Ammonia (free)	230	6000–9000
COD	4200–16 640	22 400–42 4000
BOD	1400–4800	15 500–18 500
Thiocyanates	0.5	350–887
Cyanide as CN	1.0	11.5–14.0
Sulphide	–	67–74

Table 3.4.3 Process-related pollution in petroleum refining and petrochemicals industry

Process	Pollutants
Crude processes	NH_3sulphides, oil chrloride, mercaptans, phenols
Cracking	
(a) thermal	high BOD, COD, NH_3
(b) catalytic	high alkalinity, oils
(c) hydrocracking	phenolics, cyanide, mercaptans
Hydrocarbon rebuilding	sulphides, mercaptans
Olefinic conversion	dissolved solids
Alkylation	suspended solids
Hydrotreatment	NH_3, sulphides, phenols
Asphalt production	Oil, high oxygenated phenolics

Aromatic compounds consist mainly of mono-, di-, or tri-hydroxy phenols, substituted phenols, and naphthols. Coke oven effluents contain up to 50% monohydric phenols, with catechol and resorcinol in equal proportion along with cresols, xylenols, and quinols in traces.

Processes involved in the refining of crude oil and manufacture of different petrochemicals produce a variety of pollutants. Waste waters arising from these processes contain oil, grease, sulphides, phenols, pyridines, naphthalenes, and polycyclic hydrocarbons. Water is consumed in enormous quantities and nearly half of it is discharged untreated into adjacent water bodies. Some characteristics of waste water from different refinery processes are summarized in Table 3.4.3.

Traditional fuels, such as firewood and sun-dried cattle dung, constitute a major source of fuel for cooking. Although some progress has been achieved in converting fresh cow dung into biogas, and several thousand demonstration plants have been established, by and large the tradition of using sun-dried cow-dung cakes as domestic fuel has not been supplanted. The health hazards associated with the burning of cow-dung cake or firewood or trash in poorly designed domestic stoves or cooking ovens have been brought to the fore by a recently conducted study in India (Smith *et al.*, 1983). Domestic fuel use also significantly contributes to the air pollution load of the metropolitan cities of Calcutta and Bombay. Release of polycyclic hydrocarbons and arsenic into the environmental compartments has ecotoxicological implications which have not been assessed realistically (Vohra, 1982).

3.4.5.2 Chemical Fertilizers

There has been an unprecedented use of fertilizers in tropical/agricultural countries following the success attending efforts to boost the output of food and cash crops. Fertilizer production (in terms of nutrients) in India is likely to increase from 5.25 million tonnes anticipated in 1984–85 to 19 million tonnes by the year 2000 to achieve a reasonable measure of self sufficiency. The discovery of substantial quantities of pure gas in the offshore Bassein fields in addition to the associated Bombay High Offshore gas has made it possible to plan the expansion of the nitrogen fertilizer capacity from 5.25 million tonnes to 9.9 million tonnes by 1990.

Environmental pollution problems likely to be posed by the naphtha or pure natural based fertilizer plants in India are yet to be identified. In contrast, coal-based fertilizer units have given rise to ecotoxicological problems, such as eutrophication and the toxicity associated with ammonia, nitrate, and urea (Desai and Keshavamurthy, 1982).

3.4.5.3 Petrochemicals

By 1960 the demand for various chemicals in India had far outstripped the supply of chemicals based on alcoholic fermentation, calcium carbide, and coal carbonization. At the same time, with the global expansion of petroleum refining capability, naphtha became available as a starting material for the organic chemical industry. From this basic raw material are produced ethylene, acetylene, and benzene, which in turn are converted into a variety of consumer articles, e.g. polyvinylchlorides, polystyrene, synthetic rubber, dyes, and insecticide intermediates. The commodity chemicals in their turn are chemically transformed into consumer products, such as paints, pipes, drugs, and insecticide formulations. Major petrochemicals produced now in India include ethylene, propylene, butadiene, benzene, methanol, monoethylene glycol (MEG), butyl alcohol, 2 ethyl hexanol, phthalic anhydride, formaldehyde, and ethylene oxide. The intermediates for synthetic fibres are polyester staple fibre, polyester filament yarn, nylon filament yarn, acrylic fibre, dimethyl terephthalate (DMT), monoethylene glycol, acrylonitrile, and capralactum.

3.4.5.4 Plastics

For convenience the Committee of Experts on Plastics of the Council of Europe has divided the large number of ingredients used in the plastics industry into two sections and two appendices. The list of monomers and initiators includes more than 300 items, whereas additives and polymerization aids give a total of over 1000 items. Besides 500 phthalate based plasticizers including di-2-ethyl hexyl phthalate and dioctyl phthalate, many adipates, amide esters of diethanol amine, azelates, benzoates, butane tricarboxylates, chlorinated paraffins, chlorofluoroethanes, and chlorofluoromethanes are also extensively used. Nearly a third of these chemicals rated as highly toxic include monomers, like vinyl chloride and acrylonitrile, and are classified as environmental mutagens and human carcinogens, whereas acrylamide is a powerful neurotoxin. The ecotoxic problems likely to emanate from the spread of plastic-based industries in developing countries have been outlined in an earlier review (Krishna Murti, 1982b).

3.4.5.5 Pesticides

Pesticides in common use today belong to the broad functional groups of insecticides, fungicides, herbicides, algicides, ascaricides, rodenticides, and nematocides. Chemically they belong to two groups: inorganic and organic. The inorganic pesticides include mercuric chloride, selenium compounds, lime sulphur, Paris green, lead arsenate, and calcium arsenate. Among naturally occurring organic insecticides are rotenone and the pyrethrins. The synthetic pesticides include organohalides, organophosphates, carbamates, and some minor compounds, like the quinones and phenols (Krishna Murti and Dikshith, 1982).

3.4.5.6 Dyestuffs

The main groups of synthetic dyestuffs used in the textile industry consist of sulphur dyes, vat dyes, direct dyes, azoic dyes, reactives, disperse pigments, and 1 : 2 metal complexes. Dyes, in general, are resistant to biodegradation, Dyestuffs are made essentially by batch processes involving a wide range of chemical reactions, unit operations, and unit processes. Dyestuffs of the non-azo class involve the use of solvents as reaction media, needing additional processing steps and sophisticated equipment. All these generate wastes.

The present trend in dyestuffs research worldwide does not indicate any major technological changes. In the developed countries the accent has been on increased automation and legislative controls to prevent environmental pollution. In countries like India, where the industry has gained considerable momentum in recent decades, there is broad scope for optimizing the unit processes and reducing the output of liquid effluents and introduction of catalytic hydrogenation instead of non-acid or zinc reductions, which produce considerable quantities of solid wastes. The hazardous waste problem associated with the dyestuff industry in Gujarat, India, has been dealt with in a recent pilot study conducted by the National Productivity Council.

3.4.5.7 Drugs and Pharmaceuticals

Drug and pharmaceutical industries which require environmental control are fermentation plants, synthetic organic plants, fermentation/synthesizer organic chemical plants, biological products units, and formulation plants (Subbiah, 1982). Pharmaceutical products belong to three main groups:

Medicinals: antibiotics, vitamins, anti-infective drugs, central depressants and stimulants—analgesics, antipyretics, barbiturates, gastrointestinal agents.

Biologicals: Serums, vaccines, toxoids, antigens.
Botanicals: Morphine, reserpine, quinine, curare, various alkaloids.

A majority of the pharmaceutical industries produce medicinals both by fermentation and organic synthesis. Fermentation processes produce vitamins B_2, B_{12}, and ascorbic acid; the antibiotics, penicillins, tetracyclines, streptomycin, kanamycin, erythromycin; organic acids (citric and gluconic acids); enzymes (amylases, lipases, proteases); amino acids, and a few others. The plant operations include seed production, fermentation for increased biomass and production of desired metabolites, filtration of biomass, removal of colour of broth, extraction by solvents or adsorption, evaporation, filtration, drying, and formulation (Mohan Rao *et al.*, 1970). Compositions of typical spent fermentation broths and untreated synthetic drug wastes are given in Tables 3.4.4 and 3.4.5.

Table 3.4.4 Characteristics of spent fermentation broth in antibiotic and vitamin B_{12} plants. From Subbiah (1982)

	Penicillin	Streptomycin	Erythromycin	Vitamin B_{12} extract
pH	6.5	5.7	6.5	8.0
Biological oxygen demand (mg/l)	8000–12 000	8000–12 000	4000–5000	72 000
Suspended solids (mg/l)	200–700	6000–10 000	300–500	Slurry
Solid SO_4 (mg/l)		2800–4500	–	–
Phosphate P (mg/l)	500–700	200–300	–	–
Total N_2 (mg/l)	1500–2500	1850	2050–3050	9500–10 500
Carbohydrate (mg/l)	2–2.5	5–10	5–10	5–7.5

Table 3.4.5 Characteristics of untreated synthetic drug waste

Component	Concentration (mg/litre) range
Calcium chloride	600–700
Sodium chloride	1500–2500
Ammonium sulphate	15 000–20 000
Calcium sulphate	800–21 000
Sodium sulphate	800–10 000
Aromatics drug derivation	1870–7300
Various solvents	3500–4400

3.4.5.8 Heavy Chemicals

Next to steel, caustic soda and sulphuric acid determine the level of industrialization of a country. The growth of paper, textile, aluminium, PVC, chemical, and soap industries, among others, depends on the availability of quality caustic soda and chlorine. The production and use of these two chemicals have given rise to significant levels of aquatic pollution from the mercury used in chloralkali cells.

3.4.5.9 Steel and Non-Ferrous Metals

Processing of iron ore and other minerals for the extraction of iron, conversion to steel, and the production of aluminium, zinc, copper, lead, tin, vanadium, titanium, nickel, chromium, and manganese is associated with varying problems of environmental pollution. Ore exporting countries like India also face problems of environmental degradation indirectly due to deforestation required for mining and transport of the ores to shipping harbours on the sea coast. Iron ore deposits of Kudremukh region in the Chikmaglur district of Karnataka are being exploited currently by transferring the ore in a slurry form to the sea port of Mangalore. The ecology of the grasslands under which mining is carried out has been reported to be disturbed. There are also indications of significant pollution of the Bhadra river and its estuaries by colloidal iron.

3.4.5.10 Chemicals and Products based on Processing of Natural Products

Under this category must be included distilleries which produce ethyl alcohol by fermentation of sugar-cane molasses; tanneries which process hides; and wood or paper and pulp plants which process wood, straw, bagasse, and rags to produce a variety of paper and boards, or process timber to produce raw material for construction and furniture. Of these the tanneries are in a class by themselves with chromium used in chrome tanning. Paper and pulp mills use large quantities of water and discharge effluents consisting of both biodegradable and non-degradable organic matter and mercury, where the caustic alkali used for pulping is produced captively by the mercury diaphragm cell process. Wood processing for timber requires the use of copper, arsenic, and glues and resins based on petrochemicals. Characteristics of wastes generated from distilleries, tanneries, and the weak black liquor of paper pulp plants are given in Tables 3.4.6, 3.4.7, and 3.4.8.

3.4.5.11 Natural Toxicants

The tropical humid climate promotes the proliferation of innumerable fungi which attack food and cash crops and produce a variety of mycotoxins (Tulpule *et al.*, 1982). The decay of wood and vegetation contributes to the cycling of

Table 3.4.6 Characteristics of distillery spent wash

Biological oxygen demand (BOD)	39 600 mg/l
Chemical oxygen demand (COD)	120 000 mg/l
Soluble sulphates	2312 mg/l
Total solids	144 840 mg/l
Total dissolved solids	113 200 mg/l
Suspended solids	34 240 mg/l
pH	4.3
Residual sugar	1.3%

Table 3.4.7 Some characteristics of tannery effluents. From Chakraborty *et al.* (1965) Reproduced with permission of the National Environmental Health Association

Process	Volume of waste (l/tonne of hide)	pH	Total solids (mg/l)	Suspended solids (mg/l)	BOD (mg/l)
Soaking	2112	9.3	21 200	2588	1610
Liming	2088	12.7	48 400	10 322	10 027
Deliming	1408	8.0	5870	1392	1522
Vegetable tanning	1584	5.8	31 800	3510	19 284
Chrome tanning	1892	2.9	11 550	482	1000
Composite waste*,	15 092	9.5	20 000	3170	7000

*Total chromium in the composite waste could be 27 mg/l.

Table 3.4.8 Characteristics of black liquor from paper and pulp industry*. From Subrahmanyam (1982)

pH	11.5–12.4
Colour as Klett units (420 nm)	55 000
Total solids (mg/l)	147 000
Total volatiles (mg/l)	79 435
Organic : Inorganic solids	54 : 46
COD (mg/l)	151 600
BOD_5 (mg/l)	29 150
Lignin (mg/l)	56 580
COD of lignin isolated from black liquor (mg/l)	1.95
COD due to lignin (mg/l)	110 330
BOD rate constant kg/day	0.18
Ultimate BOD (mg/l)	36 320
Organics other than acid-insoluble lignin measured as COD and which do not exert BOD in the test (mg/l)	4950

*Waste waters from unbleached kraft pulp mills contain abietic acid, dehydroabietic acid, isoprimaric acid, palustric acid, pimaric acid, and sandaracopimaric acid. In caustic extraction state waste water one encounters trichloroguaicol, tetrachloroguaicol, monochlorodehydroabietic acid, epoxy stearic acid, and dichlorostearic acid. Minor toxic components in Kraft pulp mill waste water are sodium sulphide, methylmercaptan, hydrogen sulphide, sodium thiosulphate, formaldehyde, sodium hydroxide, and sodium chloride.

Table 3.4.9 Production and consumption figures of fertilizers for some countries in the arid/semi-arid and tropical/semi-tropical regions. From FAO (1982a)

	Nitrogenous (×1000 tonnes N_2				Phosphatic (×1000 tonnes P_2O_5)			
	Production		Consumption		Production		Consumption	
	1978/79	1982/83	1978/79	1982/83	1978/79	1982/83	1978/79	1982/83
1	2	3	4	5	6	7	8	9
ARID/SEMI-ARID								
Egypt	216.0	625.9	490.4	667.8	96.6	47.9	86.7	73.7
Morocco	25.6	40.6	80.7	92.4	34.6	37.5	72.3	80.8
Afghanistan	48.3	49.6	35.5	31.4	28.6	32.0	18.1	14.1
Iran	72.3	21.8	167.2	492.8	7.8	7.4	119.1	400.0
Israel	79.5	74.9	37.6	39.1	29.5	26.7	24.2	15.7
Qatar	104.0	304.5	–	–	–	–	–	–
Saudi Arabia	120.9	161.0	7.0	59.0	–	4.2	4.5	33.1
Syria	24.9	80.0	65.0	95.9	–	53.8	33.0	54.9
TROPICAL/SEMI-TROPICAL								
Zimbabwe	50.0	84.0	55.3	73.1	–	–	34.6	45.0
Cuba	34.3	96.1	225.0	275.0	13.5	4.3	58.8	81.6
Guatamala	6.6	12.4	52.0	58.7	4.5	10.6	23.3	18.1
Mexico	593.0	1067.0	752.2	1254.6	227.0	252.5	258.7	486.0
Brazil	273.0	396.8	705.9	642.3	1185.0	1139.0	1519.0	1210.0
Venezuela	61.7	208.3	91.5	70.8	14.0	38.9	59.0	47.4
Bangladesh	134.0	236.6	227.8	306.0	28.6	32.0	100.9	130.4
Burma	57.5	51.2	67.9	114.7	–	–	14.4	43.1
China	7903.0	10 456.0	9280.0	12 210.0	1108.0	2589.0	1347.0	3191.5
India	2173.0	3429.7	2986.0	4043.0	792.0	1002.0	964.6	1200.0
Indonesia	694.0	951.0	549.0	981.0	–	256.6	137.9	356.1
Pakistan	336.7	994.4	684.3	758.0	27.3	73.5	187.9	265.2
Philippines	45.8	18.1	205.4	231.4	36.2	18.9	49.7	51.0
Malaysia	37.4	19.8	109.5	308.0	28.8	26.0	235.0	149.0
Vietnam	40.0	40.0	209.0	250.0	30.0	30.0	13.0	38.4
Sri Lanka	–	97.0	80.9	79.4	–	4.2	24.4	30.7

Table 3.4.10 Production and consumption of fertilizers in the developed and developing countries of the world

	Nitrogenous fertilizers (1000 tonnes N_2)		Phosphatic fertilizers (1000 tonnes P_2O_5)	
	1978/79	1982/83	1978/79	1982/83
Production				
Developed countries	40 367	41 106	26 146	25 113
All developing countries	15 540	22 242	5375	7876
Consumption				
Developed countries	33 578	34 634	22 740	21 078
All developing countries	20 184	26 386	7174	9755

arsenic (Krishna Murti, 1984a). There are many endemic areas of fluorosis in India due to the presence of high levels of fluoride in water bodies and in the soil (Bulusu *et al.*, 1985). The emphasis hitherto has been to explore the health effects of these toxicants in man and animals. Relatively less information is available on aquatic biotoxins in the tropical region.

3.4.6 PRODUCTION AND USE PATTERN OF CHEMICALS

3.4.6.1 Fertilizers

Realistic estimates for the global production, import, and consumption of this important group of heavy chemicals are available in the FAO yearbook of statistics (FAO, 1982a). The increasing trend of fertilizer consumption in developing countries is more than apparent from the figures given in Tables 3.4.9 and 3.4.10.

The situation in India, may be taken as a case study of the increasing trend in use of fertilizers. From a production of 151 MT (million tonnes) foodgrains in 1983–84 the level targeted for 2000 AD is 225 MT (*see* Table 3.4.11 for trends in the output of important food items). Eighty percent of this increase has come from increased use of fertilizers. The application rates statewise are revealing (*see* Table 3.4.12). From a meagre figure of 65 000 nutrient tonnes in 1950–51 fertilizer consumption has gone up to 7.9 MT (sevenfold increase in seventeen years) in 1983–84, making India the fourth largest producer of nitrogenous fertilizers in the world. The growth of the industry is reflected in the data summarized in Table 3.4.13 (Vittal, 1984).

Nearly 80% of fertilizer consumption in India is in assured irrigated areas, which cover only 30% of the cultivated land. Even in irrigated areas the consumption per hectare varies widely. It is assumed that the increase in fertilizer use will be from 91.4 kg to 120 kg/ha in irrigated areas and from 11.2 kg to 25 kg/ha in rainfed areas. The output from 33 nitrogenous and complex

Table 3.4.11 Increase in output of agricultural products over three decades in India. Values in million tonnes

	1949/1950	1978/1979
Rice	23.5	53.8
Jowar	5.9	11.4
Bajra	2.8	5.6
Maize	2.1	6.2
Wheat	6.4	35.5
All cereals	46.8	119.7
All pulses	8.2	12.2

Table 3.4.12 Fertilizer consumption in major agricultural states of India in 1982/1983. Figures in parentheses indicate total irrigation potential tapped up to 1980–81

	Fertilizer input: kg/ha cropped land (NPK)	Ultimate irrigation potential (million hectares)
Uttar Pradesh	60.6	25.7 (15.8)
Punjab	127.8	6.6 (5.3)
Andhra Pradesh	53.0	9.2 (5.1)
Maharashtra	26.3	7.3 (3.0)
Tamil Nadu	58.6	3.9 (3.1)
Karnataka	38.3	4.6 (2.2)
Gujarat	38.7	4.8 (2.5)
Average for whole country	36.6	

Table 3.4.13 Fertilizer production and consumption* in India from 1951 to 1981/82

	Nitrogenous		Phosphatic		Potassic
	Production (1000 tonnes)	Consumption (1000 tonnes)	Production (1000 tonnes)	Consumption (1000 tonnes)	Consumption (1000 tonnes)
1951/52	28.9	–	9.8	–	–
1960/61	98.0	212	52.0	53.0	29
1965/66	233.0	575	111.0	113.0	77
1975/76	1535.0	2149	320.0	467.0	278
1978/79	2170.0	3420	770.0	1106.0	592
1979/80	2224.0	3498	749.0	1151.0	606
1981/82	3143.3	4069	950.0	1322.0	–

*As of 1983, the level of consumption of all fertilizers stands at 7.2 million tonnes, recording a sevenfold increase in the past 17 years. It is also interesting to note that the percent share of various feedstocks for the production of 5.8 million tonnes of nitrogenous fertilizer in 1982 was natural gas 15, naphtha 50, fuel oil 20, and coal 10

fertilizer plants, six by-product ammonium sulphate plants, three triple superphosphate units, and 40 single superphosphate plants with an installed capacity of 5.17 MT (million tonnes) of N and 1.48 MT of P_2O_5 was 3.4 MT of N and 0.98 MT of P_2O_5 for 1982–83. The gaps will have to be met by imports or by development of biofertilizers (Sohbti, 1983). The land use pattern of a number of countries in the arid and tropical regions indicates a shift towards use of forest land for agricultural use. In all these countries there are efforts to bring more arable land under irrigation and hence one can predict an increasing use of synthetic fertilizers in the coming decades. (Table 3.4.14).

3.4.6.2 Thermal Power Generation

Oil is the principal source of energy and its share in world energy production is nearly 50%. Natural gas and coal account for 19.2% and 30% respectively. One of the imperatives of rapid national development is the availability of

Table 3.4.14 Application rate of fertilizers in some countries of arid/semi-arid and tropical/semi-tropical regions (kg/ha agricultural area). From FAO (1982a)

	Nitrogenous		Phosphatic	
	1974/76	1981	1974/76	1981
World	9.2	13.0	5.6	6.7
Africa	1.2	1.9	0.8	1.3
Egypt	141.8	204.5	25.3	38.5
Morocco	3.4	3.9	3.3	3.8
Afghanistan	0.7	0.7	0.3	0.3
Iran	3.3	6.4	2.5	4.8
Israel	28.7	32.5	14.9	14.6
Qatar	3.2	15.8	–	–
Saudi Arabia	0.1	0.5	–	0.3
Zimbabwe	9.5	14.2	6.0	6.0
Cuba	28.1	53.7	10.1	14.0
Guatemala	15.6	19.4	7.1	10.3
Mexico	7.4	11.3	2.5	3.9
Brazil	5.0	5.6	2.8	3.3
Venezuela	3.2	3.3	2.1	2.2
Bangladesh	13.6	25.8	5.2	12.3
Burma	4.0	10.6	0.7	4.3
China	11.7	29.8	4.0	7.6
India	11.3	21.4	2.8	6.5
Indonesia	10.9	31.7	3.7	10.2
Pakistan	17.8	32.9	3.9	8.9
Philippines	15.6	19.1	3.8	4.7
Malaysia	28.2	39.5	–	–
Vietnam	17.4	18.2	10.1	2.6
Sri Lanka	19.7	34.3	5.4	11.5

abundant energy. With petroleum reserves insufficient to meet today's demand and other forms like nuclear, geothermal, solar, and biomass yet to be developed, India has decided to meet her energy requirements for the foreseeable future by coal. Adequate reserves of coal are known to exist in many coal basins, the most important one being the Jharia coalfield, which is the country's main source of prime coking coal. Total coal output went up from 73 MT in 1970–71 to around 125 MT in 1981–82. The share of non-coking coal increased during this period from a little over 55 MT to over 98 MT, an increase of 78%. On the other hand, production of coking coal rose from about 18 MT to 26.5 MT, recording only a 48% growth. Coking coal production is likely to be around 30 MT in 1984–85 and to rise to 40–42 MT by 1990. The existing thermal capacity is 24 656 megawatts (MW). It is proposed to add another 22 000 MW by the end of 1990. The National Thermal Power Corporation will be implementing schemes resulting in an aggregate capacity of 10 000 MW. Six super thermal stations in Farakka, Korba, Ramagundam, Rihand, Singrauli, and Vindhyachal are under implementation. There are also plans to locate two more super thermal stations at Moradnagar in Uttar Pradesh and at Kahalgaon in Bihar, each with a capacity of 840 MW. The NTPC plans to achieve a target of 28 000 MW by 2000 AD. The balance of the need of 20 000 MW will have to be met by State thermal plants.

A major impediment to the effective utilization of Indian coal in metallurgical industries and thermal power plants has been its poor quality and high ash content ranging from 30 to 45%. With current plans to mine deeper and adopt mechanization progressively, the problem of utilizing poor-quality coal has assumed multi-faceted complexities. As at present only 15% of the country's coal output (that is about 60% of coking coal output) is being beneficiated. The corresponding figures for France, UK, Korea, and Japan vary between 90 and 100%, while those for Canada, Australia, Czechoslovakia, and West Germany are around 60%.

There are 16 operating washeries in India with a total capacity of 20.5 MT/year as against an annual production of 30 MT/year of coking coal. The washeries in operation have a ratio of 50 : 40 : 10 for clean coals, middlings, and rejects. The high yield of middlings creates many problems of waste disposal and pollution. Since beneficiation of non-coking coals may yield middlings and rejects with ash content over 45% and 60% respectively, captive use for these middlings and rejects will have to be built up in order to avoid major pollution problems. By the end of the Seventh Plan (1990) Indian coal output will exceed the 200 MT mark, nearly 170 MT of which will be high-ash non-coking coals, the ash content of an estimated 120 MT of which might exceed 40%.

Technology for conversion of high-ash coal to methanol is also under examination in view of the plans to undertake coal transport by slurry pipelines. The best prospects for slurry pipelines in India appear to be over 1200 km from the Singrauli mines in Uttar Pradesh to the Mankobari power plant (Gujarat) and thence to the proposed Vapi 2000 MW station (over 1000 km). It is estimated that a slurry pipeline project in the north expected to be commissioned towards 1995 will transport 62 MT coal per year over 3897 km to power stations with an aggregate capacity of 13 500 MW. The feasibility of underground gasification of coal deposits in the oil wells of Gujarat is also under wider study (Rajan, 1984).

Problems of pollution related to fuel use in the Indian or African context cannot be divorced from the use pattern of domestic fuels (Table 3.4.15). The potential for supplying cheap domestic cooking fuels to the millions of families in the rural sector has to be sought not only in petroleum products like kerosene or gas but in biomass based on plant or animal life. A case study conducted in 1982 in Bangalore, which houses India's major aircraft factory, the biggest establishment for telephones, machine tools, and electronics as well as two of India's biggest R & D establishments, viz. the Indian Space Research Organization and the Indian Institute of Science, revealed that about 1260 tonnes of firewood are required per day for household consumption, to supply which 10 ha of tropical forests have to be cleared every day. By 1991 the projected need will be 1760 tonnes of firewood per day (Reddy and Reddy, 1982). The interest of India in investing in biomass technology is therefore understandable.

Bagasse and sugar-cane molasses, the by-products of the sugar-cane processing industry, provide respectively 1.2 MT and 0.34 MT of fuel per annum, which can yield 934 million litres of ethanol, although the estimated annual production is only 640 million litres. A total of 203 MT of agricultural residues can be obtained annually from rice, wheat, maize, cotton, jute, barley, sugar-cane, coconut, etc. (Bose *et al.*, 1983).

The common aquatic pest, water lily or water hyacinth, which was once considered as a curse has attracted attention as a source of renewable energy. It is estimated that in India nearly 200 000 ha are covered by this weed. In shallow waters with a temperature range of 38–30°C and pH range of 4–8 the weed grows at prolific rates, absorbing 80% of nitrogen and 60% of phosphorus in five days from secondary effluents, giving a yield of 148 tonnes/ha/year on

Table 3.4.15 Sector-wise and source-wise commercial energy consumption in India, 1978–79. From Statistical Outline of India (1982). Reproduced with permission

	% share of different sources of energy			% share of different sectors
	Coal	Oil	Electricity	
Household	10	71.2	18.8	13.7
Agriculture	–	61.8	38.2	10.6
Industries	44.5	7.9	47.6	38.5
Transport	13.3	83.9	2.8	31.7
Others	11.9	36.2	51.9	5.5

Table 3.4.16 Technologies proposed for using water hyacinth as source of power gas and power alcohol

	Yield (per tonne of dried plants)	Energy value
Saccharification by acid digestion followed by fermentation	56 litres ethanol, 0.2 tonnes residue	7700 Btu
Gasification air and steam with subsequent recovery of ammonia	40–56 kg ammonium sulphate, 40 000 cu ft gas	142.8 Btu H_2, methane, CO_2, CO, and N_2
Bacterial fermentation and utilization of evolved gas for power production	26 500 cu ft gas	600 Btu methane hydrogen CO_2 oxygen

a dry weight basis. Three different technologies, as summarized in Table 3.4.16, are being proposed to convert this biomass into power gas and power alcohol.

Algae, such as *Scenedesmus*, *Chlorella*, and *Spirulina*, can also yield biomass, ranging from 60 to 90 tonnes of dry matter/ha year.

Yet another traditional domestic fuel receiving renewed attention in India is cattle dung. Approximately one million tonnes of cow dung is generated every day by the 85 million livestock population, not all of which is used at present for biogas production. From pilot studies on plants working in remote rural areas it is learnt, however, that the four-day accumulation of the dung of five cows is sufficient to produce 13 cubic metres of biogas (Bose *et al.*, 1983).

3.4.6.3 Petroleum and Industrial Organic Chemicals

In order to indicate the trend in the use of petrochemicals, projected demands and capacity gaps for five items are given in Table 3.4.17. The present low per capita level of petrochemical consumption (0.5 kg) and the large potential market have raised expectations for the rapid growth of the industry in the next decade. It is, therefore, proposed to establish two gas-based petrochemical complexes, one in Maharashtra and the other in Gujarat. The Maharashtra complex will produce 300 000 tonnes of ethylene and 40 000 tonnes of propylene, 80 000 tonnes of low-density polyethylene (LDPE), 135 000 tonnes of linear low-density polyethylene (LLDPE), 5000 tonnes of ethylene oxide, 50 000 tonnes of ethylene glycol, and 60 000 tonnes of polypropylene, whereas the Gujarat complex will produce 158 000 tonnes of ethylene and 17 500 tonnes of propylene a year. In addition three aromatic complexes are being set up for benzene, toluene, and xylene.

Table 3.4.17 Projected demand by 1987–88 of petrochemicals in India. (in 1000 tonnes). From Doraiswamy (1984)

Base chemical	Current licensed capacity	Estimated demand	
		High level	Low demand
Ethylene	268	659	495
Propylene	150	262	197
Butadiene	62	110	83
Benzene	151	467	350
Xylenes	60	212	160

3.4.6.4 Plastics

The foundations for the plastics industry in India were laid during the 1950s, with the establishment of facilities for making polystyrene and LDPE based on industrial alcohol derived from molasses by fermentation. The industry registered rapid growth only after the commissioning of the petrochemical complexes. The ready availability of a large number of commodity plastics stimulated the rapid growth of the plastics processing industries during the sixties and seventies. The trend of polymer consumption during the last five years (Table 3.4.18) indicates an annual growth rate of 10%. The commissioning of petrochemical complexes and ready availability of commodity plastics resulted in the blooming of a large number of small-scale units, particularly in the extrusion sector. Thus as of 1982, 7700 units employing 81 700 workers have been registered. The plastics machinery industry is also well established. New generations of thermoplastics, such PP-homopolymer, PP-copolymer, and linear low-density polyethylene (LLDPE), have also been introduced.

3.4.6.5 Pesticides

Pesticide consumption data (for 1981) for some arid and semi-tropical countries are summarized in Table 3.4.19, indicating an increasing demand for chemicals in pest control in the developing countries. The Indian situation is typical.

Table 3.4.18 Consumption of major plastics in India (1000 tonnes). From Doraiswamy (1984)

	1979–80	1983–84
Low density polyethylene (LDPE)	73.6	115
High density polyethylene (HDPE)	62.7	75
Polyvinyl chloride (PVC)	82.0	137.5
Polypropylene (PP)	15.1	30.0
Polystyrene/High impact polystyrenes (PS/HIPS)	12.5	17.0

Table 3.4.19 Pesticide consumption for 1981 in some countries in arid and semi-tropical zone (×100 kg). Source: FAO Year Book 1982b

	DDT	HCB	Lindane	Aldrin	Toxaphen	Other chlorinated pesticides
Egypt 1974–76)	993	–	3460	10 187	10	19 157
Sudan (1974–76)	6501	–	–	–	–	1482
Israel	–	–	–	–	–	1630
Zimbabwe	2437	–	–	542	–	2702
Mexico	6000	2500	350	–	18 000	10 350
India	30 000	27 000	500	2000	–	19 000
Pakistan	1034	2062	55	30	–	3294

	Parathion	Malathion	Other OP	Pyrethrin	Other botanicals	Arsenicals
Egypt (1974–76)	397	3573	54 267	–	20	420
Sudan	–	–	6767	–	–	–
Israel	–	–	11 230	–	–	–
Zimbabwe	215	91	3018	–	–	–
Mexico	50 000	12 000	54 520	–	–	–
India	20 000	27 500	64 350	–	–	–
Pakistan	530	2381	8993	–	–	–

	Carbamate	Mineral	Others	Sulphur	Lime
Egypt	–	34 973	29 463	85 207	–
Sudan	1590	–	–	–	–
Israel	2160	13 800	7810	6240	–
Zimbabwe	4837	670	–	232	–
Mexico	21 630	–	–	12 000	–
India	32 800	13 800	–	42 000	–
Pakistan	3511	–	1139	–	–

	Copper Compounds	Dithiocarbamates	Aromatics
Egypt	5027	30	–
Israel	6100	3200	–
Zimbabwe	–	795	856
Mexico	16 500	38 350	10 600
India	49 000	23 250	–
Pakistan	912	370	–

	Other functions	Mercurials	Seed Dressing
Egypt	7210	–	–
Israel	1940	–	305
Zimbabwe	–	–	–
Mexico	22 620	1500	7640
India	4550	100	–
Pakistan	417	41	45

continued

Table 3.4.19 *continued*

	2,4-D	MCPA	2,4,5-T
Egypt	–	–	–
Zimbabwe	–44	300	–
Mexico	13 500	–	300
India	4000	–	–
Pakistan	2	–	–
	Triazines	Carbamate herbicides	Urea derivatives
Egypt	–	–	–
Zimbabwe	5537	209	315
Mexico	9200	1450	4060
India	600	–	3050
Pakistan	184	–	–
	Other herbicides	Bromides	Other fumigants
Egypt	–	–	290
Israel	6220	8270	1000
Zimbabwe	4803	–	–
Mexico	20 890	8000	3880
India	11 760	350	1200
Pakistan	346	8	252
	Anticoagulants	Rodenticides	Other pesticides
Israel	–	60	27 910
Zimbabwe	3	–	22 164
Mexico	50	400	5050
India	350	3000	1250
Pakistan	–	279	114

The relevant figures are given in Table 3.4.20. There were 96 manufacturing units in 1983–84 with an installed capacity of 96 749 tonnes but with an output of only 61 743 tonnes, of which benzene hexachloride alone accounted for 34 802 tonnes.

Insecticides have a dominant share, with over 72%, whereas fungicides account for only 21.3%, weedkillers 3.4%, fumigants 2%, rodenticides 0.7%, and acaricides 0.4%. There are 32 insecticides in use in agriculture, of which 18 are organophosphates. Since 1965 the Indigenous availability of yellow phosphorus, phosphorus pentasulphide, and trichloride has led to the establishment of plants for the manufacture of malathion, parathion, dimethoate, monocrotophos, quinalphos, phosalone, etc. Demand for organophosphates is expected to increase from the present figure of 13 000 tonnes to 20 000 tonnes in the Seventh Plan Period for the public health programme alone. For the agricultural sector the development of microencapsulated slow-release formulations of organophosphates is in progress, with the shift in emphasis from

Table 3.4.20 Capacity and production of pesticides in India. (From Rajan, 1984)

Pesticide category*	No. of Units	Installed capacity (tonnes)	Production established 1983–84 (tonnes)
Insecticides	56	82 700	53 067
Fungicides	11	2460	1070
Herbicides	5	2535	540
Weedkillers	4	685	585
Plant growth regulators	5	200	79
Rodenticides	5	1266	320
Fumigants	10	2188	986
Antibiotics	2	915	-
Total	96	96 740	61 743

*Insecticides include: BHC, DDT, Malathion, Parathion, Metasystox, Fenitrothion, Fenthion, Dimethoate, Phosphamidon, DDVP, Quinalphos, Ethion, Carbaryl, Monocrotophos, Lindane, Endosulphan, Thimet, Dicofol, Phosolone, Fenvelerate
Fungicides include: copper oxychloride, thiocarbamate, nickel chloride, organomercurials, Carbandigine
Herbicides include: 2,4-D and Isoproturon
Weedkillers include: Paraquat, Basalin, and Dalapon
Plant growth regulators include: Cycocil and alphanaphthalene
Rodenticides include: Ratafin and zinc phosphide
Fumigants include: aluminium phosphide, methyl bromide, and ethylene dibromide
Antibiotics include: Aureofungin and Streptocycline

insect to disease control. Demand for dithiocarbamate fungicides is likely to increase. With the successful use of herbicides in Punjab and Haryana, the Green Revolution belt, the use of weedkillers is also likely to spread to cereal crops from their present limited use in plantation crops, such as tea. Of the 200 herbicides registered in developed countries, only 25 are registered in India, of which only 14 are being used. Currently, 2,4-D, butachlor, thiobencarb, and fluchlorabin are used for weed control in rice.

One million hectares are under herbicide coverage and the figure may increase to 2–3 million hectares by 1990. On the basis of consumption data per hectare of total land surface available, the low level of 400 g has always been contrasted against the 10 470 g in Japan, the 1490 g in the USA, or 1870 g in Europe to press for more extensive use of pesticides. However, if the data is recast on the basis of pattern of use in areas where intensive agriculture is practised, as in Punjab, Andhra Pradesh, or Tamil Nadu, the consumption per hectare of actual cultivated area may have to be revised several factors above the oft-quoted figure of 400 g (Sarma, 1983; Rajan, 1984: Krishna Murti, 1984c).

3.4.6.6 Minerals, Metals

The data available for the present discussion are from Indian statistics. Information on the growth of mineral production for the two decades from

1960 to 1980 are given in Table 3.4.21. The country does not produce mercury and its entire requirements are imported. Indigenous production of zinc is around 60 000 tonnes, although the demand is for 120–130 000 tonnes. Nearly 70 000 tonnes of lead are consumed whereas indigenous production is only 25 000 tonnes. Requirements of copper are around 100 000 tonnes although only 25–35 000 tonnes are indigenously produced (Raghavan, 1983). It is of relevance in the context of the present discussion that there has been almost a doubling within a period of 20 years of the rate of production of materials with a potential to diffuse toxic elements in the environment.

3.4.6.7 Chemicals and Products Based on Biomass Technology

Production of alcohol has gone up from 355 million litres in 1979–80 to 534 million litres in 1982–83. For 1989–90 the estimate is 902 million litres. The installed capacity for paper from 121 paper and paper-board mills is 1.54 million tonnes per annum. In addition, there are three rayon grade mills with an output of 120 000 tonnes per annum, and one newsprint mill with 75 000 tonnes per annum. Three more paper mills (233 000 tonnes per annum) and one newsprint mill (30 000 tonnes per annum) are ready to be commissioned. The demand for paper by 1985–86 has been put at 2.5 million tonnes, requiring around 3 million tonnes per annum capacity (Subrahmanyam, 1982).

A total of 58.5 million skins are processes in India annually in about 2160 tanneries. This would mean 314 000 tonnes of leather being processed with an annual discharge of 9.4 million m^3 waste water containing 28 260 tonnes of BOD (Arora, 1982). Another industry which has grown rapidly in the last decade with a high biological pollution potential is the processing of cassava for the preparation of industrial starch.

Table 3.4.21 Mineral production in India. From Statistical Outline of India (1982). Reproduced with permission

	1960	1980
	($\times 10^3$ tonnes)	
Bauxite	387	1532
Chromite	107	305
Copper ore	448	1997
Gypsum	997	817
Lead concentrate	6.2	16.3
Manganese ore	1452	1740
Mica	29	8
Zinc concentrate	9.6	50.3
	($\times 10^6$ tonnes)	
Coal	52.6	107.0
Iron ore	16.6	39.0
Limestone	12.9	28.2
Petroleum crude	0.5	9.5

3.4.7 DISPERSION IN ENVIRONMENTAL COMPARTMENTS

Information on the dispersion of chemicals in the environmental compartments of air, water, and soil has to be compiled with the aid of inventories of the sources and use pattern of a chemical in a given location, supplemented with data on the levels of the chemical in air, water, and soil, or in agricultural and industrial products. Although some limited information has been presented in the preceding section on production and consumption of chemicals, mostly in the light of the Indian experience, the task of compiling reliable data for the entire tropical, semi-tropical, arid, and subpolar zones remains incomplete. The picture in regard to monitoring environmental pollution in countries with a tropical and semi-arid climate is even less satisfactory.

3.4.7.1 Air Quality in Selected Urban Areas

Under the WHO/UNEP Air Quality Monitoring Project, as of 1982, 40 countries with approximately 150 monitoring stations were participating in the programme to measure sulphur dioxide and suspended particulate matter (SPM) or smoke in 40 different regions of the globe (WHO, 1983). This is indeed the only project of global dimension in which a systematic attempt has been made to measure levels of important environmental pollutants in different geographical regions, representing temperate, tropical, and arctic climates. India has also instituted its own national network of air quality monitoring and valuable information has been derived for pollutant levels in urban centres (NEERI, 1983).

By far the most extensive study on air pollution is the epidemiology related investigation of the health effects of air pollutants in Greater Bombay (Kamat *et al.*, 1983). Data on the trends of air pollution levels in Bombay City over a 15-year period are summarized in Table 3.4.22. Seasonal variations in the levels of SO_2, NO_2, sulphation rate, and particulate matter in the city of Ahmedabad have been carried out by the National Institute of Occupational Health (NIOH, 1979–83). An increasing trend of pollution levels as tested by Spearman rank correlation technique was observed for the period 1978–83 (*see* Table 3.4.23). The general levels of SO_2 and NO_2 were lower than those reported for other industrial cities of the developed countries and also well below their respective Air Quality Criteria Levels (SO_2: 80 $\mu g/m^3$; NO_2: 100 $\mu g/m^3$). However, the concentration of suspended particulate matter was definitely higher than those reported from most of the urban centres of the developed countries. Levels of SO_4 (4.7–15.9 $\mu g/m^3$) were comparable with those reported for other urban centres of the world; however, the values of nitrate (5.04–22.7 $\mu g/m^3$) were found to be higher than the 3–5 $\mu g/m^3$ reported in temperate regions.

Emission of sulphur dioxide in Delhi from 360 industries in six industrial belts is estimated to be about 175 tonnes per day. Likewise the emission from

Table 3.4.22 Trends in air pollution levels in Bombay City over 15 years. All values in $\mu g/m^3$. From Shetye *et al.* (1984). Reproduced with permission

(a) 1966–75 Eastern Suburb

	1966–69	1970	1972	1973	1975
SPM	316	–	–	–	–
SO_2	65	36	36	39	44
NO_x	8	–	10	–	–
O_3	20	–	13	6	–

(b) 1971–73 All City

	South		Central		Western Suburb		Eastern Suburb	
	1971	1972–73	1971	1972–73	1971	1972–73	1971	1972–73
SPM	–	270	–	364	–	291	–	325
SO_2	61	43	65	78	29	43	54	54
NO_x	10.6	30	12.3	46	9.6	37	7.3	43
O_3	7	9	7.5	14	3.3	11	5.1	12
Sulphation rate	0.94	0.98	0.98	1.01	0.35	0.47	0.70	1.15

(c) 1978–80

	South			Central			Western Suburb			Eastern Suburb		
	1978	1979	1980	1978	1979	1980	1978	1979	1980	1978	1979	1980
SPM	–	–	–	255	270	264	204	204	231	222	238	236
SO_2	–	–	–	97	94	90	35	24	27	59	41	37
NO_x	–	–	–	22.5	12.5	25	18	9.9	16.9	27.1	12.4	22

(d) Benzopyrene levels ($\mu g/1000\ m^3$)

South				Central				Western Suburb				Eastern Suburb			
1972	73	75	80	1972	73	75	80	1972	73	75	80	1972	73	75	80
4.5	39	7.2	–	271	173	211	6.2	–	5.6	–	–	1.9	1.0	0.73	0.82

Table 3.4.23 Trend of air pollution in Ahmedabad

Pollutant†	Area‡	1978–79	1979–80	1980–81	1981–82 (G.M. ± S.G.D.)	1982–83	1983–84	Spearman rank correlation: Trend	Spearman rank correlation: Level of significance
SO_2	H.P.	29.6 ± 2.67 (77)	22.5 ± 1.62 (82)	22.4 ± 1.52 (63)	26.2 ± 2.00 (38)	22.8 ± 2.10 (23)	39.1 ± 2.31 (48)	Positive	10%
	L.P.	–	10.6 ± 1.50 (25) **	10.8 ± 1.51 (34) **	12.0 ± 2.54 (21) **	12.9 ± 1.80 (18) *	13.9 ± 1.87 (28) **	Positive	1.0%
NO_2	H.P.	10.4 ± 1.50 (77)	17.6 ± 1.48 (82)	17.2 ± 1.64 (63)	26.0 ± 1.48 (38)	30.2 ± 2.17 (23)	32.4 ± 1.61 (48)	Positive	5.0%
	L.P.	–	12.1 ± 1.81 (25) **	10.8 ± 1.90 (34) **	17.6 ± 1.65 (20) **	28.8 ± 1.91 (18) NS	20.5 ± 2.09 (29) **	Positive	5.0%
SPM	H.P.	–	608 ± 1.40 (29)	407 ± 1.88 (14)	526 ± 1.36 (17)	460 ± 2.29 (17)	521 ± 1.66 (46)	Negative	NS
	L.P.	–	331 ± 1.77 (10) **	297 ± 2.00 (9) NS	342 ± 1.15 (10) **	258 ± 2.37 (13) **	293 ± 1.58 (27) **	Negative	NS

†Levels in $\mu g/m_3$. Figures in parentheses indicate number of observations.
‡H.P. = high-pollution area;, L.P. = low-pollution area.
* = $P < 0.05$; ** = $P < 0.01$, compared to L.P. area. NS = not significant at $P > 0.10$.

Table 3.4.24 Chemical composition (%) of fly ash from different thermal stations in India. From Satapathy and Ramana Rao (1984). Reproduced with permission of the Council of Scientific and Industrial Research, New Delhi

	SiO_2	Al_2O_3	Fe_2O_3	TiO_2	P_2O_5	CaO	MgO	SO_3	Unburnt carbon
Bandel	55.84	23.85	4.39	1.05	Traces	2.14	1.28	Traces	10.10
Singareni	50.91	16.05	5.59	1.00	0.16	5.32	1.48	0.73	16.90
Kanpur (KESA)	49.91	21.00	7.50	1.03	0.21	2.84	0.94	0.24	15.60
Madras (Basin Bridge)	46.90	21.55	8.45	1.30	0.33	3.19	1.29	0.73	13.03
Rourkela Steel Plant	46.31	20.57	9.74	1.28	0.95	3.34	1.61	0.56	12.60

Table 3.4.25 Elemental composition of fly ash from Indraprastha Thermal Power Station. From Misra (1984). Reproduced with permission of the National Environmental Health Association

Element	ppm
Copper	155
Chromium	274
Manganese	253
Cadmium	2
Strontium	14
Lead	120
Arsenic	101
Nickel	150
Zinc	177
Iron	26 725
Cobalt	18
Selenium	72
Polyaromatic hydrocarbons	90–130

automobiles is estimated to be about 2 tonnes per day. In nine grids from north to south and six from east to west, each of 4 km, the area source emission was 1.0 $\mu g/m^2/s$ and from vehicular traffic 0.07 $\mu g/m^2/s$. The 24-hour maximum concentrations of SO_2 did not exceed the EPA primary standard (265 $\mu g/m^3$) but 24-hour maximum values of ground level concentration of SO_2 have exceeded secondary standards (260 $\mu g/m^3$) during pre-monsoon, post-monsoon, and winter seasons in two pockets (Gupta and Padmanabhamurty, 1984).

Any discussion of air pollution in India cannot be complete without referring to the fly ash and gaseous materials released from the thermal power stations. It is estimated that 8–10 million tonnes of fly ash are released from thermal power plants every year. The fly ash can adsorb cyanide and phenols (Satapathy and Ramana Rao, 1984). Typical composition of fly ash from five different thermal stations is summarized in Table 3.4.24. The fly ash collected from the Indraprastha Thermal Station in New Delhi has been found to contain 60–130 μg of polycyclic hydrocarbons and a variety of toxic metals as shown in Table 3.4.25 (Misra, 1984).

3.4.7.2 Aquatic Pollution

From the flow rate and characteristics of the waste waters generated, estimates of the quantity of effluents and pollutant load as BOD in the water bodies have been made in Table 3.4.26. Since human settlements and industrial units tend to be concentrated in existing urban clusters on the banks of rivers, pollution of rivers has become inevitable.

Table 3.4.26 Approximate estimates of pollutants as BOD from six industries

	Flow (m^3/day)	BOD (kg/day)
Petrochemical	1231	1739
Paper and pulp	310/tonne	50/tonne of pulp
Producer gas plant	110 m^3/day	1.7/tonne of coal
Tannery	30 m^3/tonne	378/tonne of hide
Pharmaceuticals	2200 m^3/day	6606/day
Distilleries	20 m^3/m^3 spirit	900 kg/m^3 spirit

The contamination of existing resources of ground water is being examined, along with the implementation of plans for a rational use of the resources for human, agricultural, and industrial needs. In recent years sophisticated geophysical techniques and remote sensing have been increasingly deployed for mapping water resources and assessing their quality (Pathak, 1983). Infra-red and thermal infra-red imagings are also used now in the detection of freshwater discharge to inland and coastal areas and the effect of pollution on waterways and estuarine systems (Satyanarayana, 1983).

3.4.7.3 Marine Pollution

Acenaphthene, acenaphthylene, benz(a)pyrene, fluoranthrene, methylphenanthrene, phenanthrene, and triphenylene were among the polycyclic hydrocarbons detected along with heavy metals in benthic organisms in the Upper Gulf of Thailand. Benz(a)pyrene was detected in all species at concentrations varying from 1.0 to 8.2 ng/g. No correlation was found between metal concentrations in animals and sediment, with the exception of copper; thus in polychaetes and clams the copper concentrations appeared to correlate with the copper/iron ratio of sediments (*see* Table 3.4.27). Degradation rates of aromatics using labelled chlorobenzene, phenanthrene, and chrysene were significantly lower in the waters and sediments of the Gulf of Thailand than the fast degradation rates of the water and sediment of the Chao Phraya river (Hung Spreugs *et al.*, 1984).

Residues of 2,4-D and 2,4,5-T have been detected in tissues of corals subject to massive mortality in the Gulf of Chiriqui, Panama, due to intense agricultural activity and high herbicide input (Glynn, 1983). Coral mortality was noticed in the Gulf over an area of 10 000 km^2 five to six weeks after corals lost their zooxanthellae. Besides death there was extensive bleaching of corals. Insecticides found in Panamian Pacific coral tissues included lindane, endrin, kepone, dieldrin, ethion, endosulphan, chlordecane, dimethoate, pp′ODT, op′DDT, op′DDD, and pp′DDD. Butyl benzyl phthalate and diethyl hexyl phthalate were found in the tissues. It is significant that relatively high phenoxy acid concentrations (0.01–0.02 ppm) were found in coral tissue in an isolated area with strong tidal flux and warm sea temperature. Herbicides are applied regularly

Table 3.4.27 Content of polycyclic hydrocarbons and toxic metals in benthic organisms of upper Gulf of Thailand. From Hung Spreugs *et al.*, (1984). Reprinted with permission ©1984 Pergamon Books Ltd

	Oyster	Green mussel	Scallop
Polycyclic hydrocarbons (ng/g)			
Phenanthrene	6.7	4.4	4.4
Acenaphthene	16.3	16.2	–
Acenaphthylene	–	–	–
Fluorathrene	470.0	–	–
Methylphenanthrene	3.5	–	2.9
Triphenylene	0.03	–	–
Benz(a)pyrene	3.5	1.0	8.1
Toxic metals (μg/g)			
Cd	1.6	0.41	0.58
Co	0.27	0.50	0.48
Cu	114.00	8.50	3.00
Pb	0.24	0.73	0.01
Ni	0.30	1.50	2.6

by aerial spraying at a recommended dose of 1 l/ha in a total area of 42 000 hectares. During the peak period of mortality, the equatorial surface warming anomaly exceeded 4 degrees over large areas and reached 6 degrees in some places. High levels of herbicide and an unusually prolonged warm spell could have acted synergistically (Glynn *et al.*, 1984). There has also appeared an interesting report on the accumulation of trace metals and chlorinated hydrocarbons in Ross seals collected in Antarctica. Mercury levels were 5 μg/g dry matter in Ross seals as compared to 74.7 μg/g noted in the Californian sea lion or 50 μg/g in Leopold seals. However, Cd levels in Ross seals were more than 100 μg/g as compared to 5–10 μg/g noted in the Californian sea lion. Copper and zinc levels were 80 μg and 200 μg respectively in Ross seals, as compared to 40 μg and 110 μg in Starbour seals of San Francisco. Levels of lead were detected only in traces (Meclurg, 1984). Arsenic enrichment was noticed in two marine microalgae of Phaeophyta (42.2–179 μg/g and 26.3–65.3 μg/g), relative to Rhodophyta (17.6–31.3 μg/g and 12.5–16.2 μg/g) and Chlorophyta (6.3–16.3 μg/g or 9.9–10.8 μg/g) (Maher and Clarke, 1984).

An interesting comparative study of the ecotoxicological effects of organic pollutants versus heavy metals was carried out from 1976 to 1980 using the blood clam *Anadora granosa* and Gobid mudgkept. The levels of Zn, Mn, Cu, Fe, Co, Ni, Cd, Cr, Pb, and Sr in water, sediments, and the two benthic species were found to be far below those that are known to affect adversely the life and quality of benthic communities. The concentration of these elements in various compartments neither revealed any systematic temporal or spatial fluctuation nor reflected the substantial increase in the total pollution budget for the 8–12 year period studied. The poor growth and high mortality in the clam *Anadora granosa* (from the Sewri clam bed may be due to anoxic conditions caused by organic pollutants discharged from the industrial units on Sewri Coast (Patel *et al.*, 1985).

3.4.7.4 Soil Pollution

The rate of land degradation due to human activity currently is 50 000 km^2 per year. However, it is to be noted that 36 million km^2 per year of land, which supports one sixth of the world's population, is ultimately at risk of degradation according to UN estimates (United Nations, 1977). Cultivation of terrestrial ecosystems leads to substantial losses of the major elements. Thus continuous cultivation of land in about a decade in tropical areas has led to the loss of 50% of carbon, 30–40% of sulphur, and 10–30% of phosphorus. The efforts to replenish the lost nutrients by chemical fertilizers require enormous investments. The attendant losses of essential trace metals or interaction with toxic metals by cycling of the latter can give rise to toxicological problems affecting land productivity, animal health, and eventually human welfare.

Population pressures, rapid and often unplanned industrialization without adequate attention being paid to the siting of industries, as well as the search for natural resources have all led to destructive exploitation of tropical ecosystems (Melillo and Gosz, 1983; Vitousek, 1983). One of the primary effects of fossil fuel combustion appears to be the imbalance between the rate of nitrogen fertilization and hydrogen ion loading (by acid precipitation) and heavy metals output. Extensive leaching in tropical soil can lead to losses of phosphorus as unavailable phosphate, resulting in reduced fertility of old soils. There must be innumerable problems associated with such degradation in developing countries in the tropics which remain to be identified. The following examples are given to illustrate the diversity of the pollution encountered.

Soil samples collected from Khetri, India (copper mining and refining town) and Zawar, India (lead and zinc mining town) have been found to be highly enriched with Cu, Pb, and Zn. In Khetri up to 150 ppm Cu is found in soils used for agriculture. In Zawar, Pb and Zn levels could reach levels as high as 1000 ppm (*see* Table 3.4.28) (Haque and Subramanian, 1982). The significance of soil pollution by these metals must be assessed in relation to observations on the differential effect of copper on root growth and shoot growth (Gupta and Mukherji, 1971), as well as the effects on seed germination (Mukherji and Ganguli, 1974; Mukherji and Gupta, 1972; Mukherji and Maitra, 1976). Trace element levels in some representative Indian soils are given in Table 3.4.29.

Considerable amounts of waste water from human settlements are applied to soil for irrigation purposes to recycle the nutrients. The hazards associated with using untreated sewage are evident from the data on increased accumulation of Pb and Cd, not only in the soil but also in the food and vegetable crops

Table 3.4.28 Metal concentration in soils, water, and plants from copper and lead/zinc mining areas in India

(a) Soil

	Exchange (ppm)			Plant available (ppm)			Total available (ppm)		
	Cu	Pb	Zn	Cu	Pb	Zn	Cu	Pb	Zn
	(Ranges for soils from different locations in the mining and refining sites)								
Khetri Copper Mine	0–79	0–7	0–17	0–225	2–20	1–20	20–767	5–41	6–100
Zawar Pb-Zn Mine	0–2.4	0.3–12	1–18	0–44	6–92	35–335	6–127	75–130	60–1050

(b) Water

	Total dissolved (ppm)		
	Cu	Pb	Zn
(Ranges for different locations of the sampling stations)			
Khetri Copper Mine	0–125	0–0.42	0–1.9
Zawar Pb-Zn Mine	0–0.82	0–1.31	0–6.32

(c) Plants

	Total in Plants (ppm)		
	Cu	Pb	Zn
(Ranges for different plants species and for different parts)			
Khetri Copper Mine	6–972	5–30	15–450
Zawar Pb-Zn Mine	3–105	24–951	31–766

Table 3.4.29 Trace elements (ppm) in some Indian soils. From Haque and Subramanian (1982)

Type of Soil	Mn	Cu	Zn	Co	Mo
Black	1081.6	156.0	72.2	47.0	1.5
Black	1426.0	82.5	59.6	38.1	1.84
Red loam	575.6	–	–	–	–
Laterite	805.0	28.0	66.0	36.4	1.28
Alluvial (acidic)	391.0	35.1	76.6	8.0	1.57
Alluvial (alkaline)	495.0	35.3	49.5	29.4	3.01

Table 3.4.30 Cd and Pb content of crops grown on soil irrigated by tubewell and sewage water. Data from Kansal and Singh (1983). Reproduced with permission of CEP Consultants Ltd

	Cd (μg/g dry matter)				
	Maize	Berseem	Cauliflower heads	Cauliflower leaves	Spinach
Tubewell irrigated	0.85	0.69	0.48	0.80	0.50
Sewage irrigated	1.74	1.67	1.60	2.24	2.59
	Pb (μg/g dry matter)				
Tubewell irrigated	1.98	1.93	1.27	2.69	3.29
Sewage irrigated	3.82	4.48	1.80	5.23	6.08

Table 3.4.31 DTPA extractable Cd and Pb (μg/g) in soils receiving tubewell and sewage irrigation. Data from Kansal and Singh (1983). Reproduced with permission of CEP Consultants Ltd

	Cd	Pb
Tubewell irrigation	0.05	0.10
Sewage irrigation*	0.10	1.28

*Average value for six samples from three towns in Punjab where sewage irrigation is used

grown on them. Some illustrative data from Punjab, India are given in Tables 3.4.30 and 3.4.31.

3.4.7.5 Contamination of Food

By and large one of the most serious ecotoxicological consequences of the diffusion of chemicals in the environment is the accumulation in the food chain of recalcitrant chemicals and toxic metals. With the increasing application of pest control chemicals in agriculture there is evidence of accumulation of pesticides in a variety of food and cash crops. The voluminous data in the literature pertaining to the Indian situation has been used recently to present an overview of the problems of residues in foods (Krishna Murti, 1984c). The content of cadmium in common foodstuffs grown in Punjab, India with a relatively high rate of fertilizer application ranges from 0.033 to 2.0 μg/g dry matter (Nath *et al.*, 1982). In the absence of agencies which can regularly monitor the levels of pollutants in foods, it is futile to draw inferences on the levels of contaminants and their impact on health. The problems related to interaction between diverse pollutants, such as mycotoxins, heavy metals, or recalcitrant organic chemicals in the environmental media, including food, remain to be identified.

3.4.8 PERSPECTIVES

Inventories of production, consumption, and diffusion of toxic chemicals have to be continuously updated in order to make reliable estimates of exposure to these chemicals. Although information on the production of aromatics in India for the year 1984–85 from coal carbonization (Table 3.4.32) is available, it is difficult to obtain related information on utilization and ultimate disposal. Similarly, almost the entire quantity of synthetic dyes is produced in one region but their use is spread over thousands of small fabric dyeing units situated all over the country. In regard to pesticides the problem of computing actual use in agriculture is even more acute. Inventories can only be a baseline source of information which has to be supplemented by monitoring programmes designed to give information on levels of selected chemicals in environmental compartments. Exposure can be due to both intentional use as well as by contamination due to various causes.

Table 3.4.32 Aromatics generated from coal carbonization in India, 1984–85

	(Million tonnes)
Coal carbonization on dry basis	16.19
Primary products, crude tar	0.40
Ammonium sulphate	0.15
Crude benzol	0.10
Secondary	
Naphthalene (tonnes)	18 552
Phenol Cresol	250
Anthracene	5230
Pitch	59 200
Tar oils	19 100
Coal tar fuel (million tonnes)	0.34
Benzol products	
Benzene (thousand litres)	68 011
Toluene	9746
Solvents	3397

While it may be relatively simple to compile data on production and consumption, the task of getting reliable information on disposal of wastes from industries or human settlements and their environmental fate is beset with innumerable difficulties. The present review has attempted to highlight the absence of documented data in most of the countries in the tropical or semi-tropical regions.

The examples used here are drawn mostly from the Indian situation, which may not be typical of all developing countries or of all countries in the semi-tropical region. Nevertheless the information can presumably be used to develop guidelines for more meaningful studies. Such studies have to be purposefully oriented towards the objective of helping the countries of the region to devise control measures for preventing the ecotoxicological effects of environmental chemicals. To illustrate this, one may take the example of the use of coal in steel-making in a country like India. By the year 1990, the coal requirements of the steel industry for producing metallurgical coke alone will be 40 million tonnes per annum. Waste discharge at an average rate of 0.25 m^3 per tonne of coal processed will be 10 million m^3 with an average daily discharge of 27 400 m^3. A population of four million living up to 100 km downstream along the Damodar River is likely to be exposed in the Durgapur-Asansol steel complex area, and a population of one million downstream of the Subarnamukhi by the discharge from the steel complex in Jamshedpur.

A recent press notice issued by the Ministry of Industry, Government of India, on environmental clearance of industrial licences prescribes the conditions for the issue of letters of Intent/Industrial Licence. A list of 18 industries has been prepared where it has become obligatory not only to install suitable pollution control equipment but also to identify the site and location of the project. The list includes: primary metallurgical producing industries, vis. zinc, lead, copper, aluminium, and steel; paper; pesticides/insecticides; refiners; fertilizer plants; paints; dyes; leather tanning; rayon; sodium/potassium cyanide; basic drugs; foundry; batteries; alkalis/acids; plastics; rubber; cement; asbestos.

The Central Board of Prevention and Control of Pollution has undertaken investigations on the extent of pollution of the major water bodies, in order to compile inventories of major and minor polluting industries as well as guidelines for waste water and effluent treatment. The Department of Environment, Government of India, has initiated an Integrated Environmental Programme on Heavy Metals to assess the existing levels of ten metallic elements of environmental significance. It is essential to know of the plans or progress of similar efforts in other countries where the climatic features are the same.

ACKNOWLEDGEMENTS

I have pleasure in gratefully acknowledging financial support from Department of Environment, Government of India, New Delhi.

3.4.9 REFERENCES

Arora, H. C. (1982). Biodegradation of effluents from tannery industry. In: Sundaresan, B. B. (ed.), *Proc. of National Workshop on Microbial Degradation of Industrial Wastes*, Feb 23–27, 1981, p. 154. National Environmental Engineering Research Institute, Nagpur.

Bolin, B., Crutzen, P. J., Vitousek, P. M., Woodmansee, R. G., Goldberg, E. D., and Cook, R. B. (1983). Interactions of biogeochemical cycles. In: Bolin, B., and Cook, R. B. (eds), *The Major Biogeochemical Cycles and their Interactions*, SCOPE 21, pp. 1–39. John Wiley & Sons, Chichester.

Bose, P. R., Vasisht, V. N., and Gupta, B. M. (1983). Viability of biomass as an alternative source of energy in India. *Res. and Ind.*, **28**, 195–202.

Bull, D. (1982). *A Growing Problem: Pesticides and the Third World Poor*. Oxfam, Oxford, UK.

Bulusu, K. R., Naulakhe, W. G., Kulkarni, D. N., and Vitade, S. L. (1985). Fluoride, its incidence in natural waters and defluoridation methods. In: *Silver Jubilee Commemoration Volume*, p. 92. National Environmental Engineering Research Institute, Nagpur.

Chakraborty, R. K, Singh Mohinder, Khan, A. Q., and Saxena, R. L. (1965). An industrial waste survey report. Part I, Tanneries of Uttar Pradesh. *Environment. Hlth*, **7**, 235–247.

Desai, A. G., and Keshavamurthy, G. S. (1982). Microbial degradation of waste water in the manufacture of nitrogenous and phosphatic fertilizers. In: Sundaresan, B. B. (ed.), *Proc. National Workshop on Microbial Degradation of Industrial Wastes*, Feb 23–27, 1981, pp. 97–111. National Environmental Engineering Research Institute, Nagpur.

Doraiswamy, L. K. (1984). Time for bolder planning in petrochemicals. The Hindu Survey of Indian Industry 1984. p. 29. *The Hindu*, Madras.

FAO (1981). *Agriculture Towards 2000*. Food and Agriculture Organization of the United Nations, Rome.

FAO (1982a). *FAO Fertilizer Year Book*, **32**, 33–120. FAO, Rome.

FAO (1982b). *FAO Production Year Book*, **36**, 45. Food and Agriculture Organization of the United Nations, Rome.

FAO (1984). *FAO Monthly Bulletin of Statistics*, **7**, 10. Food and Agriculture Organization of the United Nations, Rome.

Glynn, P. W. (1983). Extensive bleaching and death of reef corals on the Pacific Coast of Panama. *Environ. Conserv.*, **10**, 149–154.

Glynn, P. W., Howard, L.S., Corcoran, E., and Freay, A. D. (1984). The occurrence and toxicity of herbicides in reef building corals. *Mar. Pollution Bull.*, **15**, 370–79.

Gupta, B. D., and Mukherji, S. (1971). Effect of toxic concentration of copper on growth and metabolism of rice seedlings. *Z. Planzen. Phytol.* **64**, 131.

Gupta, R. N., and Padmanabhamurty, B. (1984). Atmospheric diffusion model for Delhi for regulatory purposes. *Mansam*, **35**, 453–458.

Haque, M. A., and Subramanian, V. (1982). Copper, lead and zinc pollution in soil environment. *CRC Critical Reviews in Environmental Control*. CRC Pres, Boca Raton, Florida.

Hung Spreugs, M., Silpipat, S., Tonapong, S., Lee, R. R., Windom, H. L., and Tenore, K. R. (1984). Heavy metals and polycyclic hydrocarbon compounds in benthic organisms of the upper Gulf of Thailand. *Mar. Pollution Bull.*, **15**, 213–218.

Kamat, S. R., Codkhindi, K. P., Shah, V. N., Bhiwankar, N. T., Palade, V. D., Tyagi, N. K., and Rashid, S. S. A. (1983). Prospective three year study of health morbidity in relation to air pollution in Bombay, India. Methodology and early results up to two years. *Lung India*, **2**, 1–20.

Kansal, B. D., and Singh, J. (1983). Influence of the municipal waste waters and soil properties on the accumulation of heavy metals on plants. *Proc. Intern. Conf. On Heavy Metals in Environment*, Heidelberg, September 1983. Vol. I. pp. 413–416.

Kostenbader, P. D., and Flecksteiner, J. W. (1969). Biological oxidation of coke plant weak ammonia liquor. *Journal WPCF*, **41**, 199.

Krishna Murti, C. R. (1982a). Integrated pest management. *Dr. T. V. Ramakrishna Ayyar Centenary Memorial Lecture II*. Entomology Research Institute, Loyola College, Madras.

Krishna Murti, C. R. (1982b). Biodegradation of wastes generated in plastic and polymer industry. In: Sundaresan, B. B. (ed.), *Proc. National Workshop on Microbial Degradation of Industrial Wastes*, pp. 291–313. National Environmental Engineering Research Institute, Nagpur.

Krishna Murti, C. R. (1984a). Cycling of arsenic, cadmium, lead and mercury in the Indian subcontinent. SCOPE Workshop on Metal Cycling, Toronto, Sep. 2–6, 1984. In: Hutchinson, T. C., and Meema, K. M. (eds), SCOPE 31 (1987): *Occurrence and Pathways of Lead, Mercury, Cadmium and Arsenic in the Environment*, pp. 315–333. John Wiley & Sons Ltd., Chichester.

Krishna Murti, C. R. (1984b). India's boom in chemicals. *World Health*, Aug/Sep 1984, pp. 18–20. World Health Organization. Geneva.

Krishna Murti, C. R. (1984c). *Pesticide Residues in Food and Biological Tissues. A Report of the Situation in India*. Indian National Science Academy, New Delhi.

Krishna Murti, C. R., and Dikshith, T. S. S. (1982). Application of biodegradable pesticides in India. In: Matsumura F., and Krishna Murti, C. R. (eds), *Biodegradable Pesticides*, pp. 257–305. Plenum Press, New York.

Lanly, J. P., and Clement, J. (1979). Present and future natural forest and plantation areas in the tropics. *Unasylvia*, **31**, 12–20.

Maher, W. A., and Clarke, S. M. (1984). The occurrence of arsenic in selected marine microalgae from two coastal areas in South Australia. *Mar. Pollut. Bull.*, **15**, 111–112.

Meclurg, T. I. (1984). Trace metals and chlorinated hydrocarbons in Ross seals from Antarctica. *Mar. Pollut. Bull.*, **15**, 384–394.

Melillo, J. M., and Gosz, J. R. (1983). Interactions of biogeochemical cycles in forest ecosystems. In: Bolin, B., and Cook, R. B. (eds), SCOPE 21: *The Major Biogeochemical Cycles and their Interactions*, pp. 177–227. John Wiley & Sons, Chichester.

Misra, U. K. (1984). Distribution of metals of fly ash in various organs of rats at various periods of exposure. *J. Envir. Hlth.*, **19**, 663–667.

Mohan Rao, G. J., Subrahmanyam, P. V. R., Deshmukh, S. B., and Saroja, S. (1970). Waste treatment at a synthetic drug factory in India. *Journal WPCF*, **42**, 1530–1543.

Mukherji, S., and Gupta, B. D. (1972). Characterization of copper toxicity in letterence seedlings. *Physiol. Plan*, **27**, 126.

Mukherji, S., and Ganguli, G. (1974). Toxic effects of mercury in germinating rice (*Oryza sativa* L.) seeds and their reversal. *Ind. J. Exptl. Biol.*, **12**, 432.

Mukherji, S., and Maitra, P. (1976). Toxic effects of lead on growth and metabolism of germinating rice (*Oryza sativa* L.) seeds and mitosis of onion (*Allium cepa* L.) root tips. *Ind. J. Exptl. Biol.*, **15**, 519.

Nath, R., Lyall, V., Chopra, R., Prasad, R., Paliwal, V., Gulati, S., Sharma, S., and Chandran, R. (1982). Assessment of environmental pollution of cadmium in North India. *Bull. Post Graduate Inst., Chandigarh*, **16**, 202–208.

NEERI (1983). *Air Quality in Selected Cities in India 1980–1981*. National Air Quality Monitoring Network. National Environmental Engineering Research Institute, Nagpur.

NIOH (1979–83). *Annual Reports of National Institute of Occupational Health, Ahmedabad.*

Oostdam, B. L. (1984). Tar pollution of beaches in the Indian Ocean, the South China Sea and South Pacific Ocean. *Mar. Pollut. Bull.*, **15**, 267–270.

Parhad, N. M., Kumaran, P., and Shivaraman, N. (1982). Microbial degradation of waste waters from the manufacture of metallurgical and domestic coke. In: Sundaresan, B. B. (ed.), *Proc. National Workshop on Microbial Degradation of Industrial Wastes, Nagpur, Feb 23–27, 1981*, pp. 78–96. National Environmental Engineering Research Institute, Nagpur.

Patel, B., Bangera, V., Patel, S., and Balani, M. C. (1985). Heavy metals in the Bombay Harbour area. *Mar. Pollut. Bull.*, **16**, 22–28.

Pathak, B. D. (1983). *Survey of Remote Sensing Applications in Ground Water Resources Development in India.* National Seminar, May 10–12, 1983. National Natural Resources Management System, Indian Space Research Organization, Bangalore pr. 11–2–1.

Pearse, A. S., and Punt, S. E. (1973). Biological treatment of liquid toxic wastes. *Effluent and Water Treatment.*, **15**, 32.

Raghavan, S. V. S. (1983). MMT's crucial role in metallurgical sector. *The Hindu Survey of Indian Industry, 1983*, p. 59. *The Hindu*, Madras.

Rajapopalan, P. K., and Das, P. K. (1984). Environmental Control of Filariasis in Pondicherry. In: Krishna Murti, C. R. (ed.), *Facets of Environmental Problems. Five Case Studies*, pp. 21–34, National Committee of SCOPE, Indian National Science Academy, New Delhi.

Rajan, T. P. S. (1984). Potential new areas of coal utilisation. *The Hindu Survey of Indian Industry. The Hindu*, Madras.

Reddy, A. K. N., and Reddy, B. S. (1982). *Energy in a stratified society. A case study of firewood in Bangalore.* Karnataka Council for Sci. and Tech. July 1982.

Sarma, P. R. (1983). Imperatives in pesticide promotion. *The Hindu Survey of Indian Industry, 1983*, p. 125. *The Hindu*, Madras.

Satapathy, B. K., and Ramana Rao, D. V. (1984). Adsorption efficiency of high carbon fly ash. *Res. and Industry*, **29**, 188–190.

Satyanarayana, B. L. (1983). *Role of Remote Sensing Technology in Water Resources. National Seminar, May 10–12, 1983.* National Natural Resources Management System. Indian Space Research Organization, Bangalore pr. 11.1.1.

Shetye, S. V., Doshi, V. P., Palade, V. D., Pipewas, V. N., Gregrab, J. K., Sonaje, A. G., and Kamat, S. R. (1984). Trends in causes of death in four zones of Bombay over 1971–1979. *Lung India*, **2**, 44–49.

Smith, K. R., Aggarwal, A. L., and Dave, R. M. (1983). Air pollution and rural biomass fuels in developing countries. A pilot village study in India and implications for research and policy energy system. *Atmosph. Env.*, **17**, 2343–2352.

Sohbti, S. G. (1983). Pressing task of stimulating fertilizer consumption. *The Hindu Survey of Indian Industry*, p. 109. *The Hindu*, Madras.Statistical Outline of India (1982). Tata Services Limited Department of Economics and Statistics, Bombay, India.

Struzuski, E. (1977). Status of waste handling and waste treatment across pharmaceutical industry and effluent limitation, 1973. *Proc. 30th Ind. Was. Conf., Purdue University.*

Subbiah, T. V. (1982). Microbial degradation of waste waters from pharmaceuticals and fermentation industries. In: Sundaresan, B. B. (ed.), *Proc. National Workshop on Microbial Degradation of Industrial Wastes, Feb 23–27, 1981*, p.197. National Environmental Engineering, Research Institute, Nagpur.

Subrahmanyam, P. V. R. (1982). Microbial degradation of waste water from pulp and paper industry. In: Sundaresan, B. B. (ed.), *Proc. National Workshop on Microbial Degradation of Industrial Wastes. Feb 23–27, 1981*, pp. 53–77. National Environmental Engineering Research Institute, Nagpur.

Tripathy, R. N., Dube, N. K., and Dixat, S. V. (1983). Atmospheric biopollutants of Gorakhpur. *Water Air Soil Pollut.*, **19**, 237–246.

Tulpule, P. G., Nagarajan, V., and Bhatt, R. V. (1982). *Environmental causes of good contamination*. Environmental India Resources Series 1. Department of Environment, Government of India, New Delhi.

UNIDO (1974). *Industrial Development Survey*, 74/11/B14. United Nations Industrial Development Organization, UN, New York.

UNIDO (1979). *Conceptual and Policy Framework for Appropriate Industrial Technology*. Monographs on Appropriate Industrial Technology No. 1, United Nations Industrial Development Organization, UN, New York.

United Nations (1977). *Desertification — its Causes and Consequences*. UN Conference on Desertification, Nairobi, Kenya, 29 Aug–9 Sep. Pergamon Press, Oxford.

Vaidyeswaran, R. (1977). Low temperature carbonization and processing byproducts. *Indian Chem.*, **14**, 35.

Vitousek, P. M. (1983). The effects of deforestation on air, soil and water. In: Bolin, B., and Cook, R. B. (eds), SCOPE 21: *The Major Biogeochemical Cycles and their Interactions*, pp. 223–245. John Wiley & Sons, Chichester.

Vittal, N. (1984). Fertilizer Unit's efforts to raise productivity. *The Hundu Survey of Indian Industry, 1984*, p. 125. *The Hindu*, Madras.

Vohra, K. G. (1982). Rural and urban energy scenario of the developing countries and related health assessment. In: *Health Impacts of Different Sources of Energy*, pp. 79–97. Proc. of Symposium, Nashville, 22–26 Jan, 1982. WHO, UNEP, and IEAS, Vienna.

WHO (1976). *Air Quality in Selected Urban Areas, 1973–1974*. WHO Offset Publication No. 30. World Health Organization, Geneva.

WHO (1978). *Air Quality in Selected Urban Areas, 1975–1976*. WHO Offset Publication No. 41. World Health Organization, Geneva.

WHO (1980). *Air Quality in Selected Urban Areas, 1977–1978*. WHO Offset Publication No. 57. World Health Organization, Geneva.

WHO (1983). *Air Quality in Selected Urban Areas, 1979–1980*. WHO Offset Publication No. 76. World Health Organization, Geneva.

Chapter 4

Effects of Environmental Chemicals on Biota and Ecosystems in Tropical, Arid and Cold Regions

Ecotoxicology and Climate
Edited by P. Bourdeau, J. A. Haines, W. Klein and C. R. Krishna Murti

4.1 Effects of Temperature and Humidity on Ecotoxicology of Chemicals

P. N. VISWANATHAN AND C. R. KRISHNA MURTI

4.1.1 INTRODUCTION

Quantification of the influence of temperature and humidity on xenobiotics and environmentally significant chemicals in general will greatly help in prescribing safety limits for community and occupational exposure in countries in the tropical region. The formulation of guidelines aimed at the protection of ecosystems from pollutants could also be more effective. Since eco-epidemiological data pertaining to this subject are limited, inferences have to be drawn from simulated experimental studies and case reports. The basis of the present overview is the information available from literature on related temperature and humidity effects and on the toxicity and environmental fate of pollutants.

The physiological effects of thermal and vapour pressure stresses have been very well studied under a wide variety of conditions (Broulia *et al.*, 1960; Hertig, 1975; Hunt, 1979; Gill, 1980; Alpaugh, 1982). However, it is not clear whether exposure to higher temperatures, with and without high humidity, affects the toxicity of environmental chemicals. Similarly, the influence of ambient temperature on environmental toxicology is not understood, in spite of a large number of reports on the physiological effects of temperature stress (Bhatia *et al.*, 1975).

4.1.2 GENERAL CONSIDERATIONS

The effects of temperature and humidity on air quality and visibility have been well studied. Air parcels with high humidity and temperature, which represent conditions conducive to the formation of sulphates, tend to produce lower visibilities (Sloane, 1983). Increased levels of SO_2 and SPM (suspended particulate matter) are implicated in this phenomenon. There is more interference of visibility by pollutants in summer than in winter (Linak and Peterson, 1983). The data on monitoring of air quality in Indian cities under the national monitoring and GEMS (Global Environmental Monitoring System) programmes indicated high particulate content (SPM) with high levels of SO_2 characteristic of both Bombay and Calcutta. Data are also available on the content of polycyclic hydrocarbons in samples of ambient air from Ahmedabad and Bombay. SPM could provide a matrix for adsorption of SO_2 and NO_x and for complex photochemical reactions, the nature of which needs elucidation in hot humid climates. Seasonal variations, with higher SPM in summer, could alter toxic effects. The release of terpenes into the air of a pine forest was found to be higher in summer than in winter (Yokuchi *et al.*, 1983). Depending on temperature and humidity, these organics interact with atmospheric chemical substances and affect ozone levels.

Seasonal variations in precipitation chemistry, with lower sulphur in winter, have been reported. Nitrate and H^+ were, however, unaffected (Pratt and Krupa, 1983). These results make the study of climatic influences in ecotoxicology a priority area in tropical countries.

The foliar uptake of pesticide aerosols as tested with labelled 2,4,5-T was higher at 37°C than at 21°C or 30°C. Similarly, humidity influences the rate of translocation. From a study of the influence of 32°C and 20°C and relative humidity of 100% and 40% it was found that washable residues of glyphosate on leaves of *Cynodon dactylon* after 48 hours were least at low temperatures and low humidity. The high- temperature, high-humidity effect was mostly through enhanced translocation (Hartley and Graham-Bryce, 1980). However, the response varied with different chemicals and features of the plant. The effect of humidity and temperature on pesticide uptake by roots has also been reviewed by the above authors. The pattern was similar to that shown by leaves with the addition that soil characteristics also play a part. It may be pointed out that very little systematic work is available in published literature on this subject.

Temperature also affects the stability of insecticides. Carbamates persist in cold water for longer periods than in warm water, due to less hydrolysis (Aly and ElDib, 1972). Some information and a few mathematical models are available regarding the influence of temperature and humidity on the fate of air pollutants, their transport, and secondary transformation. Similarly, these aspects have received attention in aquatic environments as well as soil matrices. Since they are beyond the scope of this section, they are not covered here. It should be pointed out, however, that from mathematical models it is possible to extrapolate to specific conditions relevant to a particular habitat.

Aerial spraying of pesticides is carried out only in very limited areas. Most of the application is done manually using simple hand operated sprayers. A variety of chemicals are used and often the application is made repeatedly under conditions showing wide variations in temperature and humidity. Application of weedkillers to a plantation crop, like tea, presents its own problems related to downpour of rain in the tea-growing region. Data on residues have been documented to a certain extent, but practically nothing is known about the dynamics of the chemicals in the ecosystem, especially the role of climate on biodegradation and biomagnification.

4.1.3 EFFECT OF TEMPERATURE AND HUMIDITY ON TOXICITY

Brown (1980) concluded that the relationship between ambient temperature and humidity and the toxic effects of pesticides in vertebrate animals is not linear. However, for man, it is presumed that there is an increased risk in hot climates due more to disinclination to wear protective garments than to the actual effects of higher intrinsic toxicity. The LD_{50} (intraperitoneal, mice) for parathion, carbaryl, and DDT at 1 °C was 16.5, 263 and 750 mg/kg, respectively, whereas at 38 °C the values were 11.3, 112, and 875. But at 27 °C, the values were 29, 588, and 1175, showing the least toxicity at this temperature. Doull (1972) also suggested that ambient temperature may affect pesticide toxicity.

Cummings (1969) showed that penetration of N octylamine through skin is greater at higher temperatures. Craig *et al.* (1977) considered that the toxicity of cholinesterase inhibitors is higher in cold conditions, because depot formation is less at higher temperatures. In such cases, where temperature alters toxicity, Cornwall and Bull (1967) have suggested that body surface area rather than weight is the ideal basis for calculation of doses. Several organophosphates and carbamates show less *in vivo* inhibition of cholinesterase at higher ambient temperatures (Brown, 1980).

Even though very little systematic work has been done on pollutants one could draw analogy from the exhaustive work on drug toxicity versus temperature (Fuhrman and Fuhrman, 1961; Weihe, 1973). LD_{50} variations with temperature in animals could also be useful. Acute toxicity of these agents, which induces, among other signs, hyperthermy, increases with ambient temperature. Sympathomimetic amines, phenothiazines, and salicylate above and below a thermoneutral zone, show increased toxicity. Procaine and caffeine cause higher toxicity at higher temperatures, but toxicity is unchanged below thermoneutrality. The effects of atropine are more severe in hot humid zones. Thermal adaptation to environment (acclimatization) can also alter the toxic effects of drugs (Zbinden, 1973), suggesting the need for more detailed studies on the influence of geographical factors on toxicity.

One major effort in the above direction was the symposium 'Toxicology in the Tropics' at Ibadan, Nigeria (Smith and Bababunmi, 1980). The relatively high toxicity of chemotherapeutic agents used in large amounts for the treatment of parasitic diseases, under conditions of high humidity and temperature, was identified as a major problem in countries in the tropical belt. People in the tropics are more prone to toxicity by food toxins, like cyanogenetic glycosides, mycotoxins, nitrosamines, and lathyrogens.

People living in the humid tropical and semi-tropical countries have coexisted with a variety of myco- and phytotoxins. How far the traditional processing of food by indigenous methods of fermentation (Koji fermentation in Japan/Korea/China; Idli/Kedli fermentation in Indonesia/Philippines/India) or dehydration modulate the properties and chemistry of these chemicals is not understood at the elementary level of investigation let alone as far as molecular mechanisms are concerned. One expects the half-lives of mycotoxins to be within time intervals which may not be consequential from the point of view of overall exposure. In contrast, some processing techniques might lead to adduct formations and acquisition of the property of recalcitrance. Ethnic variations in detoxification processes may also affect toxicity.

That humidity and temperature affect the toxic potential of environmental pollutants like lead, parathion, and antimony has been suggested by Baetjer (1968). Some information is available regarding the effect of temperature variations on the toxic response of cold-blooded animals. Rehwoldt *et al.* (1972) reported that the toxicity of mercurous ion in fish increased threefold when water temperature rose from 15°C to 28°C. However, the toxic effects of Cu, Zn, Ni, Cd, and Cr remained unchanged with an increase in ambient temperature. In an exhaustive review, Cairns *et al.* (1975) suggested that temperature affects toxicants through altered metabolism, increased diffusion, altered oxygen levels, or actual interaction with toxic process. With ammonia and several chemicals, although not all, toxicity increases with temperature. Hg accumulation in fish is also greater at higher temperatures. With pesticides, the data do not show a regular pattern. Higher temperature enhanced endrin toxicity to fish while that of DDT was decreased. With detergents, a rise in temperature increases toxicity. Thus temperature influence depends on the nature of the toxicant, the organism, and on water quality.

Suskind (1977) reported that higher ambient temperatures enhance the percutaneous absorption of chemicals. In the case of miners, the effect of varied conditions of temperature on physiological processes, health, and productivity have been exhaustively studied (Wyndham, 1970), but how far higher ambient temperature has an impact on the effects of simultaneous chemical stress is not clear. Also, the ethnic variations in heat stress deserve more detailed study in tropical countries.

Casarett and Doull (1975) pointed out that any investigation on the interrelationship between temperature and toxic response should include the study of the effect of the toxic agent on temperature regulation as well as the study of the environmental temperature on drug response. Generally, it can be said that the response of a biological system to a drug would be decreased with a decrease in environmental temperature. This is not always true with toxicants. Parathion toxicity is increased in hyperthermia but malathion is potentiated by cold exposure.

Keplinger *et al.* (1959) studied the acute toxicity of 58 compounds in rats at different ambient temperatures. Warfarin was less toxic at 26°C than at 8°C or 36°C. DNOC and pentachlorophenol became less toxic as the temperature became progressively lower. DDT had the same effect at 8°C and 26°C but was more toxic at 36°C. Strychnine was less toxic at 26°C than 8°C and 36°C, which elicited similar responses. LD_{50} of the rodenticide ANTU to rats was 1.9, 2.9, 4.0, and 1.2 mg/kg at 37, 48, 72, and 89°F (Meyer and Karel, 1948).

Variations in temperature and intensity, quality, and duration of light are well-known modulators in plant physiology. There is evidence to show that higher ambient temperature influences the response of plants to air pollutant toxicity (Heck and Dunning, 1967).

Climatic factors also affect toxic residue dissipation in plant tissues, as evident from the study on parathion in citrus orchards (Gunther *et al.*, 1977). With a temperature range of 50–100°F, it was found that residue in leaves dissipated faster at higher ambient temperatures. Similarly, leaf and soil residues in ethion sprayed grape crops decreased more rapidly at higher temperatures. Seasonal and diurnal variations in temperature also affected urinary *p*-nitrophenol in parathion sprayers. Higher temperatures enhance the excretion of dermally exposed parathion, in spite of higher absorption.

Another instance of temperature affecting toxicity is the recovery of grasshoppers exposed to DDT when the temperature rises from 20°C to 25°C. Also, ppb levels of DDT enhance the lower lethal temperature for certain fish, changing their cold resistance (Holdgate, 1979). The lethality of caffeine to rats maintained at room temperature of 84–89°F as compared to 74–79°F, caused hyperpyrexia in addition to the usual caffeine toxicity symptoms (Boyd, 1972).

Wolfe *et al.* (1961) observed temperature dependence in the toxicity of dinitro-*o*-cresol(DNOC) which also increases body temperature of exposed humans at ambient temperatures above 22°C and decreases it below 26°C (Hayes, 1963). The toxicity of alphachloralose to birds and rodents also showed variation with ambient temperature (Brown, 1980).

Many toxicological studies on fish indicate increased toxicity with temperature. Studies at the sublethal and lethal levels have shown that the rate of uptake for water-borne substances, such as lead, mercury, and zinc, will increase with temperature and that the time of death will be advanced (Somero *et al.*, 1977; MacLeod and Pessah, 1973). The bioconcentration factor or amount accumulated has been shown to increase with temperature for mercury and DDT (Cember *et al.*, 1978; Boudou *et al.*, 1980; Reinert *et al.*, 1974). In contrast, studies on other organic contaminants have shown toxicity to decrease with increasing temperatures (Kumaraguru and Beamish, 1981; Brown *et al.*, 1967). An assessment of the influence of temperature on contaminant toxicity to fish indicates that this response will increase over a wide temperature range; but there are some exceptions for different substances, particularly organic contaminants, which can vary among species.

Acute toxicities of environmental pollutants, such as organic solvents, heavy metals, and agricultural chemicals, are known to be aggravated at higher or lower environmental temperatures (Nomiyama *et al.*, 1980a). Environmental temperature has been known to modify acute toxicities of environmental pollutants. Parathion, lead, and 1,1,1-trichloroethane have been reported to be more toxic under high environmental temperatures (Baetjer and Smith, 1956; Baetjer *et al.*, 1960; Horiguchi and Horiguchi, 1966; Horiguchi *et al.*, 1979). Acute toxicities of benzene, trichloroethylene, mercuric chloride, cadmium chloride, fratol, methyl parathion, and dieldrin increased markedly under low environmental temperatures and were also enhanced at a high temperature (Nomiyama and Nomiyama, 1976; Nomiyama *et al.*, 1980a). However, acute toxicities of toluene, copper sulphate, and chromium trioxide were enhanced at a high temperature of 38°C. Toxicity of beryllium was also enhanced under environmental temperatures of 8°C and 38°C. Higher temperatures, furthermore, aggravated cadmium induced testicular haemorrhage (Matsui and Nomiyama, 1979). Acute toxicity of methyl mercury was aggravated under the environmental temperature above 80°F. The acute toxicity of methyl mercury at 38°C was higher than at 22°C (Nomiyama *et al.*, 1980b). Yamanouchi *et al.* (1967) reported that the body temperature of mice, which were fixed on a board, decreased in a cold room at 10°C but that the body temperature of five mice kept in one cage remained unchanged because they warmed each other. Yamaguchi *et al.* (1984) studied the effects of environmental temperatures on the toxicity of methyl mercury in rats and observed that high temperatures can result in increased mortality and neurotoxicity, and even low temperatures may result in some increases in mortality in comparison to room temperature. Neurotoxicity due to mercury poisoning can be responsible for further dysfunction due to its effect on food and ingestion during heat and cold stress.

The environmental temperature can influence the actions of drugs and chemicals in warm-blooded animals (Farris and Griffith, 1949).

A considerable amount of work has been done in the USSR on the effect of ambient temperature, humidity, and other factors (Filov *et al.*, 1978). Temperature affects toxicity, through impaired heat regulation, water loss, respiratory/circulatory disturbances, basal metabolism changes, and altered individual reactions. Increase or decrease in toxicity with temperature varies with the chemical nature of the toxin. The toxicity of narcotics, nitrogen oxides, mercury, petroleum solvents, trichlorfor, methyl styrene, CO, thiotic poisons, etc. is affected by higher temperature. Ambient temperature may affect toxicokinetics also.

Filov *et al.* (1978) include a 20 page table in Russian, 'Combined action of industrial poisons and elevated ambient temperature'. Generally, higher temperature enhances toxicity and accelerates onset. The exception was silicosis, which is alleviated in animals at higher temperature. Aniline inhalation was more toxic at temperatures above 35°C in rats but not in dogs. The effect of higher temperature on CO toxicity was different for guinea pigs, rats, and rabbits. Temperature effect was more marked in chronic toxicity.

The combined effects of heat and toxicant can be taken as a mutual aggravation syndrome. Such effects have to be taken into account while fixing

MAC (Maximum Allowable Concentration) values. It was suggested that MAC for pesticides should be reduced by a factor of 5–10 in hot climates. Also, the temperature range of minimum toxicity varies for different substances and can be worked out. Type I narcotics are generally hydrophilic and their toxicity decreases at higher temperature, while the reverse is the case with hydrophobic type II narcotics.

According to Jahnke (1957), there is no drug whose actions in man are more influenced by climatic conditions than atropine.

4.1.4 EFFECT OF TEMPERATURE ON PLANTS

Indirect injury may occur at a temperature just below the point of denaturation (Langridge, 1963). As the temperature increases, the reaction rates and metabolic activity increase proportionately. As temperatures approach 30°C, the metabolic rate can be very high. Cellular damage at high temperatures may also result from the formation of toxic substances in certain cells exposed to a high localized temperature. The toxic material may subsequently be translocated to other parts of the plant and cause widespread injury (Yarwood, 1961).

Beevers and Cooper (1964) grew rye plants under various temperature regimes. They found that plants growing continuously at 12°C had a higher carbohydrate and nitrogen content than similar plants grown in warmer regimes. The relatively high carbohydrate level was considered to result from the slower degradation of carbohydrates caused by decreased respiration at lower temperatures. The young stem developing early in the growing season may grow best at one temperature while later growth stages and reproduction require different temperatures (Went, 1953).

4.1.5 HUMIDITY

High ambient humidity causes swelling of the stratum corneum so that penetration of chemicals through skin is enhanced (Suskind, 1977). Lindquist *et al.* (1982) reported that the interconversion of nitrogen oxides in the atmosphere is influenced also by humidity and temperature, which may be significant in photochemical smog.

Without moisture in the atmosphere, there would be no corrosion of materials, even in the most heavily polluted air (MacCormick and Holzworth, 1976). In the formation of secondary pollutants, like sulphate, which are more toxic, moisture plays a role (Wagman *et al.*, 1967).

Humidity also affects toxicity by forming more irritant products from the pollutants, for example nitric acid from NO_2, and HCl or Cl_2 from labile chlorinated organics.

The effects of exposure to cold and hot occupational environments have been exhaustively reviewed by Horvath (1979). Relative humidity and soil moisture have been noted to exert a marked effect upon the sensitivity of plants to phytotoxic air pollutants, plants grown under drought conditions being less sensitive (Heck *et al.*, 1965).

Environmental conditions causing full opening of stomata cause more severe air pollution phytotoxicity. A mixture of SO_2 and O_3 was more toxic in high humidity than low (Carlson, 1979). At 55–90% relative humidity, rates of photosynthesis by maple and white ash leaves were reduced by 67% and 58% in 1 day, by 50 pphm O_3 and 50 pphm SO_2. The corresponding figures for 20–50% humidity were 26% and 60%. Subsequent periods of fumigation were also different. Thus different plants may vary in their response.

The influence of humidity on nicotine toxicity on i.p. (intraperitoneal exposure indicates factors other than absorption (Brown, 1980).

4.1.6 COMBINED EFFECTS

It is well known that higher temperature and humidity have a role in enhancing the release and affecting the transport of aeroallergens and other viable particulates (Jacobson and Morris, 1976). This is of great concern for health in tropical countries due to the abundance of sources of biological pollution and human, plant, and animal pathogens.

Baetjer (1968) has reviewed the effect of climatic factors on toxicity of chemicals. Since the bulk of lead poisoning in children takes place in summer, lead toxicity of rats was studied after ip/iv (intraperitoneal or intravenous) injections. Death rate was higher at 95°F than at 72°F. In the absence of sweat glands, humidity was indirectly tested by withdrawal of water, which increased mortality. Lowering temperature at night reduced the effects. Similar results were also obtained with mice. Urinary lead decreased at higher temperatures indicating higher retention. The temperature effect on Pb was independent of its effect on normal physiology. With parathion also, death rate was higher and survival time shorter in mice at 96°F than 73°F. The effect of temperature variation was even more rapid than with lead. On the basis of these data and of those on antimony and benzol, it has been concluded that high environmental temperature generally increases susceptibility to toxic chemicals. In the case of inhaled chemicals, environmental temperature and humidity could affect the ciliomucous clearance mechanism, thereby enhancing toxicity. This was established from the clearance of intratracheally injected ^{131}I in chicks exposed to variation from 40°F to 95°F and 3 mm to 39 mm Hg vapour pressure. Higher temperature may also stimulate cutaneous blood flow and also enhance skin reaction to irritant chemicals. Excessive sweating, when not evaporated due to high ambient humidity, could lead to hydration of skin thereby increasing penetration of chemicals. Sweat may also dissolve many chemicals settled on skin. Further, high temperature may enhance the vapour pressure of chemicals, increasing the risk of exposure. Thus higher temperature and humidity could enhance chemical toxicity (Baetjer, 1968).

By measuring urinary *p*-nitrophenol in humans exposed to parathion, Funkes *et al.* (1963) showed that dermal absorption increased directly with ambient temperature, from 14.4 to 40.5°C. Other cases of temperature and toxicity of pesticides are listed by Brown (1980).

Crayfish can withstand mercury in water better at lower than higher temperatures due to less intake and to altered metabolic activity (Heit and Fingerman, 1977).

4.1.7 THEORETICAL CONSIDERATIONS

In the absence of adequate experimental studies, the mechanisms governing the alteration in toxic response as a function of temperature and humidity can best be understood from conceptual logic based on theoretical considerations. In most cases higher temperature enhances and accelerates toxicity. Filov *et al.* (1978) considered that the temperature component of the environment acted indirectly, modifying the functional state of the organism. Water loss, impaired thermoregulation, and higher breathing and blood flow may lead to larger amounts of toxicant entering and being transported to target loci. The mutual aggravation syndrome with temperature and toxicants, implies that MAC for pesticides in hot climates should be 10–20% of that elsewhere. Temperature controls the physico-chemical properties of the toxicants. Ballard (1974) has considered the theoretical aspects of temperature variation on toxicokinetics and stated that temperature could also weaken protein binding of toxicants. How far other views based on experimental animal studies are applicable to ecotoxicology is not clear. Similarly, high humidity can enhance irritancy of chemicals, such as nitrogen and sulphur oxides, forming the acids.

The influence of thermal stress on ecosystems has been well studied, mostly from the pollution by heated effluents (Laws, 1981). When toxicant stress is superimposed, the combined effect could vary. The median tolerance level of fish to copper is known to be lower at higher temperatures (EPA, 1971). The response could be attributed to faster metabolic rate, toxicant mobilization, and lesser oxygen availability at higher temperatures so that toxicity of a particular dose becomes more serious (Metelev *et al.*, 1971). The toxicity of cyanide, metals, and phenol becomes faster and more drastic at higher temperatures. The bioaccumulation of mercury in fish is also enhanced by higher ambient temperatures (Reinert *et al.*, 1974; Cember *et al.*, 1978; Kumaraguru and Beamish, 1981). The enhanced toxicity to fish of a xenobiotic at higher temperature is also indicated by a behavioural response with SO_2 at higher temperatures (Burton *et al.*, 1978). Alterations in biotransformation by liver of polychlorinated benzene and naphthoflavon exposed fish were caused by variations in temperature (Foertin *et al.*, 1983). As in the case of fauna, the toxicity of mercury to algae was also found to be higher at elevated temperatures (Huisman *et al.*, 1980). Also, in the case of insects, there are a few reports suggesting greater toxicity at higher temperature, DDT being a notable exception (Sun, 1963).

4.1.8 QUANTITATIVE STRUCTURE ACTIVITY RELATIONSHIP (QSAR)

The QSAR approach has been found to be a useful tool in environmental toxicology, as well as in occupational toxicology and pharmacology (Kaiser, 1983). Since the octanol-water partition coefficient, which is an important parameter in QSAR ecotoxicology, is likely to be influenced by temperature, once the effect of temperature of one compound is understood it may be possible to calculate it for others. Similarly, by suitable models, it may be possible to extrapolate effects from one temperature to another. Recently, the QSAR approach has been applied in predicting responses in environmental toxicology (Birge and Cassidy, 1983; Black *et al.* 1983). In cold-blooded animals the influence of ambient temperature variation on toxicokinetics can be extrapolated from experimental studies more directly than in warm-blooded animals. As such, some of the models suggested for aquatic species (Hermens *et al.*, 1985; Lipnick *et al.*, 1985; Bobra *et al.*, 1985) can be modified to include the temperature component.

4.1.9 TEMPERATURE, CARBON DIOXIDE AND PHOTOSYNTHESIS

Considerable information is available on the possible climatic changes caused by increasing levels of CO_2 due to uncontrolled fossil fuel burning (Clark, 1982). How far increased temperature affects the phytotoxicity of xenobiotics is not fully understood. In the case of acid deposition hotter and wetter climates promote soil acidification and oxygen deficiency leading to root injury (Ulrich, 1983).

One finding of significance could be the differences in the effect of temperature on the CO_2 assimilation capacity of C_3 and C_4 plants. At usual levels of CO_2, C_4 plants are less affected than C_3 plants by higher temperature. But on a threefold increase in CO_2, the pattern becomes similar (Cooper, 1982). It may be that at higher temperatures there is better tolerance to higher CO_2 concentrations, or vice versa.

4.1.10 TEMPERATURE AND ECOTOXICOLOGY OF POLLUTANTS

The effects of elevated temperatures on ecosystems and individual species have been fairly well understood (Connell and Miller, 1984). The decrease in dissolved oxygen, apart from direct effect on species, could also influence their response to toxicants. The biophysical aspects of the influence of temperature on various species have been discussed by Gates (1981). Even though the influence of high temperature on insect physiology is well understood, data relating to pesticidal response are limited. However, such information is vital in biological monitoring of pollutants and their effects. The solubility of several organochlorine insecticides in water increases with ambient water temperature, leading to greater uptake and possibly higher toxicity to aquatic biota (Phillips, 1980). Studies with DDT and mosquito fish, and rainbow trout support this. Similarly, the uptake of several toxic metals by different higher and lower aquatic fauna was found to be increased with temperature. Cairns *et al.* (1975) have reviewed the information on the effect of temperature on the toxicity of various pollutants to aquatic biota. Whereas *Daphnia magna* had similar responses to phenol, chlorobenzene, diethanol, amine, and ethylene glycol at 20°C and 24°C, *Ceriodaphnia* was distinctly more sensitive at the higher temperature (Cowgill *et al.*, 1985). Cooney *et al.* (1983) also observed temperature to influence toxicity in lower fauna.

4.1.11 ENVIRONMENTAL SIGNIFICANCE

The influence of ambient temperature and humidity on the toxicity of an environmental pollutant may have diverse practical significance. Hot and humid climates favour breeding of insects, parasites, rodents, and noxious weeds so that an effective strategy for control is needed. This has to include any superimposed effect of climate on toxicity to both target and non-target species, while designing the nature, dose, mode, and time of applying biocides. The effect of climatic extremes on toxic response may vary from species to species, at least quantitatively if not qualitatively. In that case, if any non-target species is likely to be specifically sensitive, such situations could be anticipated and avoided. Further, if higher temperature and humidity influence the safety of a plant protection chemical, a severe situation could arise during situations of drought. In the context of polyclonal natural forest ecosystems, the interrelationship between climate and xenobiotics has to be such that the natural balance is not altered. The same situation could also apply to other ecosystems when under chemical stress. Where the non-target species is mostly grown in monoculture, as in commercial agriculture, sylviculture, or pisciculture, ideal situations can be developed using seasonal variation as a means to reduce sensitivity to the chemical. The pattern of use of pesticides, fertilizers, and growth parameters can be guided by data on climatic influence. Since there is a paucity of experimental and field studies on this, it is a priority area for ecotoxicologists.

4.1.12 CONCLUDING REMARKS

Epidemiological and experimental data are very limited on any superimposed effect of high temperature or humidity, as prevalent in tropical climates, on the toxicity of occupational and environmental xenobiotics. From available information it is clear that there are chances of increased risk. Therefore, in-depth research in this direction is needed for the assessment and abatement of problems of environmental toxicology in tropical countries.

4.1.13 REFERENCES

Alpaugh, E. (1982). Temperature extremes. In: Olishifski, J. B. (ed.), *Fundamentals of Industrial Hygiene*, 4th ed., pp. 371–400. National Safety Council, U.S.A.

Aly, O. M., and ElDib, M. A. (1972). Studies of the persistence of some carbamate insecticides in the aquatic environment. In: *Fate of Organic Pesticides in the Aquatic Environment*, Gould, R. F. (ed.), pp. 210–243. Advances in Chemistry Series 111. American Chemical Society, Washington.

Baetjer, A. M. (1968). Role of environmental temperature and humidity in susceptibility to disease. *Archs. Envir. Hlth.*, **16**, 565–570.

Baetjer, A. M., and Smith, R. (1956). Effect of environmental temperature on reaction of mice to parathion and anticholine esterase agent. *Amer. J. Physiol.*, **186**, 39–46.

Baetjer, A. M., Joardar, S. N. D., and McQuary, W. A. (1960). Effects of environmental temperature and humidity on lead poisoning in animals. *Archs. Envir. Hlth.*, **1**, 463–477.

Ballard, B. E. (1974). Pharmacokinetics and temperature. *J. Pharm. Sci.*, **63**, 1345–1358.

Beevers, L., and Cooper, J. P. (1964). Influence of temperature on growth and metabolism of rye grass seedlings. II. Variation in metabolites. *Crop Sci*, **4**, 143–146.

Bhatia, B., Chhina, G. S., and Singh, B. (1975). *Selected Topics in Environmental Biology*. Interprint Publishers, New Delhi.

Birge, J. W., and Cassidy, R. A. (1983). Structure-activity relationship in aquatic fauna. *Fund. Appl. Toxicol.*, **3**, 359–368.

Black, J. A., Birge, W. J., Westerman, A. G., and Francis, P. C. (1983). Comparative aquatic toxicology of aromatic hydrocarbons. *Fund. Appl. Toxicol.*, **3**, 353–358.

Bobra, A., Yingshin, W., and Mackay, D. (1985). QSAR for the acute toxicity of chlorobenzene to *Daphnia magna*. *Environ. Toxicol. Chem.*, **41**, 297–305.

Boudou, A., Ribeyre, F., Delachre, A., and Marty, R. (1980). Bioaccumulation et bioamplification des dérivés du mercure par un consommateur de troisième ordre: *Salmo gairdneri*—incidences du facteur température. *Wat. Res.*, **14**, 61–65.

Boyd, E. M. (1972). The human, animal and physical environmental elements. In: *Predictive Toxicometrics*, pp. 193–202. Scientechnica Publishers, Bristol.

Broulia, L., Smith, P. E., and Stopps, G. J. (1960). The physical environment and the industrial worker. In: Fleming, A. J., and Alonzo, C. A. (eds.), *Modern Occupational Medicine*, pp. 137–180. Lea & Febiger, Philadelphia.

Brown, V. K. (1980). Test animals. In: *Acute Toxicity: Theory and Practice*, pp. 33–67, 117. John Wiley & Sons, Chichester.

Brown, V. M., Jordan, D. H. M., and Tiller, B. A. (1967). The effect of temperature on the acute toxicity of phenol to rainbow trout in hard water. *Wat. Res.*, **1**, 587–594.

Burton, D. T., Graves, W. C., and Margrey, S. L. (1978). Behavioural modification of estuarine fish exposed to sulphur dioxide. *J. Toxicol. Environ. Hlth.*, **13**, 969–978.

Cairns, J., Heath, A. G., and Parker, B. C. (1975). Temperature influence on chemical toxicity of aquatic organisms. *WPCF Journal*, **47**, 267–280.

Carlson, R. W. (1979). Reduction in the photosynthetic rate of *Acer*, *Quercus* and *Fraxinus* species caused by sulphur dioxide and ozone. *Environ. Pollut.*, **18**, 159–170.

Casarett, L. J., and Doull, J. (1975). Factors influencing toxicity. In: *Toxicology. The Basic Sciences of Poisons*, pp. 133–147. Macmillan Publishing Co. Inc., New York.

Cember, H., Curtis, E. H., and Blaylock, B. G. (1978). Mercury biconcentration in fish: temperature and concentration effects. *Environ. Pollution*, **17**, 311–319.

Clark, W. C. (1982). *Carbon Dioxide Review*. Clarendon Press, Oxford, New York.

Connell, D. W., and Miller, G. J. (eds) (1984). Thermal pollution. In: *Chemistry and Ecotoxicology of Pollution*, pp. 371–387. Wiley Interscience, New York.

Cooney, J. D., Beauchamp, J. J., and Gehrs, C. W. (1983). Effect of temperature and nutritional state on the acute toxicity of acridine to calanoid copepod *Diaptomus clavipes* Schacht. *Environ. Toxic. Chem.*, **2**, 431–439.

Cooper, C. F. (1982). Food and fibre in a world of increasing carbon dioxide. In: Clark W. C. (ed.), *Carbon Dioxide Review*, pp. 299–333. Clarendon Press, Oxford, New York.

Cornwall, P. B., and Bull, J. O. (1967). Alphakill—a new rodenticide for mouse control. *Pest Control*, **35**, 31–32.

Cowgill, U. M., Takahashi, I. T., and Applegate, S. L. (1985). A comparison of the effect of four bench mark chemicals on *Daphnia magna* and *Ceriodaphnia* tested at two different temperatures. *Environ. Toxic. Chem.*, **4**, 415–422.

Craig, F. N., Cummings, E. G., and Sim, V. M. (1977). Environmental temperature and the precutaneous absorption of a cholinesterase inhibitor. *J. Invest. Derm.*, **68**, 357–361.

Cummings, E. G. (1969). Temperature and concentration effects on penetration of N-octylamine through human skin *in situ*. *J. Invest. Derm.*, **53**, 64–70.

Doull, J. (1972). The effect of physical environmental factors on drug response. In: Hayes, W. J., Jr. (ed.), *Essays in Toxicology*, Vol. 3. Academic Press, N.Y.

EPA (1971). *Water Quality Criteria No. R3.73.033*. Environmental Protection Agency, Washington D.C. 549–551.

Farris, E. J., and Griffith, J. Q., Jr. (1949). *The Rat in Laboratory Investigation*, 2nd edn, pp. 303–314. Lippincott, Philadelphia.

Filov, O. A., Goluber, A. A., Liublinea, E. I., and Tolokonsten, N. A. (1978). The toxic effect as a result of interaction between the poison and the living organism. In: *Quantitative Toxicology. Selected Topics*, pp. 1–22. Wiley Interscience, New York.

Foertin, L., Anderson, T., Koivusare, U., and Hansen, T. (1983). Influence of biological and environmental factors on hepatic steroid and xenobiotic metabolism in fish. Interaction with PCB and B-naphthoflavone. *Marine Environ. Res*, **14**, 1–4.

Fuhrman, G. J., and Fuhrman, F. A. (1961). Effects of temperature on the action of drugs. *Toxic. Appl. Pharmac.* , **1**, 65–78.

Funkes, A. J., Hayes, G. R., and Hartwell, W. O. (1963). Urinary excretion of *p*-nitrophenol by volunteers following dermal exposure to parathion at different ambient temperatures. *J. Agr. Fd Chem.*, **11**, 455–457.

Gates, D. M. (ed.) (1981). Temperature and organisms. In: *Biophysical Ecology*, 527–569. Springer Verlag, New York.

Gill, F. S. (1980). Heat. In: Waldron, H. A., and Harrington, J. M. (eds.), *Occupational Hygiene*, pp. 225–256. Blackwell Scientific Publications, Oxford.

Gunther, E. A., Iwata, Y., Carman, G. E., and Smith, C. A. (1977). The citrus re-entry problem. Research on its causes and effects and approaches to its minimization. *Residue Revs.*, **67**, 1–132.

Hartley, G. S., and Graham-Bryce, I. J. (eds. 1st edition) (1980). Penetration of pesticides into higher plants. In: *Physical Principles of Pesticide Behaviour*, Vol. 2, pp. 545–657. Academic Press, New York.

Hayes, W. J. (1963). *Clinical Handbook on Economic Poisons*. Public Health Service Publication No. 476, Environmental Protection Agency, Washington D.C.

Heck, W. W., and Dunning, J. A. (1967). Effect of O_3 on tobacco and pinto beams as conditioned by several ecological factors. *J. Air. Poll. Contr. Assoc.*, **17**, 112–114.

Heit, M., and Fingerman, M. (1977). The influence of size, sex and temperature on the toxicity of mercury on two species of cray fishes. *Bull. Environ. Cont. Toxicol.*, **18**, 572–580.

Hermens, J., Konemann, H., Leeuwangh, P., and Musch, A. (1985). QSAR in aquatic toxicity studies of chemicals and complex mixtures of chemicals. *Environ. Toxicol. Chem.*, **4**, 273–279.

Hertig, B.A. (1975). Work in hot environments. Threshold Limit Values and proposed standards. In: Cralley, L.V., and Atkins, P.R. (eds), *Industrial Environmental Health*, pp. 219–231. Academic Press, New York.

Holdgate, M.W. (1979). Effect of chemical pollutants on animals. In: *A Perspective of Environmental Pollution*, pp. 116–125. Cambridge University Press, Cambridge.

Horiguchi, S., and Horiguchi, K. (1966). Effect of environmental temperature on the toxicity of 1,1,1-trichloroethane in mice. *J. Ind. Hlth.* (Japan), **13**, 290–291.

Horiguchi, S., Kasahara, A., Morioka, S., Utsunomiya, T., and Shinagawa, K. (1979). Experimental study on the effect of hot environment on the manifestation of lead poisoning in rabbits. *Sumitomo Sangyo Eisei*, **15**, 122–128.

Horvath, S.M. (1979). Evaluation of exposures to hot and cold environments. In: *Patty's Industrial Hygiene and Toxicology*, Vol. III, Cralley, L.V., and Cralley, L.J. (eds) 1st edition, pp. 447–464. John Wiley & Sons Inc., New York.

Huisman, J., Hoopen, H.J.G., and Fuchs, A. (1980). The effect of temperature upon the toxicity of mercuric chloride to *Scenedesmus acutus*. *Environ. Pollut.*, (A), **22**, 133–148.

Hunt, V.R. (1979). The physical environment. In: *Work and the Health of Women.*, pp. 61–96. CRC Press, Boca Raton, Florida.

Jacobson, A.R., and Morris, S.C. (1976). The primary air pollutants. Viable particulates, their occurrences, sources and effects. In: Stern, A.C. (ed.), *Air Pollution*, 3rd edn, vol 1, pp. 169–196. Academic Press, New York.

Jahnke, W. (1957). Antropinvergiftugen in heissen Klima. *Archs. Toxicol.* **16**, 243–247.

Kaiser, K.L.E. (1983). *QSAR in Environmental Toxicology*. D. Reidel Publishing Co., Dordrecht.

Keplinger, M.L., Lamer, G.E., and Deichman, W.B. (1959). Effects of environmental temperatures on the acute toxicity of a number of compounds in rats. *Tox. Appl. Pharmac.*, **1**, 156–595.

Kumaraguru, A.K., and Beamish, F.W.H. (1981). Lethal toxicity of permethrin to rainbow trouth (*Salmo gairdneri*) in relation to body weight and water temperature *Wat. Res.*, **15**, 503–505.

Langridge, J. (1963). Biochemical aspects of temperature response. *A. Rev. Plant Physiol.*, **14**, 441–462.

Laws, E.A. (1981). Thermal pollution and power plants. In: *Aquatic Pollution*, pp. 266–300. John Wiley & Sons, New York.

Linak, W.P., and Peterson, T.W. (1983). Visibility: pollutant relationships in southern Arizona. II. A. Winter/summer field study. *Atmos. Environ.*, **17**, 1811–1823.

Lindquist, O., Ljungstrom, E., and Svensson, R. (1982). Low temperature thermal oxidation of nitric oxide in polluted air. *Atmos. Environ.*, **16**, 1957–1972.

Lipnick, R.L., Johnson, D.E., Gilford, J.H., Bickings, C.K., and Newsome, L.D. (1985). Comparison of fresh toxicity screening data for 55 alcohols with QSAR predictions of minimum toxicity for non-reactive non-electrolytic organic compounds. *Environ. Toxicol. Chem.*, **4**, 281–296.

MacCormick, R.A., and Holzworth, A.C. (1976). Air Pollution. In: Stern, A.C. (ed.), *Climatology in Air Pollution*, 3rd edn., vol. I, pp. 643–700. Academic Press, New York.

MacLeod, J. C., and Pessah (1973). Temperature effects on mercury accumulation toxicity and metabolic rate in rainbow trout (*Salmo gairdneri*). *J. Fish Res. Bd. Canada*, **30**, 485–492.

Matsui, K., and Nomiyama, K. (1979). Effects of environmental temperatures on the cadmium induced testicular injury. *J. Hyg.* (Japan), **34**, 620–623.

Metelev, V. V., Kanaev, A. I., and Dzabokhova, N. G. (1971). Effects of ecological factors on resistance to fish toxicants. In: *Water Toxicology*, pp. 25–37. Amerind Publishings, New Delhi.

Meyer, B. J., and Karel, L. (1948). The effect of environmental temperature on naphthyl thiourea toxicity to rats. *J. Pharmac. Exp. Ther.*, **93**, 420–422.

Nomiyama, K., and Nomiyama, H. (1976). Effects of environmental temperature—the acute toxicity of Cd in mice. *Kankyo Hoken Rep.* **38**, 153–155.

Nomiyama, K., Matsui, K., and Nomiyama, H. (1980a). Environmental temperature, a factor modifying acute toxicities of organic solvents, heavy metals and agricultural chemicals. *Toxicol. Letters.* **6**, 67–70.

Nomiyama, K., Matsui, K., and Nomiyama, H. (1980b). Effects of temperature and other factors on the toxicity of methyl mercury in mice. *Toxicol. Appl. Pharmacol.*, **56**, 392–398.

Phillips, D. J. H. (1980). *Quantitative Aquatic Biological Indicators*, pp. 185–187. Applied Science Publishers, London.

Pratt, G. C., and Krupa, S. V. (1983). Seasonal trends in precipitation chemistry. *Atmos. Environ.*, **17**, 1845–1847.

Rehwoldt, R., Menapace, L. W., Merrie, B., and Alessandrello, D. (1972). The effect of increased temperature upon the acute toxicity of some heavy metal ions. *Bull. Environ. Cont. Toxicol.*, **8**, 91–96.

Reinert, R. E., Stone, L. J., and Willford, W. A. (1974). Effects of temperature on accumulation of methyl mercuric chloride and pp′-DDT by rainbow trout (*Salmo gairdneri*). *J. Fish. Res. Bd.* Canada, **31**, 1649–1652.

Sloane, C. (1983). Summertime visibility decline: meteorological influences. *Atmos. Environ.*, **17**, 763–774.

Smith, R. N., and Bababunmi, E. A. (eds) (1980). *Toxicology in the Tropics*. Taylor and Francis, London.

Somero, G. N., Chow, T. J., Yancey, P. H., and Snyder, C. B. (1977). Lead accumulation rates in tissues of the estuarine teleost fish (*Gillichthys mirabilis*); salinity and temperature effects. *Archs. Environ. Contam. Toxicol.*, **6**, 337–348.

Sun, Y. (1963). Bioassay, insects. In: Zweis, H. (ed.), *Analytical Methods for Pesticides, Plant Growth Regulation and Food Additives*. vol. I, p. 40. Academic Press, New York.

Suskind, R. R. (1977). Environment and the skin. *Environ. Hlth. Persp.*, **20**, 27–37.

Ulrich, B. (1983). A concept of forest ecosystem stability and of acid deposition as driving force for destabilization. In: Ulrich, B., and Pankvalls, J. (eds), *Effect of accumulation of air pollutants in forest ecosystem*, pp. 1–19. D. Riedel Publishing Co., Holland.

Wagman, J., Lee, J. R., and Axt, C. J. (1967). Influence of some atmospheric variables on the concentration and particle size distribution of sulfate. *Atmos. Environ.*, **1**, 479.

Weihe, W. H. (1973). The effect of temperature on the action of drugs. *A. Rev. Pharmac.*, **13**, 409–425.

Went, F. W. (1953). Effect of temperature on plant growth. *A. Rev. Plant. Physiol.*, **4**, 347–362.

Wolfe, H. P., Durham, W. F., and Batchelor, G. S. (1961). Health hazards of some climatic compounds. *Archs. Envir. Hlth.*, **3**, 104–111.

Wyndham, C. H. (1970). Adaptation to heat and cold. In: Lee, D. H. K., and Minand, O. (eds), *Physiology, Environment and Man*, pp. 177–205. Academic Press, New York.

Yamaguchi, S., Shimajo, N., Sano, K., Kano, K., Hirota, Y., and Saisho, A. (1984). Effects of environmental temperatures on the toxicity of methyl mercury in rats. *Bull. Environ. Contam. Toxicol.*, **32**, 543–549.

Yamanouchi, C., Takahashi, H., Ando, M., Imaishi, N., and Nomura, T. (1967). Effect of environmental temperature on the acute toxicity of drugs. *Exp. Anim.*, **16**, 31–38.

Yarwood, C. E. (1961). Translocated heat injury. *Plant Physiol.*, **36**, 721–726.

Yokuchi, Y., Okanowa, M., Ambe, Y., and Furwa, K. (1983). Seasonal variations of monoterpenes in the atmosphere of a pine forest. *Atmos. Environ.*, **17**, 743–740.

Zbinden, G. (1973). Comparative toxicology. In: *Progress in Toxicology*, vol. 3, pp. 38–45. Springer Verlag, Berlin.

Ecotoxicology and Climate
Edited by P. Bourdeau, J. A. Haines, W. Klein and C. R. Krishna Murti

4.2 Effects in Arid Regions

A. S. AND R. Y. PERRY

4.2.1 INTRODUCTION

An agro-ecosystem is a complex entity comprising a number of elements which interact with one another to form and stabilize the system.

An arid ecosystem differs in many respects from one of tropical or temperate zones. Unlike temperate zones, an arid zone is characterized as an area of low precipitation, high temperature, and high rate of evaporation. Frequency of rainfall does not describe accurately the type of the arid zone nor does the amount of precipitation govern the type of vegetation. Nevertheless, precipitation serves as a criterion for the subdividing of arid zones. In the broad sense, 57–300 mm mean annual rainfall is typical of an arid zone, while 300–550 mm is that of a semi-arid zone (Thornthwaite, 1948; *see* Figure 4.2.1).

The dry climate of the world occurs in five great geographical areas lying between 15° and 35° latitude. The largest of the five, known as the African-Eurasian Dry Zone, includes the Sahara desert from the Atlantic coast, extending eastward to the Arabian Peninsula, Pakistan and India; northward, it includes Iran and southern Russia, and still farther north, it includes Chinese Turkestan and Mongolia. The southern margin includes the semi-arid zone, the so-called 'Sahel' (Meigs, 1953). This vast region includes also the Middle Eastern arid zone and that of Israel.

Typically, the arid zone of Israel is characterized as a desert with a very low rainfall, low relative humidity, high solar radiation, and high potential for evaporation (Evanari *et al.*, 1971). In considering the complexity of the arid ecosystem of the land, one must take into consideration physical, chemical, and biological parameters such as air, water, soil, biota, and vegetation.

4.2.2 THE ARID ECOSYSTEM OF ISRAEL

The arid zone comprises mainly the Negev and the Arava regions extending north to the Jordan Valley and leading to the southern edge of Lake Kinneret (The Sea of Galilee).

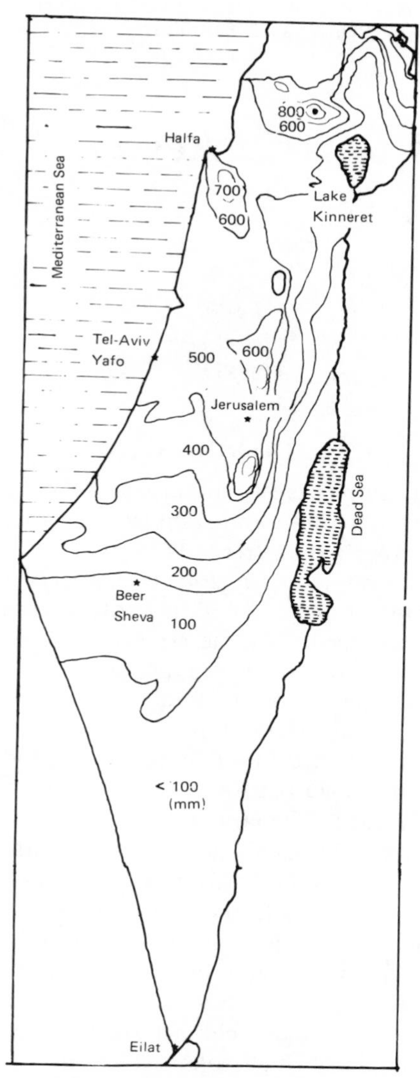

Figure 4.2.1 Climatic regions of Israel. Mean annual rainfall (mm). From Atlas of Israel. Reprinted with permission from Survey of Israel

The Negev comprises a total area of 1 186 000 ha, the Arava a total area of 193 400 ha, or 56.4% and 9.22%, respectively, of the total area of Israel.

4.2.2.1 The Climate

The desert of Israel extending south of Be'er Sheva to the city of Eilat consists

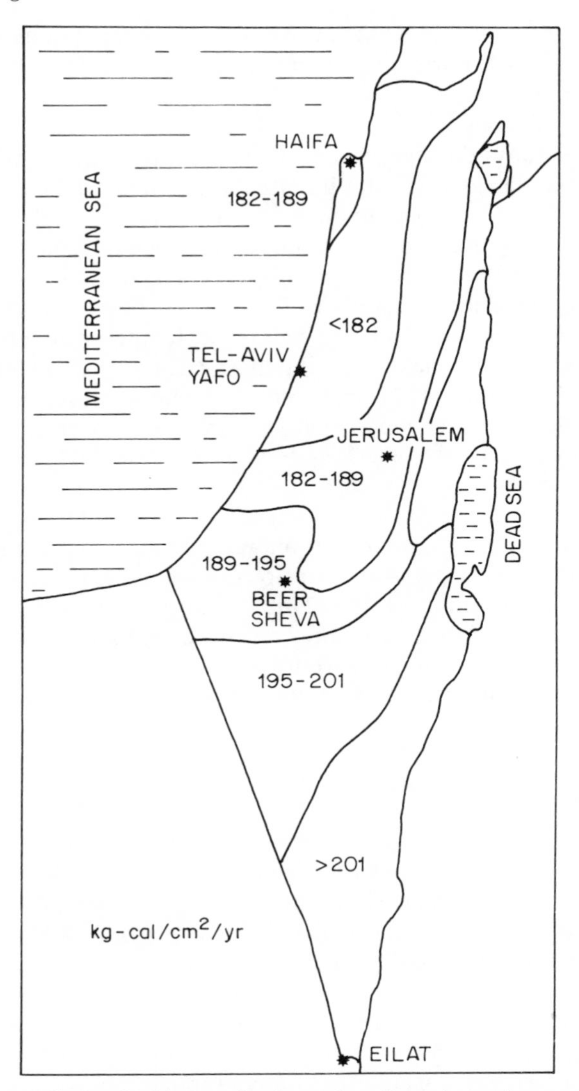

Figure 4.2.2 Solar radiation incident on horizontal surface. Annual mean (kg-calories/ cm^2/year). From Atlas of Israel (1970). Reprinted with permission from Survey of Israel

of two distinct climatic ecosystems, i.e. the Negev and the Arava. These areas are characterized by high temperatures exceeding 30°C for a good part of the year, e.g. at Eilat 206 days of the year register maximum daily temperatures exceeding 30°C. This is defined as 206 tropical days (Atlas of Israel, 1970).

Relative humidities average 50% in January (the coldest month) to 28% in July (the hottest month). The average yearly rainfall in the central Negev ranges between 28 and 168 mm, with an average of 86 mm (Atlas of Israel, 1970: Evenari *et al.*, 1968). Further south in the Arava desert, the yearly rainfall rarely exceeds 100 mm.

The desert of Israel is one of the sunniest areas of the world, having 200 or more cloudless days during the year. Consequently, solar radiation is received in an even manner and is quite intense (*see* Figure 4.2.2).

Strong north winds abound in the southern Negev during the summer months. Primarily in spring but also in autumn, easterly winds from the Sahara Desert move toward the most part of Israel but are felt more strongly in the desert area. These winds, called 'sharav' or 'hamsin', are hot dry desert winds carrying sand and dust particles and can raise the temperature up to 42°C. The Meteorological Service classifies light sharav as a day on which mean relative humidity is lower than 45% irrespective of temperature. A heavy sharav is defined as a day on which mean relative humidity is lower than 20% and mean daytime temperature is above its long-term monthly mean (Lomas and Shashoua, 1974). The number of sharav days in a year varies from a few days to 34 days in the wheat-growing region of the northern Negev. These winds have been reported to have a damaging effect on agricultural production in this region.

Concerning vapour pressure, the lowest values occur in the mountains, decreasing to a minimum of 10–12 mb on an annual average. In the central and southern parts of the Negev, the values are 13.6 and 12.8 mb, respectively. The annual run of vapour pressure forms a simple wave more or less parallel to that of temperature, i.e., high values in July–August, low values in January–February (Atlas of Israel, 1970).

4.2.2.2 Soil and Water

The soil in the arid zone is characterized by neutral or high pH (7.0–8.7), high calcium carbonate content, and low organic matter. Several soil types occur in the desert region, some of which can be utilized for agriculture. One of the obstacles to utilization of such soils is the high salt content (generally 0.15–2.0% or greater), mainly in the form of NaCl but also $CaSO_4$. To obviate such an obstacle, the need for irrigation water becomes paramount. Hence, the development of agricultural enterprises in the arid zone is primarily a function of the availability of water for irrigation. A portion of the water supplied by the National Water Carrier is relatively high in chloride content (about 250 mg/litre). This, coupled with low rainfall and the lack of leaching due to water shortage all contribute to the accumulation of salts in the soil profile (Bielorai and Levy, 1971). Detailed information on soil types and other parameters is given in the Atlas of Israel (1970).

4.2.2.3 Biota

Very few vertebrate species are found throughout the entire country. The species which occur under desert and semi-desert conditions include jirds and gerbils, the

Levant vole (*Microtus guntheri*), Tristram's jird (*Meriones tristrami*), and the ancestor of the house mouse, *Mus musculus praeraxtus*. These rodents periodically undergo explosive population outbreaks and could become agricultural pests. Indeed, the house mouse (*Mus musculus*) has recently become a major pest of green peppers in the desert region (Zook-Rimon, 1985).

Other mammalian, reptilian, and avian species include Baluchistan's gerbil, Cairo spiny mouse, fat jird, Negev jird, Ethiopian hedgehog, Egyptian dabb-lizard, Turkish gecko, Sinai agama, tropical cat-snake, crown-necked raven, Sinai rock partridge, and Desert partridge. Of seven species of poisonous snakes in Israel, six inhabit the desert and semi-desert areas. *Aspis vipera* occurs in the Arava (Atlas of Israel, 1970).

4.2.2.4 Vegetation

In the desert areas where mean annual rainfall is 150–300 mm, there exists almost no arboreal climax community. The vegetational landscape is characterized by poor but more or less continuous grey dwarf-shrub formations. In the areas where mean annual rainfall scarcely reaches 150 mm, vegetation is rare and patchy, mainly being limited to seasonal stream beds.

Trees occur only near permanent water bodies or in otherwise favourable habitats. Native desert vegetation includes principally the genera *Tamarix*, *Iris*, *Acacia*, and *Pistacia*. The genus *Tamarix* is represented in Israel by 11 indigenous species. All species require large amounts of moisture, which they obtain through their exceedingly long roots. Tamarisks are very useful in dune fixation and afforestation of deserts and salt pans, and as windbreaks and sea spray interceptors. The genus *Acacia* includes four species, of which three are found in the desert areas. In the Arava valley, the most impressive feature of the landscape is the acacia pseudosavannah, comprising mainly *Acacia raddiana* and *A. tortilis*. One of the largest tropical oases of the northeastern corner of the Arava displays no less than 40 tropical species of plants. Halophilous shrubs dominate the vicinity of the Dead Sea and the Eilat-Timna segment, where extensive salt flats occur (Atlas of Israel, 1970). A complete and beautifully illustrated book on the flora of the Negev and Sinai desert, including edible native vegetation, has been published by Danin (1983).

4.2.3 AGRO-ECOSYSTEM

Agricultural development in arid zones must take into consideration, in addition to everything else, the shortage of natural resources. One of the typical features of the arid or semi-arid zones is the lack of geographical proximity between agriculture's two main natural factors of production—land and water. In Israel, most of the arid land and all of the water resources are state-owned. This facilitates water development from an economic viewpoint. The National Water

Carrier, whose source is Lake Kinneret (The Sea of Galilee), is responsible for supplying water to the farmers in the arid zone at relatively low cost. In addition to the supply of natural water, waste water can be used as a source of irrigation water in arid areas. Sewage water contains a considerable quantity of nutrient elements for plant growth, including a high nitrogen content, in addition to large quantities of organic matter and of trace elements. These waters represent a valuable nutrient source and can improve the structural and physical properties of soils in the arid region (Yaron *et al.*, 1983). Agricultural production for irrigated areas in the arid zone can attain full self-sufficiency in fruits and vegetables and other agricultural commodities. Obviously, this would entail substantial capital investment.

4.2.3.1 Effect of Irrigation

Introduction of water into an arid land inevitably will convert that area into an agro-ecosystem typical of the particular crop to be grown. Concurrently, changes will occur in the diversity of insect fauna as well as in the ecology of the insect dwellers of the arid zone. A typical example is the creation of an environment favourable for *Schistocerca gregaria*, where irrigation satisfies the moisture requirements for oviposition while vegetation becomes more abundant for supporting larger nymphal populations (Shulov, 1952; Popov, 1958; Stower *et al.*, 1958). Another example occurs during the period of the dry desert winds in springtime, when a mass migration of *Thrips tabaci* (Lindeman) takes place from the withering uncultivated vegetation to the irrigated fields of peanuts and cotton. Later in the summer the insects disappear from these crops (Rivnay, 1964). Bishara (1932) in Egypt pointed out that heavier outbreaks of *Agrotis ypsilon* (Rottenburg) took place in irrigated rather than non-irrigated fields. Likewise, greater infestations in irrigated fields by *Prodenia litura* (Fabricus) and by *Spodoptera exigua* (Hubner) in alfalfa, cotton, and peanut fields in Israel have been reported (Rivnay, 1964). Other insects, such as *Psallus seriatus* (Reuter), *Adelphocoris rapidus* (Say), *Lygus lineolaris* (Palisot de Beauvois), and *Tetranychus telarius* (Linnaeus), as well as leafhoppers on alfalfa caused more damage in irrigated than in non-irrigated plants (Adkisson, 1957). *Hylemia cilicrura* (Rondani) ordinarily oviposits in humid ploughed earth. The eggs and larvae are not capable of living in an arid soil and this is the reason why the fly does not exist in its larval stage in summer in Israel (Yathom, 1961). However, as a result of irrigation, several crops, such as peanuts, cotton, and water melon, may become infested.

Irrigation may also interfere with the normal life history of subterranean insects. In Israel, moisture causes the termination of diapause of scarabaeid larvae of the genera *Phyllopertha* and *Anisoplia*. In regions where rainfall is below 300 mm diapause is usually terminated toward the end of December and early January. The recent practice of early irrigation of winter crops in this region in November has caused an earlier diapause termination of the larvae with a consequence of serious economic damage (Rivnay, 1964).

The type of irrigation may also create different ecological conditions for certain insects. For example, the eggs of *Raphidopalpa foveicollis* (Lucas) are laid in the ground around the stem of the plant and they need contact with moisture for successful development. Ditch irrigation does not provide such favourable conditions for the eggs as overhead irrigation does. Thus, in the latter type the beetles become more numerous. However, in this case, the excessive moisture around the plant is not conducive to healthy larval development, which normally takes place in the soil. Hence, the larvae migrate to the top surface where they attack the fruit with devastating results (Rivnay, 1964). In Israel, drip irrigation is now being practised for many irrigated crops. This new type of irrigation consists of partial wetting of the soil aiming particularly at the root zone only. Systemic pesticides can also be applied with this system, thus minimizing the amount of pesticide used and the extent of leaching. Irrigation water for the Negev comes from the north, but the western region and the Arava contain an abundance of underground water which is saline. Using such water for irrigation, although causing no damage to the plant itself, can be a potential pollutant to the arid zone. This is due primarily to the high Na/Ca ratio which has an adverse effect on soil properties (Yaron *et al.*, 1985). Under desert conditions such irrigation, although favourable from the standpoint of water conservation, may accentuate a change in the ecological conditions of insect life, resulting in adverse effects on crop production and necessitating perhaps more extensive applications of chemicals for pest control.

4.2.3.2 Effect of Other Agricultural Practices

As new virgin land is being exploited for agriculture it is anticipated that many changes will take place in the fauna and flora of that ecosystem. With the introduction of new crops not indigenous to the area, various insect pests may migrate to such territories and cause considerable damage. The abundance and continuity of food brought about by the application of agrotechnical methods inevitably will create favourable ecological conditions for insect development, resulting in outbreaks of pests. Many such examples are known (Rivnay, 1964). Soil temperature has a definite influence on the microclimatic conditions affecting the development of soil-dwelling insects. In semi-arid or more temperate zones the tobacco white fly, *Bemisia tabaci* (Gennadius), is a severe pest of tomato seedlings, cotton, and many other crops in the summer months. The adult insects can withstand temperatures of up to 45°C, but if the soil temperature reaches above 46°C, they will succumb to the intense radiation.

Mulching the soil may have the same effect on insect survival. An excellent example of this agrotechnical practice is thermal sterilization of the soil against soil insects and weed germination. This is accomplished by watering the soil and then covering the area with transparent polyethylene sheets. After several days this raises the soil temperature to 70°C or higher, thus burning the seeds that had germinated and killing soil pests, such as fungi (Katan *et al.*, 1976b). This type of treatment obviates the use of methyl bromide, which is used extensively for soil sterilization in the north, in the temperate zones as well as in semi-arid and arid agro-ecosystems.

The thermal sterilization of the soil as outlined above has some severe limitations with regard to incorporation in most crop rotations due to the long period of solarization needed to obtain effective control, thus delaying the growth period of the crop itself. However, a more important reason is the cost of treatment. Solarization of the soil for control of soil-borne pests costs approximately $1500 per hectare. Treatment with metham-sodium, which provides good control of the soil-borne pathogens, costs only $210–420 per hectare, and methyl bromide treatments, such as those used extensively by most growers, cost only $350 per hectare (Krikun *et al.*, 1982).

4.2.3.3 Environmental Effects on Plant Growth

In general, plant growth in arid zones is subjected to damage more from natural sources than from introduced contaminants due to man's activities.

(a) Injury due to natural sources

In planning agricultural enterprises in the desert region, one must select plants that can withstand certain natural environmental conditions, such as sandstorms. Many plants are susceptible to injury by sandstorms, especially leafy vegetables such as curcurbits. The injury is mainly due to strong winds and sand particles which cause abrasion of the leafy surfaces of plants. Lack of vegetation in the desert is one of the primary causes of sandstorms: this lack results in the destabilization of the soil surface, which then becomes windblown. In the northern Negev, the amount of suspended particles in the air can reach three to four times the normal amount in the air. Most of these particles are larger than 100 μm. The problem becomes more acute further south (Lerman, 1985).

(b) Injury due to introduced industrial contaminants

Introduced contaminants may arise from the establishment or relocation of chemical industries from highly populated areas to desert regions. Pollution emanating from fossil fuel, photo-oxidants in the atmosphere, by-products from pesticide industries, flying ash from power plants, waste-water effluents from industrial plants, etc., may cause significant injury to plant growth. In the arid zone, acid rain is not a serious problem due to the alkalinity of the soil. However, flying ash with a high sulphur content (4% sulphur) produced by an electrical plant in the desert area has caused corky tissue of citrus fruit. Flying ash can also have a detrimental effect on cotton fibre (Lerman, 1985). Moreover, flying ash can contain as much as 12% of coal as residue as well as 12% of heavy metals such as molybdenum and selenium. With a high pH in an arid environment, movement of heavy metals is deterred, thus the metals remain more or less confined to the soil surface. The significance of heavy metals in flying ash may become more acute in agricultural areas when residues which normally remain on the soil surface are ploughed in and become available to plant roots (Yaron *et al.*, 1985).

The concentration of heavy metals in domestic effluent is fairly low and is usually within a pH range of 7.0–7.5. It is anticipated, therefore, that under such conditions no detrimental effects would result. However, in light and coarse soils, which are characteristic of arid land, the metals in the effluent might move more freely and penetrate into deeper layers of the soil, thus becoming more available to the plant (Yaron *et al.*, 1985).

Excessive build-up of bromine occurs in native vegetation as a result of emission from a nearby pesticide industry. Theoretically, desert plants, such as *Tamarix* spp. and *Eucalyptus* spp., seem to be more tolerant to air pollution. However, a potential danger exists in the unwariness of the chemical industries spewing man-made pollutants into the desert atmosphere.

Suffice it to say that man-made pollution is not characteristic of desert regions but is typical of man's intervention in other ecosystem types.

4.2.3.4 Pesticides in the Desert Agro-ecosystem of Israel

The limited availability of agricultural land and the need for irrigation necessitate the use of intensive agricultural practices for crop production in Israel. Such practices are conducive to creating favourable conditions for soil-borne pathogens, which can cause great damage to crops if left uncontrolled. Pathogenic fungi, such as *Verticilium dahliae* and *Macrophomina phaseolina*, and the root-cortex-invader nematode, *Pratylenchus thornei*, are especially injurious to many crops (Krikun *et al.*, 1982).

Of particular interest is the influence of *Pratylenchus thornei* on wheat growing in the Negev desert region. Winter-sown wheat is grown in this area to utilize the sparse rainfall (200–250 mm) during the winter months. In cases where soil treatments were made with vaporized methyl bromide or with sprinkler irrigation-applied metham-sodium, which effectively controlled the nematode, the plants had higher leaf water potentials, increased water use efficiency, increased nitrogen uptake, and highly increased yields (Amir *et al.*, 1982). This led to the conclusion that under nematode attack the roots were unable to take up the scantily available water as efficiently as the non-parasitized roots, leading to a significant decrease in yield. In non-treated soils the pathogen builds up rapidly due to the low organic matter and relatively dry soil conditions which,

Table 4.2.1 Effect of metham-sodium application on crop yield and on pathogen eradication. From Krikun *et al.* (1982). Reproduced with permission of the Agriculturist Research Organization

Crop	Pathogen	Concentration of metham-sodium (ppb) μg/l a.i.	% Pathogen eradication	% Increase in yield	Absolute increase in yield (tonne/hectare)
Wheat	*Pratylenchus thornei*	160	100	48	1.76
Onion	*Pratylenchus thornei*	320	100	23	12.17
Potato	*Pratylenchus thornei*	160	85	25–29	7.7–12.2
Onion	*Pratylenchus terrestris*	320	84	>3.5 cm diam/45	5.5
				<3.5 cm diam/328	34.8
Peanut	*Pratylenchus myriotylum*	320	–	69	2.46

in turn, lead to low microbial activity and reduced antagonism. Another important factor is the poor mycorrhizal development in the affected plants, which leads to root cortical necrosis (Krikun *et al.*, 1982). Control of the nematode has made it possible to extend the wheat growing area further south to regions of lower soil moisture content. The effect of metham-sodium application on crop yield and on pathogen eradication is shown in Table 4.2.1.

4.2.3.5 Fate of Pesticides in the Environment

The disappearance of chemicals from the environment is due directly or indirectly to chemical, photochemical, physical, microbiological, and to higher plant and animal metabolic degradations. Chemical degradation of pesticides in soils is governed by a variety of factors, such as pH, presence of water, and the presence of various catalysts and reagents capable of attacking reactive compounds.

(a) Photochemical decomposition of pesticides

It is well known that light in the ultraviolet (UV) region of the spectrum contains energy capable of inducing chemical transformation in a variety of compounds able to absorb energy. Many insecticides, such as chlorinated hydrocarbons, organophosphorus esters, carbamates, pyrethroids, rotenone, and others have been shown to undergo photoreactions to form products that are either more toxic or less toxic than the parent compounds (Rosen *et al.*, 1966; Rosen, 1972; Robinson *et al.*, 1966; Chen and Casida, 1969; Crosby, 1972a, 1972b;

Khan *et al.*, 1974). Ordinarily, photolysis takes place only on the surface and is more pronounced on water surfaces. In the deepest layers of soil, photolysis is not a major process. However, the high surface temperatures of arid soils may cause a higher rate of volatilization of pesticides. Most of these photolytic reactions have been studied in the laboratory under simulated conditions. Hence, extrapolation of these data to actual field conditions should be done with reservation.

Under conditions of an arid environment where summer temperatures reach 35°C and higher it is anticipated that photoreactions will take place with greater intensity in comparison with temperate regions. In the arid regions of Israel more than 200 days per year are quite cloudless (15–26% cloudness) and are of high-intensity radiation. It is likely that photodecomposition of pesticides will take place more rapidly under these conditions but such data have not been documented by field experimentation. The toxicity of certain compounds may be enhanced by solar radiation. A recent example is the solar radiation-induced toxicity of anthracene to *Daphne pulex* (Allred and Giesey, 1985). Several classes of herbicides, too, undergo photodecomposition (Crosby and Li, 1969; Yaron *et al.*, 1984). Loss of herbicidal activity of organoborates under arid conditions has also been linked to sunlight (Rake, 1961). The difficulty in demonstrating clearly that many herbicides as well as insecticides undergo photodecomposition in the field is due to the fact that many of the products formed are identical with those produced by metabolic processes of plants and microorganisms. However, there is little doubt that photodecomposition of pesticides takes place in the field and that this phenomenon could play an important role in the disappearance of such chemicals from the environment (Kilgore, 1975). In this respect pesticide residues might be of less consequence in arid environments than in temperate regions.

(b) Physical factors

Volatilization is not necessarily a factor directly responsible for the disappearance of pesticides from the environment, but it is a means of transporting a pesticide from a region of inactivity to an environment more conducive to degradation (Kilgore, 1975).

Moisture lowers the persistence of pesticides in the soil, either by aiding their evaporation, by hydrolysis, by catalyzing photolytic reactions, by participating in the reduction of pesticides, or by enhancing their decomposition by microorganisms (Bowman *et al.*, 1965; Harris and Lichtenstein, 1961; Lichtenstein and Schulz, 1960, 1964; Crosby, 1972a, 1972b; Matsumura and Benezet, 1978; Cripps and Roberts, 1978; Woodcock, 1978). In submerged soils decomposition of pesticides proceeds at a faster rate (Sethunathan, 1980).

The soil of the Negev desert, though poor in organic matter shows a high level of microbiological activity, at least during the rainy season in winter, in spite of a decrease in the average air temperature, which falls to 10–12°C. Beginning March-April there is a drastic fall in microbiological activity which lasts throughout the summer months. This fall in activity occurs over the entire soil profile excluding the lowest part where microbial activity is minimal throughout the year (Buyanovsky *et al.*, 1982). Disregarding other factors, decomposition of pesticides in arid soils with low moisture content will take place at a slower rate. On the other hand, high soil temperatures increase the tendency of pesticides to vaporize or decompose. The interplay between these two parameters, i.e. soil temperature and soil moisture, most likely will determine the rate at which pesticides will disappear from arid soils or will decompose to less bioactive compounds.

4.2.3.6 Arid Zone as an Effective Storage Environment for Cereal Grains

The high temperature and low humidity of the desert climate provide an excellent environment for the airtight storage of grains. Utilizing the principle of airtight or hermetic storage of grain (Hyde *et al.*, 1973), experiments were carried out in the northern Negev to determine the feasibility of storing grain in bulk under conditions of high temperature and low humidity. Wheat of 11.4% moisture was stored in airtight PVC-covered bunkers formed by a polyethylene liner at its base and a UV-resistant PVC sheet over the surface of the grain. Within three months the O_2 concentration fell to 6% and the CO_2 concentration rose to 9%. Under these conditions insects were controlled before they could cause significant economic damage to the grain. Wheat damage attributed to insects was estimated at 0.15% and that to moulds at 0.06% (Navarro *et al.*, 1984). It appears that the desert with its long hot dry summers, mild winters, and low rainfall, is an excellent environment for the temporary and long-term storage of grain.

4.2.3.7 A Typical Example of a Desert Crop—The Date Palm

(a) Insect pests of date palm trees and their control

The green leafhopper is known as one of the main pests of date palms in the Near East and North Africa. The insect causes damage to the tree by sucking the sap and by exuding large amounts of honeydew. The leafhopper identified as *Ommatissus binotatus* (Fieb.) var. *lybicus* (De Berg) of the family Tropiduchidae was first detected in Israel in 1981 and by 1983 it had spread throughout the entire region of the Arava desert. There was no sooty mould development on the honeydew and there is no evidence to date of direct damage to the tree, except for the accumulation of large quantities of sugars emanating from the trees in response to the insect's attack.

Control experiments with systemic insecticides, such as Temik (Aldicarb), butocarboxime, Rogor (dimethoate), and deltamethrin, applied to the soil following drip irrigation showed that the leafhopper can be effectively controlled for a period of six months or more with a single spring application of the first three compounds. Deltamethrin was ineffective. Control plots showed the presence of large populations of the leafhopper except in May when there was a drastic fall in population density due to extensive damage to the insects caused by sandstorms (Aharonson *et al.*, 1984). Chemical analyses showed that Aldicarb and butocarboxime are effectively absorbed by the roots and are translocated in substantial quantities to the leaf blades and fruit where they persist for relatively long periods of time (Tables 4.2.2. and 4.2.3). Only after 150 days from the start of treatment was a noticeable reduction in residues obtained, approximating to 1–3 ppm. By analogy with the persistence of Aldicarb and butocarboxime in cotton, citrus, peaches, etc., the rate of disappearance of these compounds from date palm is slow. Dimethoate might be less persistent but this has to be ascertained (Aharonson *et al.*, 1984). Butocarboxime has an advantage over Aldicarb in that its toxicity to warm-blooded animals is much lower (200 mg/kg vs. <1.0 mg/kg, respectively). Consequently, its allowable residue in the fruit could be higher.

Table 4.2.2 Persistence and disappearance of the systemic insecticides Aldicarb and butocarboxime in the leaf blades of the date palm following their application to the soil close to the drip irrigation. Data from Aharonson *et al.* (1984). Reproduced with permission

Days after application	Aldicarb (ppm)		Days after application	Butocarboxime (ppm)
	30 g a.i./tree	60 g a.i./tree		100 g a.i./tree
24	9.8 ± 4.8	8.1 ± 2.1	32	30.1 ± 12.0
54	6.3 ± 0.5	5.7 ± 1.8	63	40.0 ± 14.0
85	2.0 ± 0.1	3.4 ± 1.0	123	23.3 ± 13.0
145	1.5 ± 0.2	2.6 ± 0.2	150	2.1 ± 1.2
172	0.5 ± 0.1	0.8 ± 0.1		

Table 4.2.3 Persistence and disappearance of Aldicarb and butocarboxime in the fruit of the date palm following their application to the soil close to the drip irrigation. Data from Aharonson *et al.*. (1984). Reproduced with permission

	Aldicarb (ppm)			Butocarboxime (ppm)	
Days after application	30 g a.i./ tree	60 g a.i./ tree	Days after application	100 g a.i./tree (experim. plot)	100 g a.i./tree (commerc. plot)
145	1.0 ± 0.1	–	42	–	3.0 ± 0.1
172	1.1 ± 0.1	1.9 ± 0.6	57	–	4.9 ± 0.2
			123	2.7 ± 0.5	–
			150	1.9 ± 0.3	–

One of the most serious scale insect pests of date palms is the Parlatoria date scale, *Parlatoria blanchardi* (Targ.). Damage is reflected in leaf withering, tree stunting, and infestation and disfigurement of the fruit, which greatly reduces its quality. Severe population explosions of this insect occurred following extensive spraying of the arid zone with chemicals for control of locusts and other pests, thus upsetting the biological equilibrium. Chemical control with 0.5% Rogor WP alone or in combination with 2% oil emulsion has given satisfactory results when used as an adjunct to a pest management programme (Kehat *et al.*, 1974). The main approach is biological control with two parasite species which keep the scale populations under control.

Other scale insects of lesser importance include *Asterolecanium phoenicis* (Green), *Saissetia privigna* (De Lotto), and *Aonidiella orientalis* (Newstead).

The pineapple mealybug, *Dysmicoccus brevipes* (Cockerell), a serious pest of pineapple, has recently been found on date palm trees in the Arava Valley. Populations of this insect occur throughout the year on the adventitious roots at the base of the trunk but causing no apparent damage. Occasionally, in late summer the mealybugs migrate upward to the ripening bunches, infesting the dates and causing great damage (Ben-Dov, 1985).

The raisin moth, *Cadra figulilella* (Gregs.), is one of the most important pests of dates in Israel. It inflicts considerable damage to the fruit from the beginning of ripening to picking time. Damage also occurs in storage if fruit is not fumigated. The insect is successfully controlled by several insecticides with the exception of cryolite (Table 4.2.4). Mechanical prevention of insect attack by covering the bunches with wire netting gives excellent protection against infestation. DDVP-treated strips (25% a.i.) were most effective and gave better protection than lindane, naphthalene, DBCP, and EDB (Kehat *et al.*, 1969).

The date stone beetle, *Coccotrypes dactyliperda*, and the sap beetle, *Carpophilus* spp., are considered to be economic pests of dates in the arid Jordan Valley but not in the arid Arava (Kehat *et al.*, 1974). Despite the establishment

Table 4.2.4 Percentage of infested date fruits following spray treatments with various insecticides

Treatment	Variety and site	
	Zahidi, Eilat	Deglet Noor, Yotvata
None (Control)	63.6	49.2
Cotnion 20 E.C. (azinphosmethyl) 0.5%	3.0	2.1
diazinon 25 E.C. 0.5%	11.0	0.8
malathion 50 E.C. 0.2%	19.0	1.6
Matacil 75 W.P. (aminocarb) 0.2%	–	3.7
Lebaycid 50 E.C. (fenthion) 0.15%	–	5.6
Dipterex 80 S.P. (trichlorfon) 0.3%	14.4	8.0
Rogor 40 E.C. (dimethoate) 0.3%	15.0	–
cryolite (44%) 1%	53.5	37.0

of an integrated pest management programme control of these insects requires four to five insecticide treatments during the season, a practice which interferes with biological control of other pests (Blumberg and Kehat, 1982).

(b) Mesurol residues in dates

Mesurol (Methiocarb) is licensed in Israel for use as a bird repellent in flowers, vegetables, beets, etc., and for the control of molluscs in flowers, ornamentals, citrus fruits, subtropical fruits, etc.

Palm trees received an application of 1% Mesurol WP at a rate of 5 gallons/tree for bird repellency six weeks prior to harvest. Chemical analysis by HPLC of six samples of fruit at harvest time showed concentrations of 1.85–7.3 ppm of the parent compound. These concentrations at harvest time were considered too high; therefore, the applications were discontinued (Adato, 1984).

The persistence of Mesurol under these conditions might be due to its high rate of absorption, its low rate of volatilization, and its high resistance to photolytic decomposition.

4.2.3.8 Insects Attacking Mango Trees

The mango thrips, *Scirtothrips mangiferae* (Priesner), has become established in Israel since 1975 and is considered the most damaging pest of mangoes in the arid Arava Valley. The thrips damage the new shoots of the trees causing the young leaves to curl along the midrib, thus distorting their shape and leading to their premature drop. No damage to fruit has yet been reported. Good control for a period of two to three months was achieved using 0.2% tartar emetic fortified with 0.3% sugar as a full-coverage bait spray. Equally good control has been achieved with cypermethrin. Control of the thrips with thionex (endosulfan), phosphamidon, and fenthion gave less satisfactory results owing to their short duration of action and the apparent development of resistance to these compounds after several treatments (Venezian and Ben-Dov, 1982; Ben-David *et al.*, 1985).

Additional host plants for the pest, including pomelos and grapevines, have recently been observed in the Arava Valley.

Other insects attacking mango trees but causing little damage include the Oriental red scale, *Aonidiella orientalis* (Newstead), a cosmopolitan and polyphagous diaspidid common in mango orchards along the Arava Valley, generally controlled biologically by *Habrolepsis aspidiotti* (Compere and Annecke), *Saissetia privigna* (De Lotto), and *Asterolecanium phoenicis* (Green) (Ben-Dov, 1985). Chemical control of these insects is being avoided for fear of potentially serious consequences to their natural enemies. The latter have, so far, maintained a good biological balance and have kept the scale populations under control.

4.2.3.9 Polychlorinated Biphenyls (PCBs)

Considerable amounts of PCBs are carried to the Mediterranean by aeolian transport (Elder *et al.*, 1976). PCBs are highly stable chemically, being resistant to thermal degradation and hydrolysis but are susceptible to photolysis (Herring *et al.*, 1972; Hutzinger *et al.*, 1972).

In areas of high water temperatures or high winds, surface evaporation will increase resulting in lower levels of PCBs in sea water (Elder and Villeneuve, 1977). However, wind-borne PCBs may be carried to desert areas and contaminate the fauna and flora of these regions.

4.2.3.10 Pesticide Usage in the Arava Region

Mention has been made of the fact that the introduction of new agricultural crops in the desert agro-ecosystem inevitably will require the use of large quantities of pesticide chemicals. Many such crops, e.g. watermelon, several varieties of melon, corn, cauliflower, tomato, eggplant, seed onion, onion, and green pepper, have been introduced in the past few years. The total acreage of the above crops is given in Table 4.2.5.

Obviously, the success of growing these crops, as in other climates, depends on using large amounts of pesticides and fertilizers. A wide variety of herbicides, insecticides, and fungicides are being used on these crops. A list of pesticides in use is presented in Table 4.2.6.

In some instances several fungicides are used together in cocktail form. For example, tomatoes are sprayed 5–6 times during the season with a cocktail consisting of triadimefon, chinomethionat, triforine, and fenarimol. Another example is an insecticide mixture consisting of cypermethrin, methamidophos, and chlorpyrifos which is used on green pepper and eggplant.

Table 4.2.5 Types and total acreage of crops grown in the Arava region. Dayan (1987). Reproduced with permission

Crop	Number of hectares Spring cultivation	Full cultivation
Watermelon	201	–
Melons	280	327
Corn	22	–
Cauliflower	12	–
Tomato	27	218
Onion	11	220
Pepper	–	257
Eggplant	–	56
Others*	12	35
Total hectares	565	1113
Grand total (ha)	1678	

*Cucumber, squash, garlic, sweet potato, asparagus, seed onion

Table 4.2.6 Types of herbicides, fungicides and insecticides used on various crops in the Arava region. (Adler (1987). Reproduced with permission

Herbicides	Fungicides	Insecticides
Metribuzin	Maneb	Cypermethrin
Isopropalin	Mancozeb	Methamidophos
Oxadiazon	Triadimefon	Chlorpyrifos
Ethalfuralin	Chinomethionat	Propargite
Diphenamid	Triforine	Cyhexatin
Nitrofen	Fenarimol	Dicofol
Chlorthal dimethyl	Iprodione	Fenpropathrin
Pendimethalin	Vinchlozolin	Endosulfan
Alachlor	Dimethrimol	Chlorobenzilate
	Imazalil	DDVP
	Pyrazophos	Methomyl
	Tridemorph	Phosphamidon
	Benomyl	Ethiofencarb

As yet no major environmental problems have arisen in the desert from the usage of such large quantities of pesticides. However, there is no way of predicting future adverse effect on the environmental quality of the desert as more and more crops are introduced and larger quantities of chemicals are used.

A problem of major importance in the Arava is the existence of large numbers of the rodent, *Mus musculus*, as mentioned earlier the ancestor of the house mouse. This causes extensive damage to green pepper. At present, the rodent is successfully controlled with 5–6% coumatetralyl placed in plastic pipes at the rate of 10 kg a.i./hectare. This rodenticide also eliminated the problem of secondary poisoning of wild animals and birds of prey, which was prevalent with the use of other rodenticides (Zook-Rimon, 1985). Changes that occurred in distribution and density of wildlife in the desert resulted mainly from changes in agrotechnical practices in arid lands.

4.2.4 GLOBAL ASPECTS OF AGRO-ECOSYSTEMS IN ARID LANDS

There is no general agreement as to what constitutes a desert. Many parameters are involved, such as rainfall, temperature, types of vegetation and fauna, soil type, salinity, geomorphological characteristics, and others. Hence, the desert as a boundary marker may sometimes be defined in different terms by various authors. The definition by Meigs (1953) seems to be the most accepted. According to Meigs the essential trait and the one upon which all others depend, is the lack of precipitation, and some sort of ratio involving precipitation and temperature.

Thirteen major desert areas of the world are recognized: Kalahari-Namib, Sahara, Somali-Chalbi, Arabian, Iranian, Thar, Turkestan, Takla-Makan,

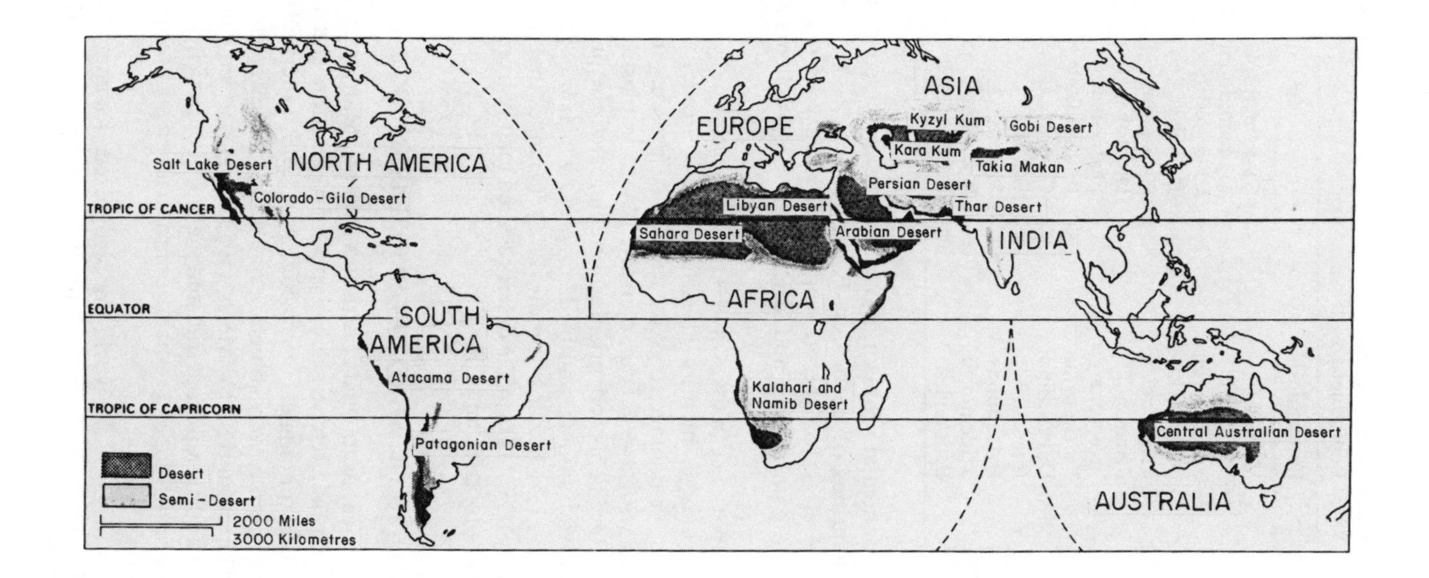

Figure 4.2.3 The deserts of the world. From Simons, 1968. Reproduced with permission from Oxford University Press

Gobi, Australian, Monte-Patagonian, Atacama-Peruvian, and North American (Figure 4.2.3).

More than a third of all the land in the world is desert or semidesert, empty and barren. During the day a fierce sun in a cloudless sky sends temperatures soaring, yet bitter winds may blow, whipping up grit and sand, while the night can be freezingly cold (Simons, 1967).

Every continent except Europe has deserts and many experts claim that most of the desert land will remain useless forever. However, as the population of the world increases more and more rapidly and the total fertile land becomes exceedingly overcrowded, it is realized that great efforts have to be made to conquer the deserts. The salinity of deserts is one of the most important and most difficult problems facing anyone who hopes to make use of such areas (Simons, 1967).

4.2.4.1 Desert Vegetation

There are two main types of desert plants. First, the *ephemerals*, which have seeds or fruits that can survive drought even though the mature plant itself cannot. Examples of such plants are the grasses *Boorhaavia rapens* and *Aristida funiculata*. These have short but widespread root systems capable of obtaining water from the shallow layers of the soil after rain.

Second, the *perennials*, which are established plants remaining in the same place for long periods. Examples, of this type are the various succulent species of cacti, the tamarisk, and acacia bushes. The last two are deep-rooted so that they can tap the water well below the surface. Because such water is somewhat saline, these plants have to tolerate a certain amount of salt. The date palm, for instance, has a high salt tolerance and thus can provide food where most other fruit trees or crops will not grow.

4.2.4.2 Desert Fauna

Animals living in the desert must also overcome the shortage of water and intense heat. Animals such as camels, goats, sheep, and small rodents have adapted well to a desert environment. Reptiles, snakes, and tortoises are also common due to their adaptation to a variable blood temperature.

Insects and arachnids are undoubtedly the best adapted to desert conditions. They require little water, can tolerate high temperatures, and are able to minimize water loss due to evaporation.

Man is not well adapted to desert life. In the past, human dwellers of the desert have been nomadic, moving about with their sheep and goats searching for pasture. Due to overgrazing, many areas which once supported good pasture became valueless scrub.

4.2.4.3 Agro-ecosystems with Irrigation

(a) Sources of water

Water for irrigation of arid lands comes from various sources. Many desert oases are fed by deep underlying water of 'fossil water'. Others are fed by shallow groundwater tables under the sand dunes. The scanty rains (50–100 mm per year) falling on the dunes seep quickly into the sand and the water accumulates on the impermeable underlying rock layers forming a shallow groundwater table which can be tapped by digging shallow wells. Fossil water has been trapped below ground for long periods of geological time. During its long stay below ground it has dissolved large quantities of salt and other minerals from the surrounding rocks. This type of water may form large aquifers, such as those found in the Sinai-Negev desert between Egypt and Israel (Issar, 1985) and the large aquifer existing under the Sahara desert (Ambroggi, 1966). Hydrogeologists now calculate that the Nubian sandstone aquifer under the Sinai-Negev desert, called Marah (*mar* meaning 'bitter' in Hebrew) or Ayun Musa (springs of Moses in Arabic), holds 200 billion m^3 of water.

By a recently developed method of isotope analysis measuring the amount of heavy hydrogen (deuterium) and heavy oxygen (oxygen-18) in water, Issar and colleagues (1985) determined the age of this palaeowater from Ayun Musa to be 30 000 years old. These brackish waters have a high content of $CaSO_4$ and $MgSO_4$ (epsomite) and are known for their bitter taste from biblical times. Agricultural settlements in the Negev utilize some of this water for irrigation. Apparently, the water is low enough in salt content to be usable for irrigation. It is anticipated that in future years more and more of this water will be tapped and will be used to irrigate new agricultural crops.

Drip irrigation has been practised in Israel for many years. Under traditional irrigation technologies large amounts of water are applied in a short period of time. On land with a low water-holding capacity, such as sandy soil or an uneven slope, a significant amount of this water is lost because of percolation beyond the root zone and runoff. Low-volume irrigation, especially drip irrigation, applies smaller amounts of water per unit time. This reduces percolation and runoff and increases the portion of applied water available to the crop (Caswell *et al.*, 1984). The relative advantage of drip irrigation over traditional methods is greater for arid and semi-arid lands and for lower quality soils. Drip irrigation was introduced into California in 1969. Other sources of water, such as saline water and blowdown water, will be dealt with under the heading 'Salinization'.

(b) India

Construction of the Rajasthan canal will service an arid area of over 2 million hectares of which 1.3 million hectares will be cultivable (Roy and Shetty, 1978). Conversion of arid land into cultivable fields and the tremendous increase in population will have a great impact on the natural vegetation in this part of the desert. The Ganganagar district, which is a part of the Great Indian Desert, was a typical desert prior to the construction of the Gang canal. At present, parts of this district are irrigated by three major canal systems. This has resulted in the transformation of a vast inhospitable stretch of arid land into a fertile land famed for its agricultural produce. On the other hand, the natural vegetation (especially xerophytic plants) has been considerably destroyed by irrigation practices.

The non-irrigated soils, consisting of sandy plains and sand dunes, are poor in organic carbon, are neither saline nor alkaline but are able to support a wide variety of desert species of vegetation.

The Chaggar alluvial plain has slightly higher soil organic carbon but also greater problems of salinity and alkalinity. Natural vegetation is sparse but includes many species of shrubs and grasses. The Chaggar river bed is used for rice cultivation during the rainy season, while wheat, barley, and gram are grown as winter crops. Accordingly, weeds of cultivation are abundant in this region.

The extensive irrigation facilities provided by these three major canals have transformed the area into a fertile agro-ecosystem. The soil is slightly alkaline, the pH ranging from 8.0 to 8.5. This district supports 50% of the cotton and 35% of the food grain produced in Rajasthan. Other crops are also being grown, such as cereals and millets, rice, barley, maize, bajra (*Pennisetum typhoides*), and jowar (*Sorghum vulgare*); pulses, such as moth (*Phaseolus aconitifolius*), urd (*P. mungo*), mung (*P. aureus*), and arhar (*Cajanus cajan*); cotton, sugarcane, tobacco, chillies, rapeseed, mustard, brown sarson, til (*Sesamum indicum*), groundnut (*Arachis hypogaeae*), linseed (*Linum usitatissimum*); fodder crops, such as guar (*Cyamopsis tetragonoloba*), different types of vegetables, etc. Fruit trees, such as citrus, guava, and pomegranates, are also grown. As in other irrigated lands many species of weeds invade the fields and orchards of this region.

Field trials on the response of mung bean (*Vigna radiata*) to irrigation and anti-transpirants in the Rajasthan district (Singh *et al.*, 1981) indicated that three irrigations applied at 0.75 bar until flowering time and at 5 bar thereafter till maturity, resulted in significant improvement in plant growth, yield attributes, and seed yield over no post-planting irrigation. Anti-transpirants had no effect on yield but straw mulch increased pod production.

The changes that have taken place in Ganganagar due to increased irrigation facilities are bound to expand to other districts along the Rajasthan canal where extensive cultivation will take place and a tremendous increase in weed flora will be evident concomitant with the destruction of natural vegetation.

There is only scant information available on the impact of pest control measures on the expanding agricultural practices in this desert region. The red hairy caterpillar, *Amsacta moorei* (Butler), is a serious pest of kharif (pearl millet) crops in the arid and semi-arid zones of Rajasthan. The most effective control chemicals were methyl parathion, carbaryl, quinalphos, and the antifeedant compound triphenyltin acetate (TPTA). BHC, malathion, and triphenyltin hydroxide (TPTH) were not effective (Verma, 1981). Most likely, many problems will arise as more pesticides are introduced into the irrigated arid zone. The judicious use of such chemicals and a thorough understanding of pesticide ecology will undoubtedly obviate much chemical pollution and environmental hazard. In this respect, the practice of integrated pest management should be encouraged and there is reason to believe that it would be successful in a desert environment.

4.2.4.4 Agro-ecosystems without Irrigation

The problems of crop production in the Indian arid zone stem from an acute ecological imbalance of the components of productivity. Harsh and unfavourable climatic conditions, wind erosion, and poor soils with low moisture retention capacity all contribute to poor crop yields.

The importance of dry farming research in the arid zone becomes prominent when it is realized that 90% of the cultivated area of this region will continue to be rainfed for many years to come (Singh, 1978). Hence, ecological and economic considerations make dry farming research more of a necessity than a choice.

Lack of moisture is further aggravated by depletion of natural vegetation due to uncontrolled grazing, and cutting of trees and shrubs for fuel, leading to large-scale loss of soil and nutrients caused by wind erosion. Pressure due to the population explosion also leads to extended use of marginal and sub-marginal lands for cultivation. Where the annual rainfall does not exceed 300 mm, large-scale afforestation with suitable species of trees and a grass cover with adapted species of grasses supporting animal husbandry should be the mainstay to stabilization of these areas. Certain types of arid-adapted vegetation with economic potential, such as jojoba bushes (*Simmondsia chinensis*), buffalo gourd (*Curcurbita foetidissima*), and others should be evaluated for cropping in non-irrigated arid regions.

Until recently, land in the Sahelian arid zone in Africa has been left fallow for many years before recropping. Thus, the land was not over-exploited. With the introduction of cash crops, such as cotton and groundnut, marginal areas were brought into cultivation to provide subsistence for the increasing population. The strain of the intensive agriculture resulted in a decline of soil fertility and in lower productivity (Cloudsley-Thompson, 1977, 1978). These adverse effects can be remedied by selecting crops whose requirements of water are compatible with the normal precipitation in the particular area.

Needless to say the introduction of cash crops in dryland farming will bring with it pest and pesticide problems, albeit of a different nature from those encountered in temperate zones.

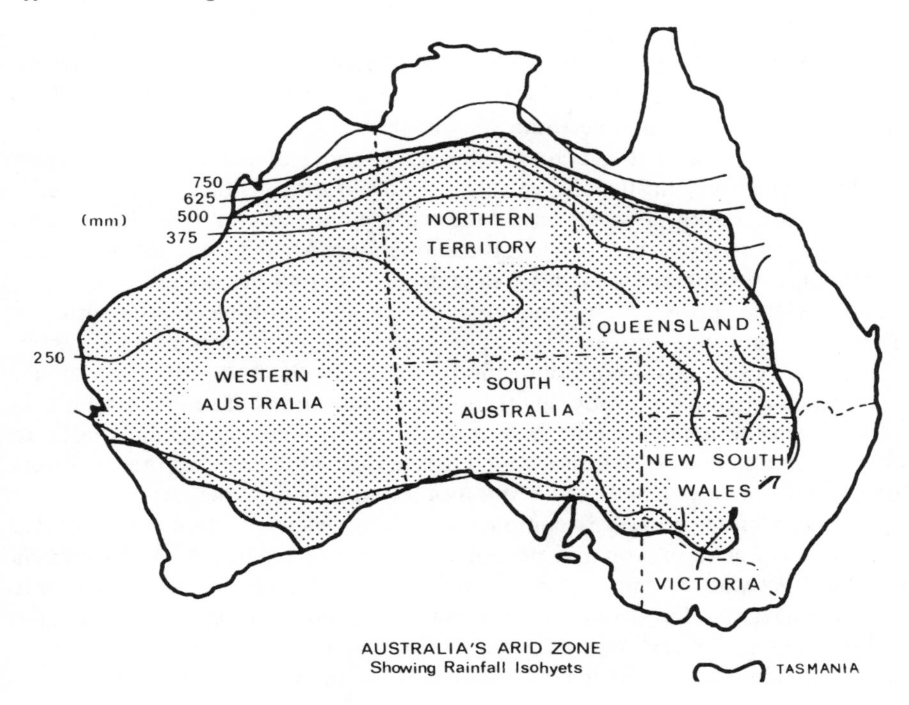

Figure 4.2.4 Australia's arid zone. From Arid Zone Newsletter, 1981, CSIRO. Reproduced with permission

4.2.4.5 Rangelands

Australia

In Australia, the arid zone is defined as that area which receives insufficient rainfall to allow pasture improvement or cropping without irrigation.

The arid zone of Australia (Figure 4.2.4) is bound to the north by the 250 mm rainfall isohyet, to the east by the 500 mm rainfall isohyet, and to the south by the 750 mm isohyet. Approximately 70% of Australia is either arid or semi-arid, and 26% of this area is unoccupied. Another 9% comprises aboriginal reserves and national parks. The remaining 65% is grazed by sheep or cattle.

Rangeland extension workers place more emphasis on the condition of vegetation than on livestock condition. It is believed that by the time a decline in livestock condition is observed considerable damage to pastures would already have been done. This condition, undoubtedly, arising from overgrazing. Progressive degrees of protection from overgrazing have been imposed by the Australian government.

The perennial grasses *Aristida leptopoda* (white spear grass) and *Astrebla lappacea* (curly Mitchell grass) are the main species of the grasslands of semi-arid Queensland in Australia. Long-term studies by Williams and Roe (1975) have clearly shown that severe mortality of seedlings of these and other grasses is due mainly to climatic factors. Drought survival of perennial range-grass seedlings is greatly influenced by soil moisture (Oppenheimer, 1960; Mueller and Weaver, 1942). However, soil temperature and nutrient supplies are also important factors (Christie, 1979). At a temperature of 30°C, and low soil-water potential, curly Mitchell grass has greater capacity to produce nodal axes quickly, higher relative growth rate, and greater root extension than other grasses. Drought endurance appears to be due mainly to differences in root system characteristics and to shoot/root ratio (Christie, 1979). It appears that surface root extension and deep root penetration are adaptations of arid and semi-arid plant species to drought endurance and survival, unlike many plant species in the temperate and humid zones which derive their moisture from upper layers of the soil; hence, their limited root system and shallow penetration.

Williams and Roe (1975) demonstrated that severe mortality of *Astrebla* seedlings as well as other native perennial grasses growing in semi-arid Australia can be directly attributed to climatic factors. The influence of soil water supply on drought survival of perennial range-grass seedlings has been well documented (Mueller and Weaver, 1942). Drought survival is closely related to species differences in the development of nodal axes and in rate of extension and depth of penetration of the root system.

Control of weeds and other undesirable plants is an ongoing process. *Parkinsonia aculeata*—a tall thorny shrub which forms dense thickets around bores, dams, or creeks—has created problems in Australia. This shrub is effectively controlled by spraying with 10% Picloram and with 40% 2,4,5-T diluted in diesel oil (Christmas *et al.*, 1980).

The sheep blow fly, *Lucilia cuprina*, has been a stumbling block in the wool industry in Australia. A wide variety of insecticides have been used to control the fly but many have failed because of the fly's ability to develop resistance to these chemicals. Resistance to dieldrin in 1957, to diazinon in 1965, and to carbaryl in 1967 has been reported (Harrison, 1969; Shanahan and Roxburgh, 1974a,b,c,d). Recently, experiments have been conducted with sheep showing resistance to the fly's attack, and with the use of $ZnSO_4$ given orally or in the drinking water (Perkins, 1980).

The most important problem for livestock in Australia has been that associated with ticks. The cattle tick, *Boophilus microplus*, a vector of piroplasmosis, had developed resistance to most insecticides used for its control, including arsenic dips (Brown and Pal, 1971; Drummond, 1977). Control measures with various acaricides as well as novel types of compounds are still being practised, so it seems inevitable that some degree of pesticide pollution will occur in rangeland areas.

California

More than 40 million of California's 100 million acres are rangeland. The forest, grassland, and rangeland environments occupy approximately two-thirds of the land area of the State, of which 50 million acres are grazed. The desert saltbush (*Atriplex* spp.), an abundant shrubby inhabitant of some of California's driest, saltiest rangelands, is one of many salt-tolerant plant species that have become adapted to growing in arid and saline lands. These salt-tolerant plants (halophytes) provide forage for livestock and wildlife throughout the West. Moreover, they are adaptable to genetic manipulation by selection and breeding (Kelley, 1984). Workers at the University of California at Davis and Riverside and at other universities have been able to improve the palatability and yield of saltbush and sagebrush. Bermuda grass and wheatgrass have been used in gene-transfer experiments to improve their salt tolerance. Native saltbush species are salt-, drought-, and heat-tolerant, as well as insect-, disease-, and fire-resistant, are easily cultivated, highly productive, nutritious, and palatable. *Atriplex* species have been found to contain up to 25% dry weight of crude protein, and one particular species produced 9 tons of forage per acre. Revegetation of marginal soils with these halophytes proved very successful.

Care must be taken in the use of halophytes for forage. Some palatable saltbush species can be toxic to livestock if consumed in excess, due to the accumulation of a toxic concentration of oxalates. *Halogeton* is a well-known toxic halophyte. The potential benefits of halophytes are still being explored. They represent important resources of rangelands, areas that provide grazing lands for livestock, habitats for wildlife, and repositories for genetic material.

4.2.4.6 Salinization—California as an Example

All waters and soils contain salt. Even the non-saline irrigation waters of the Sacramento-San Joaquin Delta in California contain enough salt to create hazard to crops if drainage is insufficient. With adequate sub-surface drainage and an average annual rainfall of 350–400 mm neither salinity nor a shallow water table becomes a problem (Oster *et al.*, 1984).

Saline and sodic soils occur naturally in arid and semi-arid regions, and as water development brings more land into irrigation the salinity problem expands. The condition is aggravated by poor soil drainage, improper irrigation methods, insufficient water supply for adequate leaching, and poor-quality water containing more than 300–800 mg/l total dissolved salts (Backlund and Hoppes, 1984).

Salts are usually leached below the root zone whenever the amount of infiltrated water exceeds that evapotranspired. Where rainfall is adequate salts will not normally accumulate. If rainfall is insufficient or the irrigation water is saline provisions for adequate leaching must be made.

Individual salt constituents as well as total salinity of irrigation water affect the stability of soil structure, and hence, water penetration. The major effects of salinity on soil properties are swelling of soil clays, dispersion of fine soil particles, crust formation, and decrease in water penetration (Rolston *et al.* 1984).

Salinity and drainage problems have plagued Californian agriculture since the inception of irrigation in the second half of the nineteenth century. Of the 10.1 million acres under irrigation in California, 2.9 million acres are affected by salinity levels of approximately 2500 mg/l or more.

The arid Imperial Valley, situated in what was once the desolate Colorado desert, has had problems of salinization for several decades, yet it has become one of the most productive farming regions of California and of the world. This success was made possible by providing adequate subsurface drainage which removed more salt than was brought in by irrigation water from the salty Colorado River. The salty waste water thus removed was drained by gravity into the Salton Sea (Kelley and Nye, 1984). However, the increasing salinity of the Colorado River is creating new problems and if salinity levels reach 1140–1290 mg/l as projected by various agencies for the year 2000, serious economic losses due to lower yields of crops would result for Imperial Valley farmers.

Strategies for increasing crop yields in saline soils

One strategy available to farmers with saline soils is to select salt-tolerant plants. Crop tolerance to salinity ranges widely from the very salt-sensitive bean and corn plants to the highly tolerant cotton and barley. Salt-tolerant varieties of barley can fare well with seawater if grown on coastal sandy areas. The importance of a *sandy* environment for salt water irrigation was promulgated by Boyko and Boyko (1964). On sand, salt-tolerant barley can grow with seawater because sand is relatively inert and has small electrical effects on salt ions. Unlike soil, sand has sufficient aeration even with heavy water application because water percolates fast through the root zone and the roots are not damaged. Experimental results suggested that the more frequently the barley in sand was irrigated with salt water the higher was the grain yield (Kirkham, 1978).

Molecular techniques, such as recombinant DNA technology, may ultimately have a significant effect on agriculture in the isolation and transfer of genes governing agriculturally important characteristics, such as salinity and drought resistance (Valentine, 1984). The genetic potential to improve crops already exists, but there are limitations, such as limited sources of genes for salinity tolerance and lack of rapid and precise evaluation methods, which until now have prevented the development of salt-tolerant varieties among sensitive crops, such as beans and corn.

Salt-tolerant species, however, cannot substitute for good management practices that prevent salt accumulation in the soil. They can be useful where good quality water is not available, and for cropping saline soils that are undergoing gradual reclamation (Shannon and Qualset, 1984).

In recent years it has been found that exposure of plants to environmental stress can signficicantly increase the occurrence or severity of *Phytophthora* root rots. Some of these stresses result from cycles of drought, heat, oxygen deficiency, and salinity. Experiments have shown that although salt itself caused no significant damage to the root system of chrysanthemum plants, it lowered the plant's resistance to disease (MacDonald *et al.*, 1984).

Similar results were obtained with tomato and citrus plants in both greenhouse and field experiments. In particular, an isolate of *P. parasitica* recovered from citrus soils showed the greatest salt tolerance. It survived and reproduced at salinity levels equal to or greater than those of seawater. Thus, it is expected that many *Phytophthora* species remain active in saline soils for long periods of time and can cause severe stress of crop plants (MacDonald *et al.*, 1984). Thus, apart from the propagation of salt-tolerant plants, a vital question is can fungi and other parasitic organisms acquire similar salt tolerance and would they become more resistance to chemical control? This area is worth investigating.

4.2.4.7 Desertification

According to the definition used in the assessment adopted by PACD, UNCOD (United Nations Conference on Desertification, 1977) desertification is the diminution or destruction of the biological potential of the land which can lead ultimately to desert-like conditions. It is an aspect of the widespread deterioration of ecosystems resulting in impoverishment and reduction in vegetative cover, exposing the soil surface to accelerated water and wind erosion, leading to loss of soil organic matter and nutrient content, deterioration of soil structure and hydraulic properties, crusting, compaction, salinization or alkalinization, and the accumulation of other substances toxic to plants and animals.

Desertification threatens 35% of the earth's land surface (95×10^6 km^2) and almost 20% of its population (approximately 850 million people). A good portion of the world's drylands is already affected and continues to be irretrievably lost to the desert at a rate of 60 000 km^2 per year, and land rendered economically unproductive is showing an increase at a rate of 210 000 km^2 per year (Karrar and Stiles, 1984). The areas affected by desertification include: rangelands (31 million km^2); rainfed croplands (3.35 million km^2); and irrigated lands (400 000 km^2).

It has long been suspected that regions immediately surrounding a desert area are vulnerable to encroachment by the desert. For example, much of what is now the arid eastern Sahara was once a productive savannah that catered to the herds of nomadic cattlemen until about 2700 BC. In our times, desertification is a threat to a vast number of people in North and East Africa, northern Asia, and the tropical Americas.

Within a span of 17 years (1958–1975) the Sudanese Sahara has reportedly advanced northward almost 200 km through deterioration of marginal areas. Land at the southern border of the Sahara is turning into desert at an estimated rate of 100 000 hectares per year (Anonymous, 1977).

It is well known that overgrazing is one of the major causes of desertification. Overgrazing depletes the already scant vegetation in arid lands and it favours an increase in the populations of grasshoppers (Merton, 1959; Roffey, 1970). The density of the grasshopper population was shown to be significantly higher in overgrazed areas of the Rajasthan desert in India (Parihar, 1981). This may have been due to the preference for oviposition of the egg pods and their subsequent hatching in open sandy patches rather than in dense grasslands where the anastomosing fibrous roots of the grasses hinder these processes. Similarly, Grewal and Atwal (1968) reported that *Chrotogonus trachypterus* were least abundant in fields under tall vegetation. The effect of overgrazing on an increase in the density of grasshopper populations has also been reported by Dibble (1940) in the USA and by Uvarov (1962) in South America.

Iran

The Lut desert is situated in the southeastern part of Iran and extends over 80 000 km^2. It is extremely poor in vegetation and this is aggravated by the extremely low annual rainfall of no more than 60 mm. The only river flowing through the region is the saline Birjand River. The advancement of the Lut desert can be particularly traced to: (1) destruction of native vegetation; (2) severe land erosion; (3) blowing of sands; and (4) extension of the sand dunes causing the destruction of the Ganat irrigation system as well as damage to farms and inhabited areas.

Desertification of this region, other than by natural causes, was aggravated by human intervention, such as: (a) destruction of vegetation caused by eradicating bushes, trees, and shrubs for use as firewood, as construction material, or as forage for cattle: (b) making charcoal out of logs and trunks of trees; (c) continuous and excessive grazing of native grasses by cattle; and (d) salinization due to irrigation with saline water (Kardavani, 1978).

Amelioration of the situation can come about, in part, if steps are taken to reduce the number of animals grazing on the sparse vegetation, to prevent the disproportionate and imbalanced utilization of plants for firewood, and to prevent salinization of the marginal land of the Lut desert.

Chile

The semi-arid region of Chile is characterized by low precipitation with high fluctuations from year to year. Agricultural activities are concentrated around the main river valleys and, to a lesser extent, along the coastal marine terraces. The intervalley areas are used mostly for subsistence grazing by goats, although some dry-farming is also found. The total intervalley area is one to two orders of magnitude larger than that of the relatively narrow river valleys. Because of overgrazing and excessive wood cutting this area is highly subject to desertification.

USSR

In recent years there has occurred in a number of arid regions a disturbance of the ecological balance in the environment resulting from an increase of man's influence upon nature and a rapid growth of the exploitation of natural resources, both in agriculture and in industry (Petrov, 1978).

Desertification is a combination of geographical and anthropogenous processes resulting in the destruction of arid and semi-arid ecosystems, and the degradation of their natural resources. The principal forms and degree of anthropogenous change of natural ecosystems in the USSR are: (a) degradation of range vegetation due to overgrazing; (b) deflation of light sandy soils utilized for non-irrigated agriculture, leading to destruction of the fertile soil layer; and (c) breaking up of compact sands by removal of shrubs for fuel, earthwork, road construction, industrial enterprises, settlement, etc.

Thus, in the course of time the old basic landscape, with its relatively stable vegetation cover that keeps surface sands and soils under control, changes into a landscape of shifting barkhan sands. It is anticipated that the influence of man on the nature of the arid regions of the USSR will grow rapidly. The 10th five-year plan for the development of the economy of the republics of Middle Asia and Kazakhstan indicate so. One of the important tasks that face scientists is the optimization of the landscape—meaning, the influence of man which can result in maximum productivity of the land. An important phase of this task will be improved methods of irrigation in agriculture. Secondly, salinization of soils will be prevented by supplying good drainage. These measures will result in the appearance of new agricultural oases in the deserts and semi-deserts of Middle Asia and Western Kazakhstan but only to the extent of 10% of the total area of the arid zone (Petrov, 1978). Other desert areas will be used for animal husbandry and for mining, giving impetus to accelerated urbanization, in turn increasing recreational facilities and tourism. If these measures necessary for the protection of the arid environment are not carried out there will be progressive desertification.

4.2.4.8 Effect of chemicals

Mention has been made of the high $CaSO_4$, $CaCO_3$, and $MgSO_4$ levels characteristic of arid lands; other chemicals might be found in abnormal concentrations. One of these chemicals is boron. Although boron is an essential plant nutrient, it becomes toxic to growing plants if excessive amounts are present in the root zone. Soils containing high native concentrations of boron occur primarily in arid and semi-arid environments where drainage and/or leaching are inadequate. This situation can be aggravated by use of groundwater or irrigation water containing high levels of boron (Peryea and Bingham, 1984).

Before high-boron soil can be used for farming its boron content must be reduced to non-toxic levels by leaching with low-boron water. Reclamation of such soils, however, is not absolute. Soluble boron appears to increase even after the soil has been reclaimed and phytotoxic conditions may recur. Boron concentrations of 4 ppm or higher in the saturation zone are considered potentially injurious to cotton and sorghum crops (Peryea and Bingham, 1984). Regeneration of phytotoxic boron concentrations can be prevented by the use of good quality irrigation water that produces good drainage and leaching after crop planting.

Pesticides in soils

Whether pesticides are applied as ground or aerial sprays, as dusts to foliage, or directly to soil, it is inevitable that large amounts of them will ultimately reach the soil, which acts as a reservoir for these persistent chemicals. Eventually, many of these chemicals find their way into the tissues of invertebrates, they move into the atmosphere by volatilization, they concentrate in bodies of water by precipitation and leaching, and the residues of the active materials or their metabolites may reach the organisms at the end of the food chain in both aquatic and terrestrial ecosystems.

Factors that affect the persistence of insecticides in soils include volatility, solubility, concentration, formulation, and type of soil to which they are applied. Persistence is longer in heavier soils than in light soils and in those with much organic matter; however, residues in heavier soils are less toxic to insects because they are absorbed and inactivated. Pesticides are less likely to be adsorbed in the light sandy soils of the desert. Experiments by Edwards *et al.* (1957) showed that 34 times as much lindane and 16 times as much aldrin was needed in a muck soil as in a sandy soil to kill the test insects (*Drosophila melanogaster*).

Environmental factors, such as soil moisture and soil temperature, also influence pesticide persistence. In dry soils insecticides are tightly adsorbed, whereas in wet soils they are released and are apt to be broken down or physically removed. Chemical degradation, bacterial decomposition, and volatilization are all influenced by temperature so that at lower temperatures these processes slow down and less insecticide is lost. Temperature also affects adsorption of insecticides in soils because sorption tends to be exothermic so that increased temperature decreases adsorption and releases insecticides. Solubility of pesticides also depends on temperature. More insecticide dissolves in soil moisture as the temperature increases and amounts of insecticide leached from soil also increase. However, warm soils are usually dry ones, so they hold insecticides more firmly than wet ones.

Needless to say, a wealth of information and data are available on pesticide residues in soils and their effects on various agro-ecosystems, especially those in temperate and humid zones (Edwards, 1975; Brown, 1978). However, there is a noticeable dearth of such information concerning arid lands. Brown and Brown (1970) found mean residues of 0.09 ppm DDT and related compounds in tundra of Canada; Lahser and Applegate (1966) found a mean of 1.60 ppm DDT and 0.20 ppm BHC in US desert land, and Ware *et al.* (1971) reported a maximum of 2.92 ppm DDT and related compounds (mean 0.48 ppm) in the US desert of Arizona, two years after cessation of its use.

In the past 30 years Lichtenstein and colleagues have done a tremendous amount of laboratory experimental work on the dynamics and fate of insecticides in soils. A few works summarizing these activities are: Lichtenstein, 1959, 1965, 1966, 1979; Lichtenstein and Schulz, 1959; Harris and Lichtenstein, 1961; Katan *et al.*, 1976a; Lichtenstein *et al.*, 1978; Fuhremann *et al.*, 1978.

The sandy loam soil that extends over approximately two-thirds of the state of Rajasthan in the desert area of India is an excellent environment for soil insects, which affect the growth and development of the major crop in that area, i.e. the bajra (*Pennisetum typhoides* P.). Experiments with carbaryl indicated that the insecticide persisted in the soil for more than 90 days. Applications of 20, 40, and 60 kg a.i./ha yielded initially residues of 140, 245, and 355 ppm, respectively; residues of 65, 121, and 243 ppm, respectively, were found after 30 days; and residues of 2.6, 13.1, and 17.4 ppm, respectively, after 90 days. Residues of carbaryl in bajra plants, however, were below detectable levels (Gangwar *et al.*, 1978). Hence it is suggested that carbaryl can safely be applied to sandy loam soils with no risk of pesticide uptake by the bajra plants. On the other hand, maize plants grown in carbaryl-treated soil absorbed the insecticide through the root system and translocated it to the aerial parts of the plant (Dhall and Lal, 1974).

Gupta and Rawlins (1966) also reported complete loss of carbaryl in 100 days from sand as against 75–85% from silt and muck soils. BHC also proved effective in the control of white grub populations attacking chilli crops. Soil residues of 2.41 ppm and 1.87 ppm were detected after 60 and 90 days, respectively (initial residue was 10.3 ppm), and 0.35 ppm was detected in the fruit after 90 days. The latter amount, however, is below the accepted tolerance level of 3 ppm at all stages of sampling (Pal and Kushwuha, 1977).

Insect pests of dryland crops in India

Soil insects: Under the direct influence of an arid environment the soil provides a variety of microclimates and microhabitats for the desert fauna. A number of insect species inflict heavy damage on the cultivated crops of the Rajasthan desert region.

(a) Termites: These are particularly injurious to chilli and wheat crops, to plantations, nursery plants, grasses, etc. Out of 27 species of termite, 14 have been observed to be pests of crops. The important species attacking guar (*Cyanopsis tetragonoloba*) are: *Microcerotermes baluchistanicus* (Ahmad), *Odontotermes guptai* (Roonwal and Bose); and *Microtermes obesi* (Holmgren). *Microcerotermes tenugnathus* (Holmgren) is a major pest of wheat, a crop widely distributed in the area. *Microtermes mycophagus* (Desneux) is a pest of the castor crop. Seedlings of guar and wheat plants may be completely destroyed by the termites. Successful control of termites attacking wheat was achieved by using the organochlorine insecticides aldrin (5% dust), BHC (10% dust), and heptachlor (5% dust), all at a dosage of 10 kg a.i./ha. The most effective compound proved to be aldrin, which reduced the extent of damaged plants by 75% (Parihar, 1980).

(b) White grubs: *Holotrichia consanguinea* (Blanchard) and *H. serrata* (Fabr.). The former is particularly injurious to kharif, especially bajra crops in the arid and semi-arid zones. The foraging beetles have been successfully controlled by 0.2% carbaryl, 0.50% monocrotophos, and 0.025% quinalphos (Kushwaha *et al.*, 1980). Control of grubs was accomplished by soil application of phorate at 2–3 kg a.i./ha, thiodemeton and lindane at 0.75–1.0 kg a.i./ha, and mephosfolan at 2.5 kg a.i./ha.

(c) Grasshoppers: The major species injurious to dryland crops are *Hieroglyphus nigrorepletus* (Boliver); *Chrotogonus trachyptorus* (Blanchard); and *Oedaleus senegalensis* (Krauss). Control strategies include: (a) tillage of soil for destruction of eggs; (b) avoiding the use of chemicals during the intensive activity of natural enemies present in the area; (c) judicious use of insecticides such as 5% BHC, and lower dosages in peripheral areas of grazing lands.

(d) Hairy caterpillars: *Utetheisa pulchella* (L) is a common pest of pastures in the arid zone, and *Euproctis* spp. are serious pests of ber, castor, and bajra. Control measures include: (a) monitoring moth emergence with light traps; (b) collection of egg masses; (c) dusting with 4% endosulfan or 5% carbaryl at 1 kg a.i./ha, preferably in the early instar stage.

(e) Armyworm: *Mythimna separata* populations can be controlled by 5% BHC or carbaryl dust at 1.25 kg a.i./ha before the population peaks out.

(f) Weevils: *Myllocerus discolor* (Boheman), *M. pustulatus* (Faust), and *M. dentifer* (Fabr.) infest bajra, sorghum, sunflower, green gram, sesamum, moth and ber. Applications of 5% BHC or carbaryl dust provide effective control.

(g) Tissue borers: Shootfly (*Atherigona* spp.) infest bajra and sorghum. Early sowing following the first rain, seed dressing with 50% WP carbofuran (60–100 g/kg), or presowing soil application of phorate or thiodemeton granules gives good control.

(h) Sap suckers: sugar-cane leafhopper, *Pyrilla perpusilla* (Walker), infests jowar and bajra.

As can be seen from the above many insect species are serious pests of various crops in the arid zone, and with some exceptions, do not differ greatly from those encountered in temperate and humid regions. The use of persistent organochlorine insecticides is still in practice, although most of these have been banned in temperate zones of the Western countries some years ago. An assessment of the hazard to the arid environment due to the continuing use of these chemicals has not been made; certainly, there is need of such an investigation.

4.2.5 CONCLUDING REMARKS

The desert can be made to bloom. Evidence for that can be seen in the green fields of the northern Negev desert of Israel, the Imperial Valley of California, the Rajasthan district of India, the rangelands of Australia, and many others. The most important requirement is the availability of water for irrigation. A desert agro-ecosystem without irrigation can support only a few crops, such as dates, figs, a few species of grasses, and, in some instances, also cotton and some of the newer exotic plants, e.g. jojoba, guayule, etc. The introduction of irrigation can convert the desert land suitable for agriculture into large oases supporting major crops of fruits, vegetables, and fibres. However, there is no doubt that irrigation will bring with it pest and pesticide problems such as those encountered in temperate and other zones.

A desert environment is less susceptible to injury by pests because of the existing natural balance between predator and prey. In the oases there is a lack of natural balance due to the prevalence of different species, resulting in overpopulations which are not naturally regulated.

The nature of agriculture in the desert is known as 'islands of agriculture' because of the discontinuous character of crop production. Under such conditions, biological control, the use of pheromones, cultural methods, or a combination of these and other practices known as 'integrated pest management' should be very successful in an overall programme of pest control. The use of pesticides will undoubtedly be necessary; hence it is paramount that the dynamics of pesticide behaviour in soil and on crops should be thoroughly investigated. To date such data are wanting.

Although modelling of pesticide distribution, persistence, and fate can be of great help in predicting the behaviour of pesticides in different environments, there is no substitute for *in situ* determination of the dynamics of their behaviour and fate under different climatic conditions. Hence, extrapolation of data from laboratory experiments and from those simulating field conditions, especially as they apply to arid environments, must be made with reservation.

Once irrigation is introduced to the desert pest problems might become similar to those encountered in temperate regions. However, the dissimilarity in soil and climatic conditions will undoubtedly alter the behaviour of pests. Hence, a thorough knowledge of pest ecology will be required.

This section by no means encompasses all that is known about the desert. The surface has barely been scratched. It merely points to some of the major problems facing arid lands, their characteristics, their use and misuse, and the prospects for their rehabilitation. History teaches us that much of the desert area of the world was once teeming with life, and history can repeat itself. Pessimistic attitudes toward arid lands and the notion of the uselessness of the desert can be offset by vivid examples presented herein of the productivity of these lands and their essential role in combating famine.

The desert is not dead; it is only neglected.

ACKNOWLEDGEMENTS

The writers wish to express their sincere appreciation to Dr. I. Adato, Mr. I. Adler, Dr. N. Aharonson, Dr. Y. Ben-Dov, Mr. R. Dayan, Dr. M. Kehat, Dr. S. Lerman, Dr. U. Mingelgrin, Dr. S. Navarro, Dr. S. Saltzman, Dr. B. Yaron, and Mr. Z. Zook-Rimon for many helpful discussions and suggestions and for supplying us with unpublished data of their works.

4.2.6 REFERENCES

Adato, I. (1984). Mesurol residues in dates. Unpublished report to the Division of Plant Protection, Israel Ministry of Agriculture, Bet-Dagan.

Adkisson, P.L. (1957). Influence of irrigation and fertilizer on populations of three species of Mirids attacking cotton. *F.A.O. Pl. Prot. Bull.*, **6**(3), 33–36.

Adler, I. (1987). Types of herbicides, fungicides, and insecticides used on various crops in the Arava region. (Unpublished material) Arava Regional Council, Sapir Center.

Aharonson, N., Venezian, A., Klein, M., and Oko, O. (1984). Biology, phenology and control of the green leafhopper, *Omonatissus binotatus* var. *lybicus* (Berg), on dates in the Arava desert of Israel. Unpublished report to the Division of Plant Protection, Israel Ministry of Agriculture, Bet-Dagan.

Allred, P. M., and Giesey, J. P. (1985). Solar radiation-induced toxicity of anthracene to *Daphnia pulex*. *Environ. Toxicol. and Chem.*, **4**, 219–226.

Ambroggi, R. P. (1966). Water under the Sahara. *Scientific American*, May, 1966.

Amir, J., Vanunu, E., Krikun, J., Orion, D., Penuel, Y., Satki, Y., and Lerner, E. (1981). Long-term field experiments in the Negev. A monoculture of wheat under semi-arid conditions. *Hassadeh*, **62**, 198–204 (in Hebrew, with English summary).

Amir, J., Carmi, Z., Krikun, J., Orion, D., Penuel, Y., and Lerner, E. (1982). Effect of a continuous wheat rotation on yield in the Negev (1980–1981). *Gan, Sadeh veMesheki*, **5**, 15–19 (in Hebrew).

Anonymous (1977). The spectra of desertification: editorial. *Ann. Arid Zone*, **16** (3), Sept.

Atlas of Israel (1970). Amiram, D. H. K., Rosenan, N., Kadmon, N., Elster, J., Gilead, M., and Paran, U. (eds.). Ministry of Labour, Jerusalem, and Elsevier Publishing Co., Amsterdam.

Backlund, V. L., and Hoppes, R. R. (1984). Status of soil salinity in California. *Calif. Agric.*, **38**, 8–9.
Ben-David, T., Venezian, A., and Ben-Dov, Y. (1985). Observations on phenology, biology and control of the mango thrips, *Scirtothrips mangiferae* (Priesner), in the Arava Valley of Israel. *Hassadeh*, **65**, 1826–1830 (in Hebrew with English summary).
Ben-Dov, Y. (1985). Further observations on scale insects (Homoptera, Coccoidea) of the Middle East. *Phytoparasitica*, **13**, 185–192.
Bielorai, H., and Levy, J. (1971). Irrigation regimes in a semi-arid area and their effects on grapefruit yield, water use and soil salinity. *Israel J. agr. Res.*, **21** (1), 3–12.
Bishara, I. (1932). The greasy cutworm, *Agrotis ypsilon* (Rott.), in Egypt. *Egypt. Min. Agr. Tech. Bull.*, **114**, 1–55.
Blumberg, D., and Kehat, M. (1982). Biological studies of the date stone beetle, *Coccotrypes dactyliperda. Phytoparasitica*, **10** (2), 73–78.
Bowman, M. C., Schechter, M. S., and Carter, R. L. (1965). Behaviour of chlorinated insecticides in a broad spectrum of soil types. *J. Agric. Fd. Chem.*, **13**, 360–365.
Boyko, H., and Boyko, E. (1964). Principles and experiments regarding direct irrigation with highly saline and sea water without desalinization. *Trans. N.Y. Acad. Sci.*, Ser. II, **26** (suppl. to No. 8), 1087–1102.
Brown, A. W. A. (1978). *Ecology of Pesticides*. Wiley Interscience, New York.
Brown, A. W. A., and Pal, R. (1971). *Insecticide Resistance in Arthropods*, 2nd edn. World Health Organization, Geneva.
Brown, N. J., and Brown, A. W. A. (1970). Biological fate of DDT in a subarctic environment. *J. Wildl. Mgmt.*, **34**, 929.
Buyanovsky, G., Dicke, M., and Berwick, P. (1982). Soil environment and activity of soil microflora in the Negev Desert. *J. Arid. Envir.*, **5**, 13–28.
Caswell, M., Zilberman, D., and Goldman, G. E. (1984). Economic implications of drip irrigation. *Calif. Agric.*, July–August.
Chen, Y. L., and Casida, J. E. (1969). Photodecomposition of pyrethrin I, allethrin, phthalthrin and dimethrin. Modifications in the acid moiety. *J. Agric. Fd. Chem.*, **17**, 208–215.
Christie, E. K. (1979). Eco-physiological studies of the semi-arid grasses *Aristida leptopoda* and *Astrebla lappacea. Aust. J. Ecol.*, **4**, 223–228.
Christmas, R., Shaw, K., and MacEllister, F. V. (1980). *Arid Zone Newsletter*, p. 49.
Cloudsley-Thompson, J. L. (1977). Reclamation of the Sahara. *Envir. Conserv.*, **4**, 115–119.
Cloudsley-Thompson, J. L. (1978). The future of arid environments. In: Mann, H. S. (ed.), *Proc. Int. Symp. on Arid Zone Res. and Devel.*, pp. 469–474. Scientific Publishers, Jodhpur.
Cripps, R. E., and Roberts, T. R. (1978). Microbial degradation of herbicides. In: Hill, I. R., and Wright, S. J. L. (eds), *Pesticide Microbiology*, pp. 669–720. Academic Press, New York.
Crosby, D. G. (1972a). The photodecomposition of pesticides in water. *Adv. Chem.*, Ser. III, 173–188.
Crosby, D. G. (1972b). Environmental photooxidation of pesticides. In: *Degradation of Synthetic Organic Molecules in the Biosphere*, pp. 260–278. National Academy of Sciences. Washington D.C.
Crosby, D. G., and Li, M. (1969). Herbicide photodecomposition. In: Kearney, P. C., and Kaufman, D. D. (eds), *Degradation of Herbicides*, pp. 321–363. Marcell Dekker, New York.
Danin, A. (1983). *Desert Vegetation of Israel and Sinai*, p. 148. Cana Ltd., Jerusalem, P.O.B. 1199.
Dayan, R. (1987). Types and total acreage of crops grown in the Arava region. (Unpublished material). Arava Regional Council, Sapir Center.

Dhall, P., and Lal, R. (1971). Colorimetric determination of carbaryl residues in field-treated corn. *J. Ass. Off. Anal. Chem.*, 1095–1099.

Dibble, C. B. (1940). Grasshoppers, a factor in soil erosion in Michigan. *J. Econ. Ent.*, **33**, 496–499.

Drummond, R. O. (1977). Resistance in ticks and insects of veterinary importance. In: Watson, D. L., and Brown, A. W. A. (eds), *Pesticide Management and Insecticide Resistance*, pp. 303–319. Academic Press, New York.

Edwards, C. A. (1975). *Persistent Pesticides in the Environment*, 2nd edn. CRC Press, Cleveland, Ohio.

Edwards, C. A., Beck, S. D., and Lichtenstein, E. P. (1957). Bioassay of aldrin and lindane in soil. *J. Econ. Ent.*, **50**, 622–626.

Elder, D. L., and Villeneuve, J. P. (1977). Polychlorinated biphenyls in the Mediterranean Sea. *Mar. Poll. Bull*, **8**, 19–22.

Elder, D. L., Villeneuve, G. P., Parsi, P., and Harvey, G. R. (1976). *Polychlorinated Biphenyls in Sea Water, Sediments and Over-Ocean Air of the Mediterranean*. Activities of the Int. Lab. of Marine Radioactivity, Monaco. International Atomic Energy Agency, Vienna.

Evenari, M., Shanan, L., and Tadmor, N. H. (1968). ''Runoff farming'' in the desert: I. Experiment Layout. *Agron. J.*, **60**, 29–32.

Evenari, M., Shanan, L., and Tadmor, N. H. (1971). *The Negev*. Harvard University Press, Cambridge, Massachusetts.

Fuhremann, T. W., Lichtenstein, E. P., and Katan, J. (1978). Binding and release of insecticide residues in soils. *ACS Symposium Series*, **73**, 131–140.

Gangwar, S. K., Kavadia, V. S., Gupta, H. C. L., and Srivastava, B. P. (1978). Persistence of carbaryl in sandy-loam soil and its uptake in bajra (*Pennisetum typhoides* P.). *Ann. Arid Zone*, **17**, 357–362.

Grewal, G. S., and Atwal, A. S. (1968). Development of *Chrotogonus trachypterus* in relation to different levels of temperature and humidity. *Indian J. Ent.*, **30**, 1–7.

Gupta, D. S., and Rawlins, W. A. (1966). Persistence of two systemic carbamate insecticides in three types of soil. *Indian J. Ent.*, **28**, 482–493.

Harris, C. R., and Lichtenstein, E. P. (1961). Factors affecting volatilization of insecticides from soil. *J. Econ. Ent.*, **54**, 1038–1045.

Harrison, I. R. (1969). Development of organophosphorus resistance in the Australian sheep blow fly, *Lucilia cuprina* (Wied.). *Soc. Chem. Ind. Monogr.*, **33**, 215–223.

Herring, J. L., Hannan, E. J., and Bills, D. D. (1972). Ultraviolet irradiation of Aroclor 1254. *Bull. Envir. Contam. Toxicol*, **8**, 153–157.

Hutzinger, O., Safe, S., and Zitko, V. (1972). Photochemical degradation of chloro biphenyls (PCBs). *Envir. Hlth. Perspect.*, **1**, 15–20.

Hyde, M. B., Baker, A. A., and Ross, A. C. (1973). Airtight grain storage. *F.A.O. Agr. Serv. Bull.*, **17**. FAO, Rome.

Issar, A. (1985). Fossil water under the Sinai-Negev peninsula. *Scient. Am.* **253**, July, 82–88.

Kardavani, P. (1978). Role of biotic and environmental factors in the extension of the Lut Desert in Iran. *Ann. Arid Zone*, **17**, 92–98.

Karrar, G., and Stiles, D. (1984). The global status and trend of desertification (Review). *J. Arid Envir.*, **7**, 309–312.

Katan, J., Fuhremann, T. W., and Lichtenstein, E. P. (1976a). Finding of ^{14}C-Parathion in soil: a reassessment of pesticide persistence. *Science*, **193** (4256), 891–894.

Katan, J., Greenberger, A., Alon, H., and Grinstein, A. (1976b). Solar heating by polyethylene mulching for the control of diseases caused by soil borne pathogens. *Phytopathology*, **76**, 683–688.

Kehat, M., Blumberg, D., and Greenberg, S. (1969). Experiments on the control of the raisin moth, *Cadra figulilella* Gregs (phycitidae, Pyralidae), on dates in Israel. *Isr. J. Agr. Res.*, **19**, 121–128.

Kehat, M., Swirski, E., Blumberg, D., and Greenberg, S. (1974). Integrated control of date palm pests in Israel. *Phytoparasitica*, **2**, 141–149.

Kelley, D. B. (1984). Halophytes as a rangeland source. *Calif. Agric.*, **38**, 26.

Kelley, R. L., and Nye, R. L. (1984). Historical perspective on salinity and drainage problems in California. *Calif. Agric.*, **38**, 4–6.

Khan, M. A. Q., Khan, H. M., and Sutherland, D. J. (1974). Ecological and health effects of the photolysis of insecticides. In: Khan, M. A. Q., and Bederka, J. P. Jr. (eds), *Survival in Toxic Environments*, pp. 333–356. Academic Press, New York.

Kilgore, W.W. (1975). Mechanisms responsible for the disappearance of pesticides from the environment. In: *FAO/UNEP Consultation on Impact Monitoring of Residues due to Uses of Agricultural Pesticides in Developing Countries*. Food and Agriculture Organization of the U.N., Rome.

Kirkham, M. B. (1978). Salt water irrigation frequency for barley. *Ann. Arid Zone*, **17**, 12–18.

Krikun, J., Orion, D., Nachmias, A., and Reuveni, R. (1982). The role of soil borne pathogens under conditions of intensive agriculture. *Phytoparasitica*, **10** (4), 245–258.

Kushwuha, K. S., Noor, A., and Rathore, R. R. S. (1980). Insect pest management in arid agriculture. *Ann. Arid Zone*, **19**, 503–509.

Lahser, C., and Applegate, H. G. (1966) Pesticides at Presidio. III. Soil and water. *Texas J. Sci*, **18**, 21.

Lerman, S. (1985). Personal communication. Institute for Applied Research, Ben-Gurion University of the Negev.

Lichtenstein, E. P. (1959). Factors affecting insecticide persistence in various soils. *Pest Control*, Aug, p. 40.

Lichtenstein, E. P. (1965). Problems associated with insecticidal residues in soils. In: *Research in Pesticides*, pp. 199–205. Academic Press, New York.

Lichtenstein, E. P. (1966). Persistence and degradation of pesticides in the environment. In: *Scientific Aspects of Pest Control*, pp. 221–229. Nat. Acad. Sci. Publ. No. 1402.

Lichtenstein, E. P. (1979). Fate of pesticides in a soil-plant microcosm. In: *Terrestrial Microcosms and Environmental Chemistry*. National Science Foundation Publication, April.

Lichtenstein, E. P., and Schulz, K. R. (1959). Persistence of some chlorinated hydrocarbon insecticides as influenced by soil types, rate of application and temperature. *J. Econ. Ent.*, **52**, 124–131.

Lichtenstein, E. P., and Schulz, K. R. (1960). Epoxidation of aldrin and heptachlor in soils as influenced by autoclaving, moisture and soil types. *J. Econ. Ent.*, **53**, 192–197.

Lichtenstein, E. P., and Schulz, K. R. (1964). The effect of moisture and microorganisms on the persistence and metabolism of some organophosphorus insecticides in soils, with special emphasis on parathion. *J. Econ. Ent.*, **57**, 618–627.

Lichtenstein, E. P., Liang, T. T., and Fuhremann, T. W. (1978). A compartmentalized microcosm for studying the fate of chemicals in the environment. *J. Agr. Fd.Chem.*, **26**, 948–953.

Lomas, J., and Shashoua, Y. (1974). The dependence of wheat yields and grain weight in a semi-arid region on rainfall and on the numbers of hot, dry days. *Israel J. of Agric. Res.*, **23** (3–4), 113–121.

MacDonald, J. D., Swiecki, T. J., Blaker, N. S., and Shapiro, J. D. (1984). Effect of salinity stress on the development of *Phytophthora* root rots. *Calif. Agric.*, **38**, 23–24.

Matsumura, F., and Benezet, H.J. (1978). Microbial degradation of insecticides. In: Hill, I. R., and Wright, S. J. L. (eds), *Pesticide Microbiology*, pp. 623–659. Academic Press, New York.

Meigs, P. (1953). World distribution of arid and semi-arid homoclimates. *Arid Zone Progrm.*, **1**, (Rev. of Research on Arid Zone Hydrology), pp. 203–209.

Merton, L. F. H. (1959). Studies in ecology of the Moroccan locust (*Docoistaurus maroccanus* Thunberg) in Cyprus. *Anti-locust Bull.*, **34**, 123

Mueller, I. M., and Weaver, J. E. (1942). Relative drought resistance of dominant prairie grasses. *Ecology*, **23**, 387–398.

Navarro, S., Donahaye, E., Kashanchi, Y., Pisarev, V., and Bulbul, O. (1984). Airtight storage of wheat in a PVC covered bunker. In: Ripp, B.E. (ed.), *Controlled Atmosphere and Fumigation in Grain Storages*, pp. 601–614. Proc. Int. Symposium, Perth, Western Australia, April. Elsevier, North-Holland.

Oppenheimer, H.R. (1960). Adaptation to drought. Xerophytism. In: *Plant Water Relationships in Arid and Semi-Arid Conditions*, pp. 105–138. UNESCO, Paris.

Oster, I. D., Hoffman, G. J., and Robinson, F. E. (1984). Dealing with salinity. Management alternatives; crop, water and soil. *Calif. Agric.*, **38**, 29–32.

Pal, S. K., and Kushwuha, K. S. (1977). Persistence of BHC residues in soil and its translocation in chilli fruits. *Ann. Arid Zone*, **16**, 169–170.

Parihar, D. R. (1980). Termite problem in desert plantations. *Ann. Arid Zone*, **19**, 329–334.

Parihar, D. R. (1981). Effects of overgrazing on grasshopper population in the grasslands of Rajasthan desert. *Ann. Arid Zone*, **20**, 291–293.

Perkins, I. D. (1980). New methods for control of green sheep blowfly (*Lucilia cuprina*). *Arid Zone Newsltr*, **9**, 67.

Peryea, P. J., and Bingham, F. T. (1984). Reclamation and regeneration of boron in high-boron soils. *Calif. Agric.*, **38**, 35.

Petrov, M. P. (1978). Development and protection of natural resources of deserts and semi-deserts. In: Mann, H.S. (ed.), *Proc. Int. Symp. on Arid Zone Res. and Devel.*, pp. 491–496. Scientific Publishers, Jodhpur, India.

Popov, G. B. (1958). Ecological studies of oviposition by swarms of the desert locust, *Schistocerca gregaria*. Forskal in East Africa. *Anti-locust Bull.*, **31**, 1–70.

Rake, D. W. (1961). Some studies on photochemical and soil microorganism decomposition of granular organoborate herbicides. *Weed Soc. Am. Abstr.*, pp. 48–49.

Rivnay, E. (1964). The influence of man on insect ecology in arid zone. *Annu. Rev. Entomol.*, **9**, 41–62.

Robinson, J., Richardson, A., Bush, B., and Elgar, K. E. (1966). A photo-isomerization product of dieldrin. *Bull. Envir. Contam. Toxicol.*, **1**, 27.

Roffey, J. (1970). The effect of changing land use on locusts and grasshoppers. In: *Proc. Int. Study Conf. on Current and Future problems of Acridology*, London, pp. 199–204.

Rolston, D. E., Nielsen, D. R., and Beggar, J. W. (1984). Effect of salt on soils. *Calif. Agric.*, **38**, 11–12.

Rosen, J. D. (1972). The photochemistry of several pesticides. In: Matsumura, F., Bouch, G. M., and Misato, T. (eds), *Environmental Toxicology of Pesticides*, pp. 435–447. Academic Press, New York.

Rosen, J. D., Sutherland, D. J., and Lipton, G. R. (1966). The photochemical isomerization of dieldrin and endrin and effects on toxicity. *Bull. Envir. Contam. Toxicol.*, **1**, 133.

Roy, G.P., and Shetty, B.V. (1978). The impact of canal irrigation on the flora of the Rajasthan desert. In: Mann, H. S. (ed.), *Arid Zone Research and Development*, Proc. Int. Symp. on Arid Zone Res. and Devel., pp. 183–189. Scientific Publishers, Jodhpur.

Sethunathan, N. (1980). Effects of combined pesticides application on their persistence in flooded rice soils. *Agrochemical Residue-Biota Interactions in Soil and Aquatic Ecosystems*, pp. 259–281. IAEA, Vienna.

Shanahan, G. J., and Roxburgh, N. A. (1974a). Insecticide resistance in Australian sheep blowfly. *Agr. Gaz. N.S.W.*, **85**, 4–7.

Shanahan, G. J., and Roxburgh, N. A. (1974b). A significant increase in larval resistance of *Lucilia cuprina* (Wied.) to diazinon. *Vet. Rec.*, **94**, 382.
Shanahan, G. J., and Roxburgh, N. A. (1974c). The sequential development of insecticide resistance problems in *Lucilia cuprina* (Wied.) in Australia. *PANS*, **20**, 190–202.
Shanahan, G. J., and Roxburgh, N. A. (1974d). Insecticide resistance in Australian sheep blowfly *Lucilia cuprina* (Wied.). *J. Aust. Inst. Agric. Sci.*, **40**, 249–253.
Shannon, M.C., and Qualset, C.O. (1984). Benefits and limitations in breeding salt-tolerant crops. *Calif. Agric.*, **38**, 33–34.
Shulov, A. (1952). The development of eggs of *Schistocerca gregaria* in relation to water. *Bull. Ent. Res.*, **43**, 469–476.
Simons, M. (1967). *Deserts—the Problem of Water in Arid Lands*. Oxford University Press, Oxford.
Singh, A., Ahlawat, I. P. S., and Saraf, C. S. (1981). Response of spring mung bean (*Vigna radiata*) to irrigation and anti-transpirants with and without mulching. *Ann. Arid Zone*, **20**, 28–34.
Singh, R. P. (1978). Dry farming research—its relevance and prospects for arid zone agriculture. In: Mann, H. S. (ed.), *Proc. Int. Symp. on Arid Zone Res. and Devel.*, pp. 475–481. Scientific Publications, Jodhpur.
Stower, W. J., Popov, G. B., and Greathed, D. J. (1958). Oviposition behaviour and egg mortality of the desert locust, *Schistocerca gregaria*. *Anti-locust Bull.*, **30**, 1–33.
Thornthwaite, C. W. (1948). An approach toward a rational classification of climate. *Geogrl. Rev.*, **38**, 55–94.
United Nations Conference on Desertification (1977). *Report A/Conf. 74/36*, Nairobi, 29 Aug.–9 Sept.
Uvarov, B. P. (1962). Development of arid lands and its ecological effects on their insect fauna. *Arid Zone Res.*, **28**, 235–248.
Valentine, R. C. (1984). Genetic engineering of salinity-tolerant plants. *Calif. Agric.*, **38**, 36–37.
Venezian, A., and Ben-Dov, Y. (1982). The mango thrips *Scirthothrips mangiferae* Priesner, a new pest of mango in Israel. *Hassadeh*, **62**, 1116–1117 (in Hebrew with English summary).
Verma, S. K. (1981). Field efficiency of insecticides and antifeedants against advanced stage larvae of *Amsacta moorei* Butler. *Ann. Arid Zone*, **20**, 253–257.
Ware, G. W., Estesen, B. J., and Cahill, W. P. (1971). DDT moratorium in Arizona. Agricultural residues after 2 years. *Pest. Mon. J.*, **5**, 276.
Williams, O. B., and Roe, R. (1975). Management of arid grasslands for sheep: Plant demography of six grasses in relation to climate and grazing. *Proc. Ecol. Soc. Aust.*, **9**, 142–156.
Woodcock, D. (1978). Microbial degradation of fungicides, fumigants and nematocides. In: Hill, I. R., and Wright, S. J. L. (eds), *Pesticide Microbiology*, pp. 731–799. Academic Press, New York.
Yaron, B., Vinien, A. J., Fine, P., Metzger, L., and Mingelgrin, U. (1983). The effect of solid organic components of sewage on some properties of the unsaturated zones. *Ecol. Studies*, **47**, 162–181.
Yaron, B., Gerstl, Z., and Spencer, W. F. (1984). Behaviour of herbicides in irrigated soils. *Adv. Soil Sci.*, **2**, 1–143.
Yaron, B., Mingelgrin, U., and Saltzman, S. (1985). Personal communications.
Yathom, S. (1961). Studies on the bionomics of *Hylemia cilicrura* Rond. in Israel. *Ktavim*, II, 51–55.
Zook-Rimon, Z. (1985). *Mus musculus* as a pest of green peppers in the Arava desert region of Israel. Personal communication.
Zook-Rimon, Z. (1985). Personal communication.

Ecotoxicology and Climate
Edited by P. Bourdeau, J. A. Haines, W. Klein and C. R. Krishna Murti

4.3 Effects in Marine Ecosystems

J. P. LAY AND A. ZSOLNAY

4.3.1 INTRODUCTION

In this paper, tropical marine areas are defined as those whose minimum surface temperatures never go below 20°C. They are shown in Figure 4.3.1 and are approximately the latitudes between 28°N and 28°S. The major exceptions to this are the areas of strong upwelling found along the western coasts of most continental areas. These introduce nutrient rich, but colder waters to the surface. For example, the major upwelling off Peru normally prevents tropical conditions from existing below latitude 5°S in that region (Figure 4.3.1).

The tropics, especially in coastal areas, are exposed to an ever increasing environmental impact through increased population, industrialization, and tourism. This has required an increased knowledge of the potential effects of these impacts. However, to date there has been very little work done in this geographical area, and the majority of it has been concerned with basic ecological principles rather than the effect of specific anthropogenic inputs. Of the 15 000 ecological publications published between 1979 and 1983, only 0.09% of them dealt with tropical environmental research (Cole, 1984). In fact, the trend appears to show a decrease over time! In 1979 and 1980 the relative number of articles in this area of research were 0.18 and 0.23% respectively, while during 1982 and 1983 this has decreased to 0.05 and 0.14%.

One possible approach to circumvent this lack of information is to use the data and results obtained from environmental studies carried out in temperate regions and attempt to extrapolate them to tropical environments. This, however, can only be done in a very tenuous manner. The physical environment is by definition different in the tropics from that in the temperate areas. Furthermore, the biota, as a result of diverging evolutionary processes, are naturally not the same in these two regions. Even when the same taxa exist in both the tropics and temperate zones, it is reasonable to assume that their ability to successfully respond to an anthropogenic input will also be affected by the different physical environments.

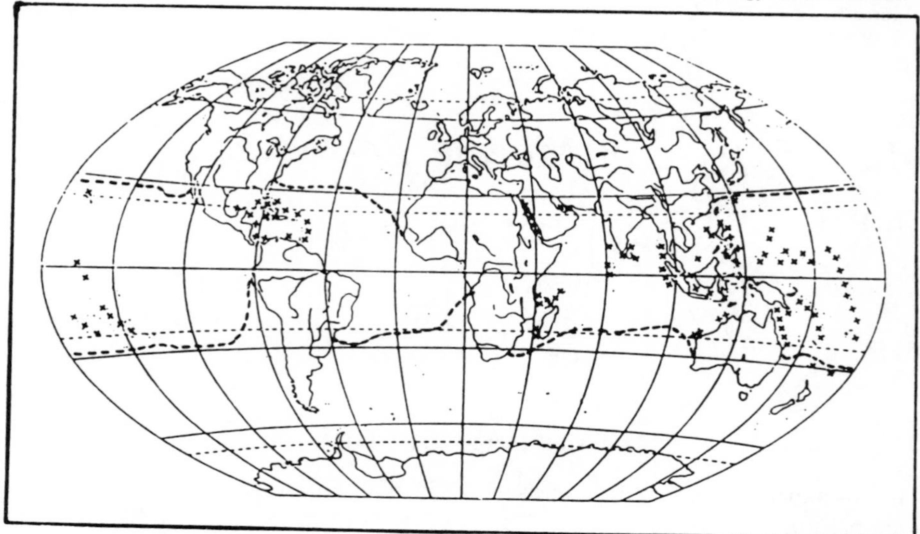

Figure 4.3.1 Distributions of corals (x) and the 20°C surface water isotherms (–)

The goal of this report is to look at these environmental differences and see in which ways they may mitigate or exacerbate the effects of anthropogenic inputs. Much of this has already been done (Johannes and Betzer, 1975), but in this report the results reported in more recent literature will be emphasized. Essentially, only marine systems will be considered. We feel that rivers and other inland waters are more complicated and that their variations within a given geographical region may be greater than the variations between geographical zones. Similar to Johannes and Betzer (1975), we have found that the predominance of tropical ecological literature is concerned with coral reefs. By necessity, our report must reflect this. Finally, we will not be considering mangroves, since this is being done elsewhere (*see* Section 5.6.3.2).

4.3.2 PRIMARY CAUSES AND ASPECTS OF THE TROPICAL ENVIRONMENT

The primary cause of tropical conditions is obviously their geographical location. This has the main effect of causing tropical areas to receive more intensive solar radiation, which in turn means that more light and higher temperatures prevail there. The latter is the chief indicator, by definition, of this region. Furthermore, the wind patterns near the equator result in a system with relatively low kinetic energy. It is unlikely that anthropogenic inputs will result in an increase of light or a decrease in the kinetic energy levels. These primary effects produce a whole series of secondary ones, both inorganic and organic, from which most of the differences between tropical and temperate regions derive (Figure 4.3.2).

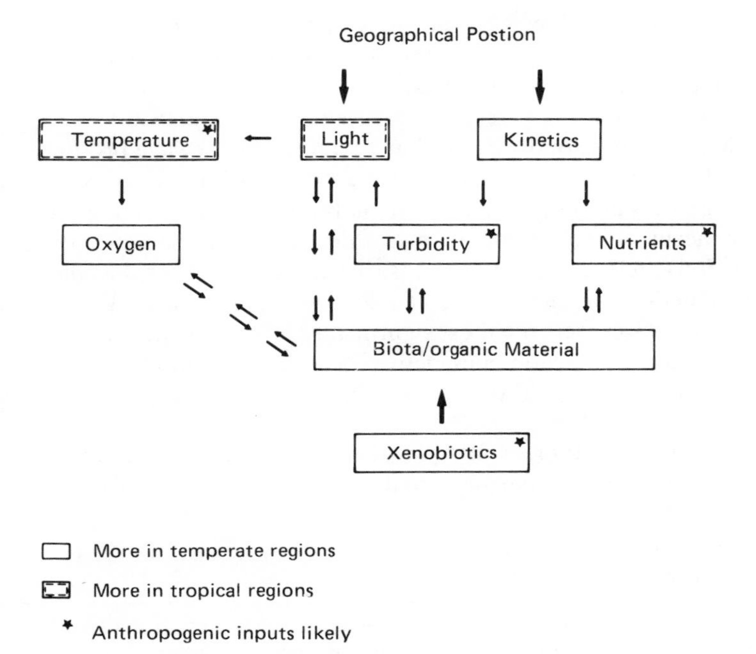

Figure 4.3.2 Primary and secondary interactions resulting in differences between tropical and temperate aquatic regions

The higher temperatures in the tropics have a direct influence on the level of oxygen saturation, which is lower there than in the higher latitudes. Power plants and similar sources of heat can have significant local effect on the water temperature. This, of course, will also reflect on the amount of oxygen that can be present in the water.

Furthermore, the lower kinetics present in the tropics result in lower turbidity levels in coastal areas. These levels, however, can be drastically increased through the influence of such activities as dredging, power boating, or sewage introduction. In addition, these lower energy levels, when combined with the tendency of surface waters to be rapidly heated, lead to a significant vertical stratification, which in turn prevents the introduction of nutrient-rich deep water into the photic zone. The result is a prevalence of nutrient-poor or oligotrophic conditions in the surface waters. This, of course, has the additional effect of keeping the biological standing stock as well as the turbidity lower in the tropics than in the temperate regions. The stronger seasonal variations in non-tropical areas lead to periods of strong mixing between the photic zone and the nutrient-rich water located below it. This results in the well-documented seasonal blooms of phytoplankton and their associated zooplankton. This phenomenon is usually absent in the tropics. However, the introduction of insufficiently treated sewage can produce a drastic increase in the amounts of nutrients available. This can then result in eutrophic conditions, which can also produce an increase in turbidity levels. The net effect is that the normal ecological system is altered into a diametrically opposite one. Furthermore, raw sewage produces a strong biological oxygen demand, causing a significant decrease in oxygen levels.

The organic material, including the biota, is usually present at considerably lower concentrations in the tropics than in the temperate zone. Sewage introduction can lead to a local increase in the amount of organic material, while the introduction of alien biota can lead to major instabilities in tropical communities. However, the most significant anthropogenic impact in this area is the introduction of xenobiotics, active chemicals introduced into the ecosystem either accidently (spills, leaks, etc.) or intentionally (crop spraying, sewage, etc.).

The main relevant topics concernng the above mentioned differences between temperate and tropical marine systems are summarized in Table 4.3.1.

Table 4.3.1 The tropical marine environment. Comparison of environmental factors between tropical and temperate zones

Environmental factor	Tropical situation compared to the temperate one	Applicable anthropogenic input	Ecological effect	Impact in tropics*
Light	More intense	Sewage	Eutrophication	+
		Xenobiotics	Photochemical alteration	+
		Turbidity from from waste	Less productivity	+
Temperature	Higher	Sewage	BOD, COD	+
		Heat	Thermal stress	+
		Xenobiotics	Secondary chemical reactions	+
		Xenobiotics	Metabolism, detoxification	?
Oxygen	Lower concentration	Sewage	Community restruc- turing	+
Nutrients	Lower concentration	Sewage	Eutrophication	+
Turbidity	Lower	Waste	Alteration of benthic biota (e.g. corals)	+
Community structure	More linear	All	Loss of diversity	+

* + = greater in tropics compared to temperate regions; ? = unclear

4.3.3 POTENTIAL ENVIRONMENTAL IMPACTS IN THE TROPICS

Three groups of pollutants are considered here: (1) hydrocarbons, including oils and chlorinated compounds; (2) metals; and (3) waste, including nutrients and turbidity-causing agents. Subsequent effects will be evaluated by giving examples of changes in light, temperature, oxygen, nutrient flux, and community structure.

4.3.3.1 Hydrocarbons

Volatile liquid hydrocarbons (VLH) are characterized by their high relative chemical reactivity, volatility (vapour pressure), and good water solubility. They make up the major constituents of anthropogenic organic material in the atmosphere (Duce, 1978) and are discharged into marine coastal waters near industrial and urban areas in environmentally significant amounts (Sauer *et al.*, 1978). They can be classified chemically mainly as alkylated aromatics, aliphatics, and alkylated cyclopentanes/hexanes. Many of them are among the most toxic components of petroleum (Anderson, 1975;McAuliffe, 1977; Baker, 1970).

Residues of these compounds in the tropical marine environment were determined, e.g. by Sauer *et al.* (1978) in the Gulf of Mexico and the Caribbean Sea from coastal, shelf, and open ocean surface waters.

Concentrations of the VLH ranged from ± 60 ng/l in open ocean non-petroleum-polluted surface water, to ± 500 ng/l in heavily polluted Louisiana shelf and coastal water. Caribbean surface samples showed very low concentrations of 30 ng/l. Subsurface VLH concentrations were only 35–40 ng/l.

Determination of the VLH load was made on the basis of their six major constituents: benzene, ethylbenzene, toluene, and *m*-, *p*- and *o*-xylene.

In the Caribbean water samples, aromatics were the only volatile compounds in seawater. Toluene was present in all samples and was suggested to be of possible geochemical origin. Heavily polluted areas contained considerable concentrations of cycloalkanes (60–110 ng/l; 20% of total VLH). Besides the apparently high concentrations of VLH due to contamination, the distribution of VLH in the regions investigated could have further origins; these include: surface currents passing through polluted water bringing VLH to the central Gulf; the atmosphere acting as a source of surface-water VLH; open ocean discharges from tankers and other ships sufficient to produce detectable VLH concentrations; and VLH present in open-ocean surface water as true residuals.

Biological effects of VLH have mainly been investigated by short-term studies of lethal effects at high dosages and have not reflected the realistic stresses that may be encountered by organisms from waters polluted by VLH (Sauer, 1980). With regard to the chronic sublethal effects of VLH on tropical marine organisms, only studies concerning behavioural or chemoreception effects at lower concentrations have been carried out. The response of snails and crabs to chemical substances that normally initiate feeding behaviour was eliminated with a 1 μg/l concentration of the water-soluble kerosene (benzene) fraction (Jacobson and Boylan, 1973; Takahashi and Kittredge, 1973; Johnson, 1977). Fertilization of macroalgae was completely inhibited by 0.2 μg/l No. 2 fuel oil (Steele, 1977). Chemoreception in marine bacteria was inhibited by 10 μg/l of benzene (Walsh and Mitchell, 1973). VLH concentrations in the microgram per litre range can be present near many coastal, urban, and industrial areas, and their steady-state concentrations are high enough to cause detrimental long-term effects to the local marine organisms present (Sauer, 1980).

Another group of hydrocarbons, the polynuclear aromatic hydrocarbons (PAHs), are of considerable concern for the marine environment. This is mainly due to their relative persistence, and carcinogenic and mutagenic potential as well as to their possible function as indicators of anthropogenic pollution. They are present, in greatly varying concentrations, in most marine environments. For example, concentrations of PAHs in the surface waters of the Baltic Sea were nine times greater than in waters from the same depth in the Sargasso Sea (Zsolnay, 1977). Aqueous solubilities of these hydrocarbons have been demonstrated to influence biota uptake, sorption processes to sediments, and transport to the marine environment (Geyer *et al.*, 1981; Mackay, 1982, Gearing *et al.*, 1980).

Whitehouse (1984) studied the solubility of six selected PAHs at various temperatures and salinities. Comparing laboratory data of PAH solubility performed under standardized conditions, he found that all six compounds tested exhibited decreasing distilled water solubility with decreasing temperature. With the possible exception of benzo(a)pyrene, they were also sensitive to small changes in temperature. At temperature ranges from 3.7 to 25.3°C and salinities from 0 to 30.7‰ phenanthrene, anthracene, 2-methylanthracene, 2-ethylanthracene, and benz(a)pyrene salted out (decreasing solubility with decreasing temperature and decreasing solubility with increasing salt concentration), whereas 1,2-benzanthracene salted-in (increased solubility upon the addition of salt). The findings were that, unlike salinity, small decreases in temperature can cause significant decreases in the solubility of four of the PAHs tested, especially in the higher temperature range. This could have significant theoretical implications and greatly complicates the attempt to predict and compare the partitioning and transport within the temperate and tropical marine systems.

The presence of polynuclear aromatic hydrocarbons is important for the photo-oxidative degradation behaviour of crude oils in the sea. Thominette and Verdu (1984) reported a photochemical study, with three different crude oils from Algeria, the Middle East, and Venezuela, under simulated natural weathering conditions (wavelength above 300 nm, temperature 16–38°C).

It was shown that, besides spreading, evaporation, dissolution, emulsification, dispersion, and sedimentation, the sunlight-induced degradation or alteration of the oils in the sea can lead to weathering processes, which modify their physicochemical properties. Photo-oxidation of the PAHs is also responsible for the stabilization of water-oil emulsions (so-called chocolate mousse).

PAHs are responsible for a great majority of photochemical initiation reactions. Studies (Thominette and Verdu, 1984) have shown that aromatic-rich oils were photo-oxidized several times faster than aromatic-poor homologues. As temperature plays an important role in the radical chain mechanism of photo-oxidation, the overall activation energies can be assumed to be higher in the tropical marine environment. A side-mechanism via the formation of singlet oxygen, which can directly give peroxides of the PAHs, is not temperature dependent but can operate simultaneously with radical chain mechanism which needs the presence of abstractable hydrogen. PAHs as well as alkyl branched aromatics were shown to be the most important compounds in the temperature-dependent radical propagation of oxidation chains.

Oil and tar

Atwood and Ferguson (1982) described a case study in which the fate of oil was investigated in a tropical environment following a major petroleum spill (PEMEX IXTOC-1 outblow in the southern Bay of Cambeche, summer 1979). This *in situ* research, combined with separate microbiological microcosm and photo-oxidation experiments, covered the chemical and microbial weathering and fate of the oil spilled.

Fiest and Boehm (1980a), who measured the water column concentrations of petroleum hydrocarbons, detected significant amounts of toxic compounds, i.e. alkylated benzenes, naphthalenes, and methyl naphthalenes, within a 20-km distance from the blowout. The concentrations ranged from 0.02 to 38.0 μg/l. The total petroleum concentration in the water column was as high as 10 000 μg/l at the wellhead but dropped off rapidly to background values of 5 μg/l. The chemical evidence for bacterial degradation of the oil was slight.

Atwood and Ferguson (1982) and Fiest and Boehm (1980b) assumed that the response of the bacterial community and the rate of bacterial weathering of spilled oils were strongly nutrient limited. Thus, even though the potential for rapid biodegradation of the oil existed, the process probably did not occur. Pfaender *et al.* (1980) and Atlas *et al.* (1980) both showed distinct effects of spilled IXTOC-1 oil on the microbial community in the water column. Although little or no increase in total bacterial cell number resulted, there were large increases in the total hydrocarbon-utilizing bacteria. The spilled oil also caused an increase in total microbial metabolic activity, but it inhibited microbial amino acid respiration. These effects were limited to areas very near the floating surface oil; at distances greater than 25 km down-plume all these parameters were at what were perceived to be natural ambient levels (Atwood and Ferguson, 1982).

The formation of 'chocolate mousse-like emulsion' on the sea surface is a photochemical as well as a microbial process in the weathering of spilled oil after long-term exposure. As both processes occur simultaneously in the marine environment when a spill occurs, separate laboratory photo-oxidation experiments and microbial studies in the dark were performed. The compounds identified from the photo-oxidation experiments (oxygenated products like alkyl benzoic, naphthoic, and phenanthroic acids) were identical to those collected from the *in situ* samples in the IXTOC-1 spill plume. The same type of conversion products were also detected after a 7-day incubation in the dark. The distribution ratios, however, were different, probably because microorganisms favoured certain isomers. The formation of the mousse was similar in both the photo-oxidation and in the microbial laboratory experiments. Collected sediment trap samples showed that significant amounts of the sedimenting oil were associated with phytoplankton material. The upper bottom sediment layer contained 100 mg/kg of substantially physically and chemically weathered oil. Only 0.5–3% of the total IXTOC-1 oil within 50 km of the well was in the sediment. As no quantification was made for each photochemical and microbial degradation process, we assume that abiotic conversion, because of the stronger solar inputs, is the dominant process in the tropical marine environment. The nutrient-limited medium restricts heterotrophic organisms from effecting a more rapid degradation of spilled hydrocarbons.

In an earlier work, Zsolnay (1978) used gas-chromatographic indices to determine that the breakdown of tar in a subtropical environment was due to physical and chemical processes rather than bacteriological ones.

The harmful effects of oil on Red Sea corals were investigated by Rinkevich and Loya (1977, 1979). They recognized that the poor coral recruitment on oil-polluted reefs near Eilat was the result not only of damage to the reproductive capacity of sexually mature corals but also of the survivorship and settlement of their planulae. The toxicity of crude oil at the concentration of 10 μg/l was evidenced by a significant reduction of settlement and survivorship of the planulae of *Stylophora pistillata*. In Figure 4.3.3 the number of planulae of *S. pistillata* and the percent of settlement following a 144-hour laboratory experiment in Petri dishes at different oil concentrations are illustrated.

The spatial and temporal variation of pelagic tar in the eastern Gulf of Mexico is reported by Van Vleet *et al.* (1983). Concentration of pelagic tar in this region collected monthly at seven stations, ranged from 0 to 26.5 mg/m^2. Using neuston tows, average concentrations of tar were found to be 1.6 mg/m^2 off the West Florida shelf and 0.05 mg/m^2 on the shelf. Pelagic tar concentrations were closely correlated with proximity to the Gulf Loop Current, with maxima in the spring and autumn of 1980.

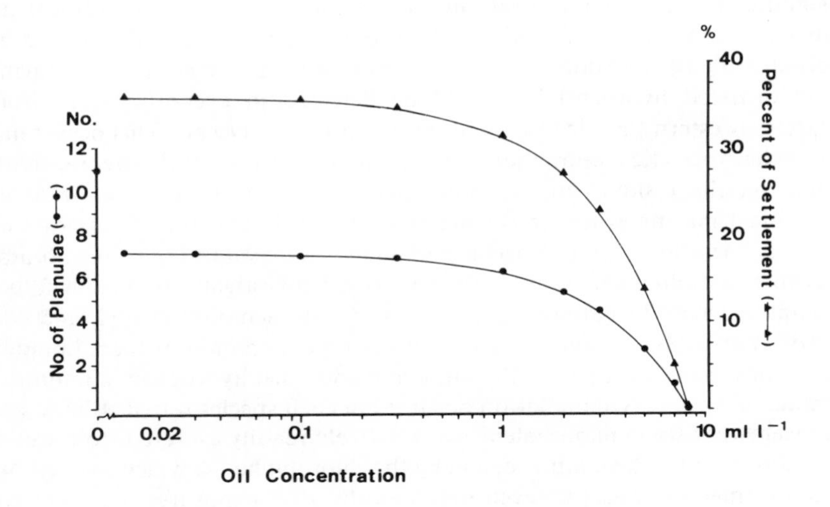

Figure 4.3.3 Number of planulae of *Stylophora pistillata* and percent of settlement under different oil concentrations. After Rinkevich and Loya (1979)

The occurrence of tar was often associated with floating *Sargassum*. The above work notwithstanding, the general impression is that more research is needed to clarify the impact of petroleum on tropical ecosystems. This is especially true in the case of chronic effects. This will require an increased need to understand natural short and long-term fluctuations (Anonymous, 1985).

Chlorinated hydrocarbons

The global distribution of the chlorinated hydrocarbons, especially in regard to the PCBs, hexachlorocyclohexane (HCH) isomers, and DDT and its metabolites, was monitored in samples collected between 1975 and 1979 by Tanabe and co-workers (Tanabe and Tatsukawa, 1980; Tanabe *et al.*, 1982). General concentration of PCBs had a range of 0.1–1.0 ng/l, DDT of 0.01–1.0 ng/l and HCHs of 1.0–10.0 ng/l in the North Pacific as well as in the Indian Ocean. High concentrations of DDT and HCH off the western coast of India and of DDT off the coast of Central America suggest that large amounts of DDT and HCH are still used in the tropical zone. The transport of these two pesticides to the ocean is predominantly via the atmosphere, and the distribution was determined by the equilibrium partitioning between air and surface waters (Tanabe and Taksukawa, 1980). HCHs were more concentrated in the northern hemisphere (especially Asia), while greater DDT concentrations have shifted from the North to the South in the last decade. PCBs were observed in significantly high concentrations in coastal regions of the tropical and subtropical zones and consisted mainly of the higher chlorinated types. Lower chlorinated PCBs were dominant in the ocean far from terrestrial environments. All chlorinated hydrocarbons could be detected in every location of the Antarctic, western Pacific Ocean, and eastern Indian Ocean. This occurrence, even in the Antarctic Ocean, indicates the global transport of these compounds.

Data on the residues of persistent organic chemicals, e.g. pesticides, in the tropical marine environment are rare. The fate and the effects of these extremely hazardous, anthropogenic compounds were studied only in laboratory experiments. Solbakken *et al.* (1984) reported investigations on the fate of ^{14}C-naphthalene, phenanthrene, 2,4,5,2′,4′,5′,-hexachlorobiphenyl, and octachlorostyrene in Bermudian corals. Uptake and elimination of these lipophilic xenobiotics were studied in 19 anthozoa and one hydrozoan common to Bermudian waters. Accumulation occurred in each species tested. Elimination rates were very slow; naphthalene was relatively readily excreted, followed by octachlorostyrene, phenanthrene, and by the chlorobiphenyl, which showed 84% retention (measured as ^{14}C) even nine months after exposure. No significant correlation was found between test species and accumulation rate.

These results underline the problem of understanding the impact in tropical waters of the highly chlorinated hydrocarbons, which are known to be persistent and possess a long-term subacute toxicity.

4.3.3.2 Metals

The occurrence, distribution, and accumulation potential of metals in the aquatic environment is known to depend on many factors, such as local enrichment, temperature, salinity and the physico-chemical properties of the metals themselves. As for hydrocarbons, most of the research on metals in tropical marine systems deals with their accumulation in the sediments or in selected organisms. Again, laboratory short-term studies try to compensate for the lack of *in situ* studies and/or of mechanistic processes. General conclusions for the tropics could not be obtained since large deviations in residual analysis and no direct comparable experimental conditions were indicated. Studies on anthropogenic metal contamination have to distinguish between natural levels and those which are enhanced from anthropogenic sources. At low levels, Fe, Cu, Zn, Co, Mn, Cr, Mo, V, Se, Ni, and Sn, as natural constituents, seem to be essential nutrients. Hg, Cd, and Pb are, besides Cu, Zn, Ag, and Cr, the most hazardous metals from the ecotoxicological as well as the public health point of view.

Heavy metal analysis is often performed in selected areas of coastal waters and sediments, when high inflows from rivers carrying a highly suspended load of municipal and industrial waste are apparent. The upper Gulf of Thailand, with five such inputs, was therefore analysed for high levels of anthropogenic sources (Hungspreugs and Yuangthong, 1983). No enrichment was found for Cr, Cu, and Zn.

Applying the ^{210}Pb method, they found that there had been strong build-ups of Cd and Pb in the upper 15 cm sediment layers during the past 30 years. Residues of Cd and Pb in the area (estuaries, upper and middle Gulf of Thailand) ranged from 0 to 6.7 mg/kg and 14 to 84.5 mg/kg respectively. Other areas in the world cited, showed similar Cd, but higher Pb, residual values (Hungspreugs and Yuangthong, 1983).

Analysis for Hg residues generally revealed that the biota (zooplankton, fish, algae) from the sites studied (tropical Pacific Ocean, Senegalese littoral and Malaysia) were relatively uncontaminated with mercury. Hirota *et al.* (1979) showed that protozoa, hydromedusae, ctenophores, chaetognaths, copepods, amphipods, euphausiids, thaliaceans, and larvae of fish and crustaceans have average Hg residues of 0.02–012 mg/kg dry weight. Methyl-Hg concentrations measured in only a few samples were generally lower than 0.02 mg/kg at different sites of the tropical Pacific Ocean.

Total bioaccumulated Hg content in algal samples (26 different species) of the Island of Penang (Malaysia) revealed 12 species with Hg residues below the detection limit and others with concentrations between 0.05 and 0.35 mg/kg (dry weight), except for one, *Phaeophycea*, with an amount of 1.03 mg/kg (Sivalingam, 1980). From 130 samples of 20 different types of fish from the Senegalese littoral it was found that Hg residues were mostly below 0.5 mg/kg, and 'mercurial pollution had not become a problem in that country at the moment' (Gras and Mondain, 1978).

The pelagic *Sargassum* community has been favoured as a useful model for the study of heavy metal–marine biosphere interactions (Johnson and Braman, 1975; Blake and Johnson, 1976). *Sargassum*, being well distributed in the North Atlantic as well as in the tropics (e.g. Caribbean Sea, Gulf of Mexico), represents a community with several trophic levels but relatively few species and is held together by a common substrate. Therefore, samples from the community possess an internal consistency. The fate of arsenic (As), germanium (Ge), and mercury (Hg) in the *Sargassum* weed and some of its associated fauna was studied by Johnson and Braman (1975). The distribution of As (III) and As (V) was on average the same in the members of the *Sargassum* community and in the surface water of the Sargasso Sea. This ration (15/85) of As III/As V in organisms and in seawater suggests the presence of a steady-state oxidation-reduction process. One percent alkyl-As was found to be produced by the *Sargassum* community *in situ*. Total As, Ge, and Hg contents were respectively 2.5–19.5 mg/kg, 0–0.28 mg/kg, and 0–0.39 mg/kg wet weight, while seawater contained respectively 1.6, 0.042, and 0.01 mg/l on average for the three metals. The species *Sargassum filipendula*, *Padina vickersiae*, and *Acanthophora spicifera* represent the important potential media for the transfer of Co, I, and Cs isotopes through food webs. *In situ* studies by Guimareas (1984) using

^{137}Cs, ^{60}Co, and ^{125}I as critical radionuclides from a local power plant, where they were released in the liquid effluent, showed a fast uptake of Co and I (equilibrium in 3–7 days) and of Cs (2–3 weeks). Bioconcentration factors were 10^1 for Cs, 10^3 for I, and 10^3–10^4 for Co. Rapid uptake, high bioconcentration factors, and relatively slow elimination by the three species tested recommended them as indicators for radioactive contamination by Co and I isotopes in the tropical sea.

Klumpp and Burdon-Jones (1982) pointed out the relationship between the concentrations of Pb, Cd, Cu, Zn, Co, Ni, and Ag in nine bivalve molluscs and their environment in tropical marine waters. *Trichomya hirsuta* (one of the nine species), representative for the tropical Australian coastal water, showed Pb and Cd accumulations linear with time and in direct proportion to the external concentrations of these metals. Following a 21-day exposure time, no loss of either metal was detected after 12 days of depuration in uncontaminated seawater. Pb and Cu in relocated *T. hirsuta* varied directly with environmental levels, while elevated levels of Zn in the environment were not indicated by this species. This organism proved to be the best indicator for elevated metal concentrations in tropical seawater, with the exception of Zn, and could be an alternative for the mytilidae (especially *Mytilus edulis*) which are used in colder waters to monitor levels of heavy metals. Accumulation of Cu by three commercially important bivalves under natural and experimental conditions was studied by Kumaraguru and Ramamoorthi (1979). The organisms, inhabiting the Vellar estuary, Bay of Bengal, had Cu levels ranging from a minimum of 7.6–11.1 mg/kg (*Anadora granosa*) to a maximum of 15.2–18.7 mg/kg (*Crassostrea madrasensis*). Experimental subjection to Cu concentration from 0.02 to 0.08 mg/l synthetic seawater showed maximum levels in gill and mantle tissues (119 and 147 mg/kg at 0.08 mg/l Cu). Toxic effects, e.g. yellowish green coagulation of the mucus and 'greening' of the gills and mantle, were observed at higher Cu concentrations (50 mg/kg).

Extensive studies on the cycling of trace metals in tropical and subtropical estuaries by the dominating seagrass *Thalassia testidinum* were undertaken by Schroeder and Thorhaug (1980). Experiments were conducted in microcosms of moving seawaters at various salinity levels and various temperatures using ^{22}Na, ^{137}Cs, ^{85}Sr, ^{54}Mn, and ^{65}Zn as model chemicals.

The uptake of the radionuclides was found to be asymptotic. Concentrations three orders of magnitude larger than in the seawater occurred in seagrass leaves and other biological compartments of the system. Uptake of the less soluble radionuclides by subsediment plant parts was much less than by leaves, although all the radionuclides were taken up by the various biological components. Significant translocation occurred in either direction, i.e. root to leaf or leaf to root, dependent on uptake site and concentration. All the studies described by Schroeder and Thorhaug (1980) showed a close relationship of uptake, as affected by temperature and time, between cations of similar valence and ionic radius. Following a single pulse, e.g. a spill, it appeared that after a rapid bioconcentration occurring almost totally through the leaves, large incorporated amounts were depurated from *Thalassia* by the ambient water. The uptake of the seagrass communities is as great as by the *Thalassia* plant itself. The subtropical and tropical seagrass communities are widespread in coastal waters and estuaries, and are, therefore, more likely to be exposed to metal pollution. Since relatively low accumulation of the metals could be found in sediments and because the main route of uptake is via leaves, these toxic elements, however, appear less potentially dangerous at the present time.

4.3.3.3 Waste

The term 'waste' includes in this context all types of anthropogenic inputs, such as sludge, nutrients, thermal load, and suspended particles. The pathways by which waste can reach the sea are manifold. Especially for countries with extensive coastlines, the possibilities of disposal to sea have obvious attractions (McIntyre, 1981). In countries with poor or no domestic plumbing systems, human faeces from coastal communities are often excreted directly onto the beach. Waste products also reach the sea via rivers by discharges into freshwaters. Finally, they may be introduced directly into the marine environment either through coastal outfall or by dumping from ships.

Most of the available data on waste disposal in the tropics result from investigations performed in estuaries, lagoons, coastal waters, and coral reefs. Biological effects were usually measured in laboratory simulation experiments rather than *in situ*. The transferability of the former to the 'real environment' is limited in most cases, mostly because of the short-term character of the studies. Laboratory studies, especially on the damage to isolated corals, have been shown to produce conflicting results due to non-harmonized testing procedures and to the inability to simulate the natural meteorological and hydrodynamic conditions/fluctuations.

Thermal pollution (heat input)

The destruction or alteration of marine communities by heated effluents from power plants and other industrial installations is greatest in the tropics (Wood and Johannes, 1975). Unlike the biota of temperate and polar regions, tropical organisms characteristically live at temperatures only a few degrees below their upper lethal limit. Laboratory experiments, indicating the narrow ranges existing between ambient temperatures and lethal temperatures for corals in Guam, the Great Barrier Reef, the Caribbean, and Samoa, have been reviewed by Wood and Johannes (1975). Earlier studies on temperature effects on Hawaiian coral species have shown that 11 out of 13 did not survive a 24-hour test with water temperature only 5–6 degrees above their normal ambient summer temperatures (Edmondson, 1928).

Sublethal heat stresses for corals (ceasing of feeding, reduction of reproductive rate, extruding of zooxanthellae, lack of settling of coral planulae) were also established. The balance of photosynthesis/respiration rates also changed; a direct relationship between temperature effect on P/R ratio and the upper thermal tolerance as well as an inverse relationship between high temperature tolerance and respiration rate were also observed (Wood and Johannes, 1975).

Thermal pollution and water movements accounted for regional differences in tropical marine zooplankton density and species composition. Youngbluth (1976) reports on a long-term study in a polluted, tropical embayment in Puerto Rico, where population dynamics were studied at the intake and outflow waters of a fossil-fuelled power plant and at the entrance of the bay. Water of about 10°C above ambient temperature is released from the power plant into a semi-enclosed cove (3900 m^3/minute). The outflow contained biocides and anti-corrosives (hypochlorite and chromate). Zooplankton fluctuations in the thermally polluted bay were related to physiological tolerances as well as to local hydrographic differences. Lower species diversity but more constant community membership at the outflow area suggests that a few species were well adapted to the pollution levels. High diversity and low abundance rates at the entrance to the bay indicated a mixing of bay and coastal populations. The consistently low densities and species richness in the thermal cove were explained by low phytoplankton biomass, high entrainment mortality, and sublethal temperature effects. Concentration of the chemicals in the effluents was not assessed but could have had an additional adverse effect on that ecosystem.

The effects of heated effluents from different power plants on *Thalassia* seagrass communities in estuaries in the subtropics and the tropics were described by Thorhaug *et al.* (1978). *Thalassia*, the dominant near-shore community in the Gulf of Mexico and the Caribbean, was shown to be denuded at sustained temperatures 5°C above the mean summer temperature of 30°C. Detectable minimal damage to *Thalassia* occurred at 1–1.5°C elevated temperatures. Plus-4°C areas showed intensive damage to biological communities at all locations. Plus-3°C effects varied from severe damage in the subtropics to 40% damage in the tropics. A legislative decision was requested that heated effluent released from power plants should never exceed 3°C above the ambient temperature with the exception of winter when sea temperatures are below 25°C.

"Tropical and subtropical ecosystems are on the brink of disaster and small increments of change by man's activities can push them beyond tolerance limits. Evidently, the marine tropics differ from marine temperate zones in their capacity to assimilate man's activities."

Thorhaug *et al.* (1978)

Sewage

The influence of environmental conditions on the biodegradation of industrial and domestic waste in a coastal tropical lagoon was investigated by Dufour (1982) using the BOD measuring technique. Between 25 and 30°C, oxygen consumption corresponds to first-order kinetics. At 20°C, an inhibition of BOD lasting for 3 days was observed. This inhibition lasted for a longer period at 30°C. The BOD was also inhibited by higher salinities (highest values tested were 35‰).

Waste waters from Abidjan City (Ivory Coast) have 330 mg/BOD. This represented 1/3 of the total BOD in the estuarine lagoon. Annual mean values for that region varied between 1.5 mg/l and more than 8 mg/l in the stagnant bays close to the city, compared to about 0.8 mg/l for water of oceanic origin.

In the Hawaiian Islands benthic nitrogen fixation was investigated by Hanson and Gundersen (1976) in a polluted coral reef ecosystem, where secondary sewage was released from two treatment plants. The range of N_2 fixation rates was 2–10 ng/g/h. Experiments demonstrated that microbial fixation by about 50% of the taxa was photosynthetically dependent. There was a significant correlation between numbers of N_2-fixing bacteria and the rates of N_2 fixation measured in the sediments.

Studies on the effects of inorganic carbon concentrations by Burris *et al.* (1983), conducted in the laboratory with isolated zooxanthellae and coral tips as well as *in situ* with whole coral colonies, indicated that inorganic carbon does not limit coral photosynthesis. The high inorganic carbon concentration in seawater appeared to be sufficient to saturate photosynthesis, while CO_2 production by coral respiration is probably not stimulatory to zooxanthellae photosynthesis. These findings are contrary to the freshwater situation, where photosynthesis is mostly limited by inorganic carbon concentration and where respired CO_2 stimulates symbiotic algal photosynthesis.

The main indicators for the presence of sewage waters are high ammonia, nitrate, and phosphate values, low O_2 concentrations, high BOD, and decrease in transparency and salinity. Several recent studies concerning these phenomena in tropical marine systems have reflected the permanent increase of high micronutrient values: in Barbados, West Indies, by Turnbull and Lewis (1981); in the Caribbean Sea of Jamaica and tropical North Atlantic by Barlow *et al.* (1968).

The effects of elevated N and P levels on coral reef growth were studies by Kinsey and Davies (1979). On the basis of a long-term experiment in a patch reef of the Great Barrier Reef at increased phosphate levels of 2 μmol and of 20 μmol urea and ammonium, they postulated that primarily the elevated phosphate level could cause coral reef communities to lose 50% of their present calcification activity, i.e. potential reef growth rate. Even if the community structure were to remain similar, the resulting algal activity would eventually progressively produce an increased dominance of the standing crop by 'soft algae', blocking hard substratum sites for the growth of calcifying organisms.

Tertiary effects, such as reducing light by algal blooms in reef communities, where the most significant calcification is locked to photosynthesis, were also assumed.

Significant increases in phytoplankton biomass by sewage discharge are apparent in polluted coastal areas. Thompson and Ho (1981) determined the chlorophyll-a content and the net phytoplankton in coastal waters of Hong Kong. They showed that a substantial increase in phytoplankton standing crops is associated with the discharge of untreated sewage into the harbour. Over a period of 20 months of observation, chlorophyll-a values rarely exceeded 2 μg/l in unpolluted coastal waters, whereas estuarine waters generally had 2–6 μg/l, and values of more than 20 μg/l were found in polluted areas.

Mining, Dredging, Drilling

The increase of suspended matter content in the tropical marine environment has caused local negative effects (reduced light and temperature) to the communities present there. For example, a comparison of coral reefs in Martinique with those in the southwest Indian Ocean by Battistini (1978) revealed a development of brown algae and calcareous coatings on the reefs, exposed to wind-generated turbidity. The outer exposed slopes showed areas where coral life was almost extinct, and where the coral heads were covered with calcareous coatings and were being colonized by sea urchins.

The potential effects of the deep-sea mining of manganese nodules in the tropical eastern North Pacific Ocean were described by Chan and Anderson (1981). Emphasis was put upon the short-term influence of mining discharges on photosynthesis rates through light reduction, discharged particulates, and by chemical intoxication caused by heavy metals from the bottom sediment. From these experiments it was predicted that in the vicinity of the mining ship a short-term reduction in primary production would occur. The reduction of light transmittance in the water column largely depends on the rates of discharges, the degree of mixing, the sedimentation speed of sediment particles out of the euphotic zone, and light-adaptation mechanisms of the phytoplankton. Effects by chemical inhibition (heavy metals or other toxic substances) were shown to be insignificant under the experimental conditions described. Extrapolation to long-term effects indicated a slight nutrient enrichment, mainly of the silicates, generating a possible increase in the diatom population in relation to other phytoplankton.

Other adverse effects are likely to occur in our opinion, i.e. drastic local damage to the bottom biota after mechanical disturbances. Resulting changes, such as those concerning the role of zooplankton within the mining area, were not investigated. This should be done for a better insight into the complexity of the total involved system.

Corals have been shown not to be dependent on a constant planktonic source of energy (Bak, 1978). Drastic reduction in light level, however, seemed to be the most important parameter in suppressing the calcification rate. Bak showed that under reduced light values (1–30% of surface illumination), coral colonies, which were inefficient sediment rejectors, lost their zooxanthellae and died. Reduction in calcification was about 33%. It was further observed that sudden deprivation of light apparently induced a calcification suppression, lasting one month, which resulted from a metabolic shock rather than from the sudden deprivation of light only.

Rogers (1979, 1983) published two papers on the effects of shading on coral reef structure and function, and on the sublethal and lethal effects of sediments on common Caribbean Reef corals. Shading of the experimental area for five weeks caused a decrease in net primary production, respiration, and the bleaching/death of several coral species. The exclusion of light was a partial simulation of one of the effects of extreme turbidity. Ten months after shading ceased, no new corals had settled on the dead corals, which being colonized by algae.

Continuous coastal development and dredging in tropical areas have increased sedimentation rates and constitute a serious threat to coral reefs (Rogers, 1983). Experiments with traps, installed for 18 months at San Cristobal Reef, Puerto Rico, to measure natural rates of sedimentation showed mean rates between 2.5 and 9.6 $mg/cm^2/day$. It was assumed that the overall normal sedimentation rates for coral reefs were of the order of 10 $mg/cm^2/day$ or less (Rogers, 1983). Additional single sediment application experiments (200 and 800 mg/m^2) indicated resistance by adult corals, with death occurring only in local areas. However, smaller doses were estimated to be sufficient to prevent establishment of coral planulae and reduction of coral growth. Even 1.1 $mg/cm^2/day$ 'normal' sedimentation rates in Discovery Bay, Jamaica, significantly reduced the average growth of *Montastraea annularis* (Dodge *et al.*, 1974).

The main effects of dredging on corals can be summarized thus: decrease of light values to marginal conditions; behavioural stress from energy-consuming sediment rejection; and the significant reduction of calcification rates.

A further experiment, simulating water column turbidity by particulate peat and subsequent effects on the behaviour and the physiology of the Jamaican reef-building coral, *Montastraea annularis*, was reported by Dallmeyer *et al.* (1982). Using an *in situ* respirometer, a large decrease in net oxygen production was measured as a result of daytime introduction of low concentrations of peat. The reduction of photosynthetic rate was emphasized by a 22% loss of chlorophyll content in the coral tissue. The effects were attributed to a loss of zooxanthellae. Similar effects were also shown as a result of elevated water temperatures.

Tin dredging and smelting activities in the vicinity of the Ko Phuket reef flat, Thailand, revealed no increased metal concentrations in coral skeletons and tissues. However, the colony size of *Porites* species was smaller, compared to unpolluted sites (Brown and Holley, 1982).

Drilling fluids or muds are highly specialized materials necessary for the drilling of offshore wells. The composition of drilling fluids is highly varied and complex, containing aliphatic and aromatic hydrocarbons, metals, clays, artificial polymers, and biocides (Trefrey *et al.*, 1981; Hudson *et al.*, 1982; Kendall *et al.*, 1983). Drilling muds can harm corals in the following ways: direct burial, increased turbidity, and chemical toxicity.

Kendall *et al.* (1983) studied the effects of used drilling fluids on the coral *Acropora cervicornis*, maintained *in situ* in plexiglass domes. A 25 mg/l concentration of the mud resulted after 24 hours in a 62% decrease in the calcification rate as compared to controls. Soluble tissue protein declined by roughly 50% compared to controls after an exposure at 100 mg/l. They also investigated the concentration of total ninhydrin-positive substance. Of these three parameters, the calcification rate was found to be the most sensitive indicator of the effect of drilling mud. Exposing the coral to pure kaolin produced considerably less drastic effects. For example, 125 mg/l kaolin reduced the calcification rate by 34% while a roughly similar concentration of drilling mud (100 mg/l) resulted in an 88% reduction. This indicated that the deleterious effects of used drilling muds is not simply due to an increase in turbidity. Figure 4.3.4 shows the average values of calcification rates in the terminal 4 cm of the branches of *A. cervicornis* at different concentrations of drilling muds.

Strong mortality (70–90%) of corals was found near the wellheads at Matinloc, Philippines, by Hudson *et al.* (1982). The effects of drilling mud on the reef-building coral *Montastraea annularis* was investigated by Dodge (1982) in a laboratory flowthrough system at concentrations of 0,1,10, and 100 μl/l drilling mud. More than six weeks continuous exposure of *M. annularis* at the

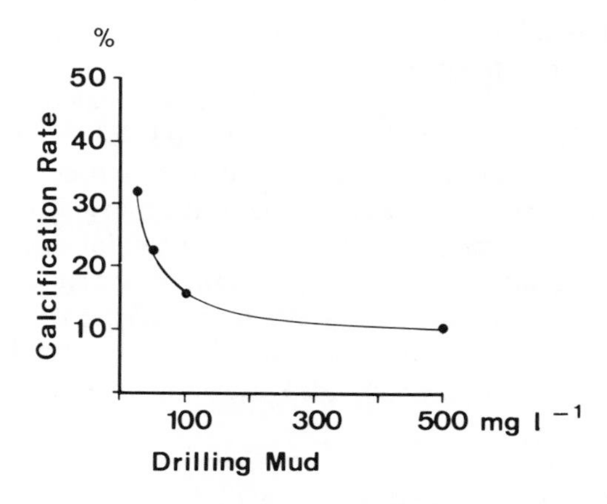

Figure 4.3.4 Calcification rates in the terminal 4 cm of the branches of *Acropora cervicornis* at different concentrations of drilling mud (means). After Kendall *et al.* (1983)

highest mud concentration significantly depressed the linear growth rate and increased the mortality of the coral. Calcification rates at 100 μl/l were more than 50% depressed after a four-week exposure. A short-term treatment of *M. annularis* with concentrated (slurry) drilling mud caused a reduced growth rate of the colony for more than six months (Hudson and Robbin, 1980). Barium levels in the coral skeleton had increased from 11 mg/kg (background) to about 1200 mg/kg after treatment. These data show that in spite of the insolubility of $BaSo_4$ in water, a considerable uptake of this toxic compound from drilling muds was possible.

In further *in situ* experiments by Thompson and Bright (1980) on selected corals, using closed aquaria at 3 m water depth, the response to drilling mud was measured as a percent of the number of retracted polyps. At 100 μl/l mud concentration a significant polyp reaction was observed for all corals, while *M. annularis* colonies did not survive at this dosage.

Acute and sublethal effects of drilling fluids to marine organisms from temperate or arctic marine regions have also been investigated. Estuarine organisms and benthic communities were subjected to a 100-day multispecies toxicity test (Rubinstein *et al.* , 1980). The fluids tested were moderately toxic to mysid shrimps at 30 and 100 mg/l. Oyster growth and lugworm survival were also significantly reduced at these concentrations.

A flowthrough system was used for a 30-day mussel toxicity test to demonstrate that the suspended drilling fluids were most toxic to larval marine organisms and much less toxic to adults. Long-term effects were observed in mussels by reduced growth rates (Gerber *et al.*, 1980).

Comparative studies of the toxicity of 18 different samples of used drilling muds from an exploratory drilling rig in a local estuary revealed that no toxic effects could be found on intermoult grass shrimps during a 96-hour test procedure at concentrations of 10 and 100 μl/l seawater. However, 30–60% mortality was found at 1000 μg/l. A mean lethal concentration of 363–739 μg/l for five of the eight mud samples was defined for the moulting grass shrimps. In a life-cycle test with the more sensitive mysids, the LC_{50} for one drilling mud was 50 μg/l (Conklin *et al.*, 1980).

Acute toxicity tests were conducted *in situ* in the Beaufort Sea, Alaska, by Tornberg *et al.* (1980). The results of the static bioassays indicated that the mean lethal concentrations were 4–70% for the drilling fluids tested, with wide variations in the responses to exposure by different species. Fish were the most sensitive organisms, followed by invertebrates, which were relatively resistant. The resistant species were found to be primarily sedentary, while sensitive species generally had the capability to migrate from the drilling fluid disposal site.

4.3.4 CONCLUSIONS

In this report, we have made an attempt to review the major known environmental impacts on tropical marine environments. We have also tried to compare the effects of these impacts with those in the temperate zone. The latter has been well covered in numerous works. However, there has been only limited research (with the possible exception of corals) done in the tropics, and much of this, for reasons of simplicity and economics, has only been done in laboratories. This makes such a comparison difficult.

It is also difficult to generalize. There are inputs with global aspects. For example, through the transport by the atmosphere and by oceanic currents, many contaminants, such as from an oil spill or from the application of pesticides, may originate in non-tropical areas but will migrate and reach significant concentrations in the tropics. In addition, there are numerous local sources of contaminants, such as sewage, construction (turbidity), thermal discharge, etc. In general, it appears that tropical systems are more sensitive and less robust in regard to many impacts than their temperate region counterparts. This is especially true for thermal and turbidity pollution. However, in some cases, such as oil spills, the greater solar radiation in the tropics does lead to a more rapid breakdown of the contaminants. But this can cause problems, since many of the resulting oxygenated products may be more toxic than their parent compounds. The net conclusion is that there is no true substitute for *in situ* tropical environmental research. It is a highly questionable practice to transfer results from either the temperate zones or from laboratory studies.

Despite this, there has been no significant increase in tropical environmental research (Cole, 1984). This is especially unfortunate, since it is precisely the tropical regions that are undergoing the greatest growth will all of its attendant problems. The problem is compounded by the fact that environmental research is expensive, requiring trained personnel, sophisticated equipment, and modern laboratories. Many of the nations in the tropical regions have, at present, only limited funds for such research. However, if this research is not done, long-term or even irreparable damage to tropical ecosystems may result, which in turn could have a significant negative economic impact.

4.3.5 REFERENCES

Anderson, J. W. (1975). Laboratory studies on the effects of oil on marine organisms: An overview. *Am. Petrol. Inst. Pub.*, **4249**, 1–70.

Anonymous (1985). *Oil in the Sea. Inputs, Fates and Effects*. National Academy Press, Washington, D.C.

Atlas, R. M., Roubal, G., Bronner, A., and Haines, J. (1980). Microbial degradation of hydrocarbons in mousse from IXTOC-1. In: *Proceedings of a Symposium of Preliminary Results from the September 1979 RESEARCHER/PIERCE IXTOC-1 Cruise*, Key Biscayne, Fl, June 9–10, pp. 441–435. (Available from NOAA Office of Marine Poll. Assess., Rockville, Md, U.S.A.).

Atwood D. K., and Ferguson, R. L. (1982). An example study of the weathering of spilled petroleum in a tropical marine environment: IXTOC-1. *Bull. Mar. Sci.*, **32** (1), 1–13.

Bak, R. P. M. (1978). Lethal and sublethal effects of dredging on reef corals. *Mar. Poll. Bull.*, **9** (1), 14–16.

Baker, J. M. (1970). The effect of oil on plants. *Environ. Pollut.*, **1**, 27–44.

Barlow, J. R., Steven, D. M., and Lewis, J. B. (1968). Primary productivity in the Caribbean Sea off Jamaica and the tropical North Atlantic off Barbados. *Mar. Sci.*, **18**, 86–104.

Battistini, R. (1978). The coral reefs of Martinique. *Cah. O.R.S.T.O.M., sér. Océanogr.*, **16** (2), 157–177.

Blake, N. J., and Johnson, D. L. (1976). Oxygen production/consumption of the pelagic *Sargassum* community in a flow-through system with arsenic additions. *Deep Sea Res.*, **23**, 773–778.

Brown, B. E., and Holley, M. C. (1982) Metal levels associated with tin dredging and smelting and their effect upon intertidal reef flats at Ko Phunket, Thailand. *Coral Reefs*, **1** (2), 131–137.

Burris, J. E., Porter, J. W., and Laing, W. A. (1983). Effects of carbon dioxide concentration on coral photosynthesis. *Mar. Biol.*, **75**, 113–116.

Chan, A. T., and Anderson, G. C. (1981). Environmental investigations of the effects of deep-sea mining on marine phytoplankton and primary productivity in the tropical eastern North Pacific Ocean. *Marine Mining*, **3** (1/2), 121–149.

Cole, N. H. A. (1984). Tropical ecology research. *Nature*, **309**, 204.

Conklin, P. J., Doughtie, D. G., and Ranga, K. (1980). Effects of barite and used drilling muds on crustaceans, with particular reference to the grass shrimp, *Palaemonetes pugio*. In: *Proc. Symp. Res. Environ. Fate Eff. Drill. Fluids Cuttings*, pp. 912–943. American Petroleum Institute, Washington, D.C.

Dallmeyer, D. G., Porter, J. W., and Smith, G. J. (1982). Effects of particulate peat on the behaviour and physiology of the Jamaican reef-building coral *Montrastraea annularis. Mar. Biol.*, **68**, 229–233.

Dodge, R. E. (1982). Effects of drilling mud on the reef-building coral *Montastraea annularis. Mar. Biol.*, **71**, 141–147.

Dodge, R. E., Allen, R. C., and Thomson, J. (1974). Coral growth related to resuspension of bottom sediments. *Nature*, **247**, 547–577.

Duce, R. A. (1978). Speculations on the budget of particulate and vapor phase non-methane organic carbon in the global troposphere. *Pure Appl. Geophys.*, **116**, 244–273.

Dufour, P. (1982). Influence des conditions de milieu sur la biodégradation des matières organiques dans une lagune tropicale. *Oceanologica Acta*, **5** (3), 355–363.

Edmondson, C. H. (1928). The ecology of an Hawaiian coral reef. *B.P. Bishop Mus., Bull.*, **45**, 64–79.

Fiest, D. L., and Boehm, P.D. (1980a). Subsurface water column, transport and weathering of petroleum hydrocarbons during the IXTOC-1 blowout in the Bay of Campeche and their relation to surface oil and microlayer composition. In: *Proceedings of a Symposium on Preliminary Results from the September 1979 RESEARCHER/PIERCE IXTOC-1 Cruise*, Key Biscayne, Fl., June 9–10, pp. 267–338. (Available from NOAA Office of Marine Poll. Assess., Rockville, Md, U.S.A.)

Fiest, D. L., and Boehm, P. D. (1980b). Subsurface distributions of petroleum from an offshore well blowout—the IXTOC-1 blowout. In: *Proceedings of a Symposium on Preliminary Results from the September 1979 RESEARCHER/PIERCE IXTOC-1 Cruise*, Key Biscayne, Fl., June 9–10, pp. 169–185. (Available from NOAA Office of Marine Poll. Assess., Rockville, Md, U.S.A.)

Gearing, P. J., Gearing, J. N., Pruell, R. J., Wade, T. L., and Quinn, J. G. (1980). Partitioning of 2 fuel oils in controlled estuarine ecosystems. Sediments and suspended particulate matter. *Environ. Sci. Technol.*, **14**, 1129–1136.

Gerber, R. P., Gilfillan, E. S., Edward, S., Page, B. T., David, S., and Hotham, J. (1980). Short and long-term effects of used drilling fluids on marine organisms. *Proc. Symp. Res. Environ. Fate Eff. Drill. Fluids Cuttings*, pp. 882–911. American Petroleum Institute, Washington, D.C.

Geyer, H., Viswanathan, R., Freitag, D., and Korte, F. (1981). Relationship between water solubility of organic chemicals and their bioaccumulation by the alga *Chlorella*. *Chemosphere*, **10** (11/12), 1307–1313.

Gras, G., and Mondain, J. (1978). Mercury content of several fish species caught off the coast of Senegal. *Rev. Int. Océanogr. Méd.*, Tomes **LI-LII**, 83–88.

Guimareas, J.R.D. (1984). Uptake by benthic algae of critical radionuclides to be released in the liquid effluent of the Angra dos Reis Nuclear Power Plant, R.J., Brazil, INIS Atomindex Report 1982, 132 pp. Rio de Janeiro, Brazil.

Hanson, R. B., and Gundersen, K. R. (1976). Bacterial nitrogen fixation in a polluted coral reef flat ecosystem in Kaneohe Bay, Oahu, Hawaiian Islands. *Pacif. Sci.*, **30** (4), 385–394.

Hirota, R., Fujiki, M., and Tajima, S. (1979). Mercury contents of zooplankton collected in the tropical Pacific Ocean. *Bull. Japan. Soc. Sci. Fish.*, **45** (11), 1449–1451.

Hudson, J. H., and Robbin, D. M. (1980). Effects of an offshore drilling fluid on selected corals. In: *Proc. Symp. Res. Environ. Fate Eff. Drill. Fluids Cuttings*, pp. 1044–1078. American Petroleum Institute, Washington, D.C.

Hudson, J. H., Shinn, E. A., and Robbin, D. M. (1982). Effects of offshore oil drilling on Philippine reef corals. *Bull. Mar. Sci.*, **32** (4), 890–908.

Hungspreugs, M., and Yuangthong, C. (1983). A history of metal pollution in the upper Gulf of Thailand. *Mar. Pollut. Bull.*, **14** (12), 465–469.

Jacobson, S. M., and Boylan, D. B. (1973). Effect of seawater soluble fraction of kerosene on chemotaxis in a marine snail, *Nassarius obsoletus*. *Nature*, **241**, 213–215.

Johannes, R. E., and Betzer, S. (1975). In: Wood, E. J. F., and Johannes, R. E. (eds), *Tropical Marine Pollution*, pp. 1–12. Elsevier Oceanography Series, **12**. Elsevier, Amsterdam.

Johnson, F. G. (1977). Sublethal biological effects of petroleum hydrocarbon exposure: Bacteria, algae and invertebrates. In: Malins, D.C. (ed.), *Effects of Petroleum on Arctic and Subarctic Marine Environments and Organisms*, vol. 2, pp. 271–318. Academic Press, New York.

Johnson, D. L., and Braman, R. S. (1975). The speciation of arsenic and the content of germanium and mercury in members of the pelagic *Sargassum* community. *Deep Sea Res.*, **22**, 503–507.

Kendall, J. J., Powell, E. N., Connor, S. J., and Bright, T. J. (1983). The effects of drilling fluids (muds) and turbidity on the growth and metabolic state of the coral *Acropora cervicornis*, with comments on methods of normalization for coral data. *Bull. Mar. Sci.*, **33** (2), 336–352.

Kinsey, D. W., and Davies, P. J. (1979). Effects of elevated nitrogen and phosphorus on coral reef growth. *Limnol. Oceanogr.*, **24** (5), 935–940.

Klumpp, D. W., and Burdon-Jones, C. (1982). Investigations of the potential of bivalve molluscs as indicators of heavy metal levels in tropical marine waters. *Aust. J. Mar. Freshwater Res.*, **33**, 285–300.

Kumaraguru, A. K., and Ramamoorthi, K. (1979). Accumulation of copper in certain bivalves of Vellar Estuary, Porto Novo, S. India in natural and experimental conditions. *Est. Coast, Mar. Sci.*, **9**, 467–475.

MacKay, D. (1982). Correlation of bioaccumulation factors. *Environ. Sci. Technol.*, **16** (5), 274–278.

McAuliffe, C. D. (1966). Solubility in water of paraffin, cycloparaffin, olefin, acetylene, cycloolefin and aromatic hydrocarbons. *J. Phys. Chem.*, **70**, 1267–1275.

McAuliffe, C. D. (1977). Dispersal and alteration of oil discharged on a water surface. In: Wolfe, D. A. (ed.) *Fate and Effects of Petroleum Hydrocarbons in Marine Organisms and Ecosystems*, pp. 19–35. Pergamon, Oxford.

McIntyre, A. D. (1981). Effects on the ecosystem of sewage sludge disposal by dumping from ships. *Wat. Sci. Tech.*, **14**, 137–143.

Pfaender, F. K., Buckley, E. N., and Ferguson, R. L. (1980). Response of the pelagic microbial community to oil from the IXTOX-1 blowout: I. *In situ* studies. In: *Proceedings of a symposium on preliminary results from the September 1979 RESEARCHER/PIERCE IXTOC-1 Cruise*, Key Biscayne, Fl. June 9–10, pp. 545–560. (Available from NOAA Office of Marine Pollut. Assess., Rockville, Md, U.S.A.)

Rinkevich, B., and Loya, Y. (1977). Harmful effects of chronic air pollution on a Red Sea coral population. In: *Proc. Third Int. Coral Reef Symp.*, pp. 585–591. University of Miami, Miami.

Rinkevich, B., and Loya, Y. (1979). Laboratory experiments on the effects of crude oil on the Red Sea coral *Stylophora pistillata. Mar. Pollut. Bull.*, **10**, 328–330.

Rogers, C. S. (1979). The effects of shading on coral reef structure and function. *J. Exp. Mar. Biol. Ecol.*, **41** (3), 269–288.

Rogers, C. S. (1983). Sublethal and lethal effects of sediments applied to common Caribbean reef corals in the field. *Mar. Poll. Bull.*, **14** (10), 378–382.

Rubinstein, N. I., Rigby, R., and D'Asaro, C. N. (1980). Acute and sublethal effects of whole used drilling fluids on representative estuarine organisms. In: *Proc. Symp. Res. Environ. Fate Eff. Drill. Fluids Cuttings*, pp. 828–846. American Petroleum Institute, Washington, D.C.

Sauer, T. C. Jr. (1980). Volatile liquid hydrocarbons in waters of the Gulf of Mexico and Caribbean Sea. *Limnol. Oceanogr.*, **25** (2), 338–351.

Sauer, T. C. Jr., Sackett, W. M., and Jeffrey, L. M. (1978). Volatile liquid carbons in the surface coastal waters of the Gulf of Mexico. *Mar. Chem.*, **7**, 1–16.

Schroeder, P. B., and Thorhaug, A. (1980). Trace metal cycling in tropical-subtropical estuaries dominated by the seagrass *Thalassia testudinum. Am. J. Bot.*, **67** (7), 1075–1088.

Sivalingham, P. M. (1980). Mercury contamination in tropical algal species of the Island of Penang, Malaysia. *Mar. Poll. Bull.*, **11**, 106–107.

Solbakken, J. E., Knap, A. H., Sleeter, T. D., Searle, C. E., and Palmork, K.H. (1984). Investigation into the fate of carbon-14-labelled xenobiotics (naphthalene, phenanthrene, 2,4,5,2',4',5'-hexachlorobiphenyl, octachlorostyrene) in Bermudian corals. *Mar. Ecol.*: Prog. Ser., **16** (1–2), 149–154.

Steele, R. L. (1977). Effects of certain petroleum products on reproduction and growth of zygotes and juvenile stages of the alga *Fucus edentatus* De la Pyl (Phaeophyceae Fucales). In: Wolfe, D. (ed.), pp. 115–128. *Fate and Effects of Petroleum Hydrocarbons in Marine Ecosystems and Organisms.* Pergamon, Oxford.

Takahashi, F. T., and Kittredge, J. S. (1973). Sublethal effects of the water soluble component of oil. Chemical communication in the marine environment. In: Ahern, D. G., Meyers, S. P. (eds), *The Microbial Degradation of Oil Pollutants*, pp. 259-264. Publ. LSU-SG-73-01, Center Wetlands Resour., Louisiana State University.

Tanabe, S., and Tatsukawa, R. (1980). Chlorinated hydrocarbons in the North Pacific and Indian Ocean. *J. Oceanogr. Soc. Japan*, **36** (4), 217–226.

Tanabe, S., Kawano, M., and Tatsukawa, R. (1982). Chlorinated hydrocarbons in the Antarctic, western Pacific Ocean and eastern Indian Ocean. *Trans. Tokyo Univ. Fish.*, **5**, 97–110.

Thominette, F., and Verdu, J. (1984). Photo-oxidative behaviour of crude oils relative to sea pollution. Part I. Comparative study of various crude oils and model systems. *Mar. Chem.*, **15**, 91–104.

Thompson, G. B., and Ho, J. (1981). Some effects of sewage discharge upon phytoplankton in Hong Kong. *Mar. Poll. Bull.*, **12** (5), 168–173.

Thompson, J. H. Jr., and Bright, Th. J. (1980). Effects of an offshore drilling fluid on selected corals. In: *Proc. Symp. Res. Environ. Fate Eff. Drill. Fluids Cuttings*, pp. 1044–1078. American Petroleum Institute, Washington, D.C.

Thorhaug, A., Blake, N., and Schroeder, P.B. (1978). The effect of heated effluents from power plants on seagrass (*Thalassia*) communities quantitatively comparing estuaries in the subtropics to the tropics. *Mar. Poll. Bull.*, **9** (7), 181–187.

Tornberg, L. D., Thielk, E. D., Nakatani, R. E., Miller, R. C., and Hillman, Sh. O. (1980). Toxicity of drilling fluids to marine organisms in the Beaufort Sea, Alaska (USA). In: *Proc. Symp. Res. Environ. Fate Eff. Dril. Fluids Cuttings*, pp. 997–1016. American Petroleum Institute, Washington, D.C.

Trefrey, J. H., Trocine, R. P., Pierce, R. H. Jr., Weichert, B. A., and Meyer, D. B. (1981). The potential impact of drilling fluids on the Texas Flower Gardens. Cited in Rogers (1982). *Mar. Biol.*, **71**, 141.

Turnbull, D. A., and Lewis, J. B. (1981). Pollution ecology of a small tropical estuary in Barbados, West Indies. 1. Water quality characteristics. *Marine Science Center Manuscript*, No. 35, pp. 1–54. McGill University, Montreal.

Van Vleet, E. S., Sackett, W. M., Weber, F. F. Jr., and Reinhardt, S. B. (1983). Spatial and temporal variation of pelagic tar in the eastern Gulf of Mexico. In: Bjoroey, M. (ed.), *Adv. Org. Geochem.* Proc. Int. Meet. 10th., pp. 362–368. 1981, John Wiley, Chichester, U.K.

Wade, B. A., Antonio, L., and Mahon, R. (1972). Increasing organic pollution in Kingston Harbour, Jamaica. *Mar. Poll. Bull.*, **3**, 106–111.

Walsh, F., and Mitchell, R. (1973). Inhibition of bacterial chemoreception by hydrocarbons. In: Ahern, D. G., and Meyers, S. P. (eds), *The microbial degradation of oil pollutants*, pp. 275–278. Publ. LSU-SG-73-01, Center Wetlands Resour., Louisiana State University.

Whitehouse, B. G. (1984). The effects of temperature and salinity on the aqueous solubility of polynuclear aromatic hydrocarbons. *Mar. Chem.*, **14**, 319–332.

Wood, E. J. F., and Johannes, R. E. (eds) (1975). *Tropical Marine Pollution*. Elsevier Oceanography Series, 12. Elsevier, Amsterdam.

Youngbluth, M. J. (1976). Zooplankton populations in a polluted, tropical embayment. *Est. Coast. Mar. Sci.*, **4**, 481–496.

Zsolnay, A. (1977). Inventory of nonvolatile fatty acids and hydrocarbons in the ocean. *Mar. Chem.*, **5**, 465–475.

Zsolnay, A. (1978). The weathering of tar on Bermuda. *Deep Sea Res.*, **25**, 1245–1252.

Ecotoxicology and Climate
Edited by P. Bourdeau, J. A. Haines, W. Klein and C. R. Krishna Murti

4.4 Effects in Arctic and Subarctic Systems

D. R. MILLER

4.4.1 COLD-ENVIRONMENT ECOSYSTEMS

The features dominating cold-climate ecosystems are primarily those of low energy fixation and harsh environmental conditions. In terms of energy, the sun is low in the horizon at the best of times, and its radiant energy strikes the earth a 'glancing blow'. In addition, the high albedo of the snow causes a good deal of solar radiation to be immediately reflected. The growing season is short, and plants must mature quickly to reproduce. Thus they are uniformly small, and productivity is low, ranging from a few per cent to less than half of one per cent of the corresponding crop in a temperate climate. Adaptation mechanisms are abundant; one finds plants growing in rosettes and cushions, storing much more of their energy in roots than their temperate counterparts, developing waxy cuticles, and so forth. Nonetheless, since they have no way to escape or hide from conditions, as hibernating animals can, the plants are generally under extreme stress and the ecosystem has practically no 'resilience' in the ecological sense.

The decreased primary production is the fundamental limitation of the cold-environment ecosystem. Such animals and birds as choose to live there generally have adaptation mechanisms that are quite effective, as long as the food supply is adequate. Some, like the bear and caribou, develop heavy lipid layers and extremely effective fur to prevent energy loss. Some simply migrate to warmer regions; some small animals live under the snow cover, which is itself rather effective insulation; and some hibernate, thus decreasing their energy requirements sharply.

There are other effects of the arctic and subarctic environment which are not nearly so well understood. For example, for a large part of the year arctic plants and animals receive sunlight that is not only in short supply, but also has a different wavelength distribution, losing energy in the blue regions because of the effectively greater thickness of the atmosphere when traversed at a slant. The magnetosphere surrounding the earth is to some extent 'open' at the poles, so that these regions receive a greater radiation burden. Also, northern regions will typically experience intense geomagnetic activity, which may have effects as yet unknown. Studies in man have shown deleterious effects of the lack of solar regularity; numerous internal biological clocks are desynchronized, with observable behaviour changes resulting; there is no reason why other inhabitants of the high-latitude environment should be spared.

In summary, we observe that cold-climate ecosystems are characterized by their low productivity, at the level of primary energy incorporation, and that they are among the most stressed ecosystems on earth. This means that any further stress, such as chemical pollution or physical disturbance, may be assumed at the outset to have profound influences on the ability of such systems to survive at all, let alone to recover; in any case, recovery will be slow.

4.4.2 CHEMICAL PERSISTENCE

Two factors interact to increase the length of time that a chemical will persist in the same chemical form, as opposed to an increased residence time, in a cold environment. Firstly, any chemical process will proceed more slowly at decreased temperatures, and secondly, bacterial and similar biological degradation processes will typically be reduced because of the generally decreased biotic activity of the microenvironment of the pollutant.

The simplest way to describe the first factor is in terms of the Q-10 rule, the handy (and surprisingly realistic) rule of thumb that a chemical reaction in general will very roughly change its rate by a factor of two when the temperature changes by ten degrees Celsius. Thus if the average ambient temperature is 10°C rather than 25°C, the rate of degradation by a pure chemical process will decrease by a factor of 2.8, which would mean that a chemical which naturally degraded with a half-life of, say, four days, would now have a half-life of 11 days approximately.

Naturally, this calculation is only a rough guide; the actual degradation process involves other factors, such as photolysis, presence of other chemicals, and so forth. Details of such calculations, taking into account these and other factors, have been published, and in fact interactive computer programs are available (NRCC, 1981; Burns *et al.*, 1981) by which numerical and graphical representations of these processes can be generated at will. The basic result does not change, namely that in a cold environment the degradation rate is significantly slowed, and this of course means that other processes which may be time-limited, such as transport, entry into a water supply or a food chain, or the process of biomagnification itself, will routinely have more time available in which to take place.

4.4.3 BIOACCUMULATION AND EXPOSURE

In general, the theme explored above recurs in this section also; in a cold climate the special considerations of the environment interact with any chemical pollutant to make its effects more serious than would be the case for the same substance in a warmer climate. This is true for uptake, food-chain concentration, and behaviour in the organism, for perhaps a series of different reasons.

4.4.3.1 Uptake

We have previously argued that cold conditions will mean that more of the polluting substance will be close to ground surface or actually carried on the emergent part of plants. This immediately means that grazing animals will pick up more of the chemical. There is a second effect, however, namely that in the North, with its lower per hectare productivity, animals tend to graze more completely, leaving very little of the above-ground plant material behind. This factor increases still further the faction of the deposited material that immediately enters the body of grazing animals. In some (albeit rare in the North) conditions, i.e. with a low wind, leafy vegetation, and recent pollutant input, uptake by grazers can be virtually complete.

To this, of course, we have to add the previous consideration that the chemical, when ingested, is more likely still to be in its original form rather than already partially degraded.

4.4.3.2 Food Chains

The main feature of cold-climate food chains is that they are less complex than their temperate counterparts. Instead of complex 'webs', with several shared or alternative food supplies at each trophic level, the food chains tend to be linear. This means that animals often have no choice of food; if the lower trophic level is somehow contaminated, even to the point of being objectionable to the grazer or predator, the choice is simply to consume it or eat nothing at all. This also, of course, makes the entire ecosystem more vulnerable since the whole structure may collapse if a single species is eliminated or even seriously depleted.

4.4.3.3 Behaviour in an Organism

We should also briefly mention a feature of cold-weather mammals that increases their vulnerability (the same general idea applies also to fishes and birds). This is connected with the abundant fat deposits many of them accumulate, both for insulation against the cold and in some cases as food supplies for hibernation. Many of the chemical pollutants in our environment are lipophilic. When ingested and distributed around the body of an animal, they will associate with the lipids and remain there, generally doing the animal relatively little harm. However, the large quantities of fatty tissue do allow the accumulation of large quantities of these chemicals.

The problem arises when the animal is forced to use its lipid reserves, either during hibernation or during a period of hunger or starvation for other reasons, sometimes environmental and sometimes behavioural, such as migration—a very energy-dependent process—or perhaps associated with the increased burden of the breeding season, e.g. nest-building or the like. During such periods, the lipids are metabolized and the chemical can be freed to come into contact with more sensitive organs. The animal, in effect, receives a major dose all in one shot; moreover, this may come at a time when the animal is already stressed by other factors. The same observation may be made of fish, which lose a great deal of their energy reserves during the hungry winter months under the ice.

We should note that this stress may also occur just when young are being raised, as in nesting birds, decreasing the ability of the parents to provide adequate food for the nestlings. In the case of bears, where the young are born and nursed during hibernation, the increased exposure to the young through milk is also obvious.

4.4.4 EFFECTS ON SPECIES AND ECOSYSTEMS

Individuals suffer from toxic chemicals rather more under cold conditions than elsewhere for several reasons. Two of these have been mentioned already, namely the fact that they are already under substantial stress because of the harsh conditions, and also because of the possibility of sudden large exposures resulting from the release into the environment of stored-up quantities of pollutant, or the sudden release of toxic agents into their own bodies when stored lipids are mobilized.

In addition, we have pointed out that the simplicity of the food webs makes for less variety in diet. This is also true, of course, for humans; people living in cold climates tend to have a much less varied diet than others. In many cases, consumption of top carnivores, such as large fish or seals, is dominant, so that bioaccumulation through the food chain has a chance to have maximum effect. Possibly the people of the North consume a larger amount of food from the top of the food chain than any other people on earth, and the exposure to such chemicals as mercury and DDT from eating such a selection of food is well known.

There are also cultural influences by which highly nutritious or otherwise preferred food is selectively consumed, sometimes by particular parts of the population. Perhaps the most outstanding example of this is the Inuit population of northern Canada, in which seal liver is selectively and traditionally reserved for consumption by pregnant women. The liver, unfortunately, contains very high concentrations of mercury, far higher than in other tissues of the same animal or in alternative foodstuffs, such as fish. (One of the intriguing questions that has not been fully resolved is why this practice does not produce more cases of mercury poisoning that it does; one current theory is that because the liver tissue also contains large quantities of selenium, known to be protective against mercury poisoning, damage is avoided. In any case, the author is not aware of any documented cases of overt mercury poisoning in this population.)

There are many other examples of how toxicity of an agent can be enhanced by low temperatures. The toxic effects of oil spills on sea birds are enhanced at low temperatures (Levy, 1980), probably because the thermal insulation efficiency of feathers is decreased. The development time of invertebrates is sharply dependent on temperature (Rosenberg and Costlow, 1976); other examples have been noted.

At the species level, interaction with cold can enhance effective toxicity, first of all, simply because the entire species is already in a stressful situation. More specifically an entire species can be wiped out because of elimination of the species (perhaps only one) below it in the food web. Alternatively, a decrease in the size of a population below the level at which predation normally takes place can increase predation pressure, as the higher organism increases the effort level to find food and ends up consuming the entire crop or population. This has been observed to happen locally with lichens, which are grazed by caribou; a decreased population of lichens, thought to be due to acidic precipitation, was almost wiped out by a too large population of caribou in the region.

Finally, at the level of the ecosystem, the entire structure is more delicate for a variety of reasons already covered. The simplicity of the food web means that elimination of one species can destroy all those above it, and even relatively small perturbations will require quite a large recovery time.

Scarcely any measure of ecosystem performance can be imagined that is not made more sensitive to additional stress by a cold environment. This is not, of course, surprising; with decreased energy input, increased requirements for basic existence, and decreased diversity, however measured, it must be obvious that the ecosystem will be less able to absorb and recover from additional insult. What is perhaps not always realized is that there is a dual effect in that transport and persistence of chemicals take place in such a way that their effects are enhanced, and environmental conditions may easily convert a low-level chronic exposure into a brief but serious dose. The message is that great care must be taken to protect cold-climate ecosystems, and information gained by the study of temperate regions is very often not relevant.

4.4.5 REFERENCES

Burns, L. A., Cline, D. M., and Lassiter, R. R. (1981). *EXAMS, An Exposure Analysis Modeling System*. Environmental Systems Branch, Environmental Research Lab., Office of Res. and Dev., U.S. Environmental Protection Agency, Athens, Georgia. (unpubl.).

Levy, E. M. (1980). Oil pollution and seabirds: Atlantic Canada 1976–1977 and some implications for northern environments. *Marine Pollut. Bull.*, **11**, 51–56.

NRCC (1981). *A Screen for the Relative Persistence of Lipophilic Organic Chemicals in Aquatic Ecosystems—an Analysis of the Role of a Simple Computer Model in Screening*. Environmental Secretariat, National Research Council of Canada, Publication No. NRCC 18570 (ISSN 0316-0114).

Rosenberg, R., and Costlow, J. D. Jr. (1976). Synergistic effects of cadmium and salinity combined with constant and cycling temperatures on the larval development of two estuarine crab species. *Marine Biol.*, **38**, 291–303.

Ecotoxicology and Climate
Edited by P. Bourdeau, J. A. Haines, W. Klein and C. R. Krishna Murti

4.5 *Effects on Domestic Animals*

R. K. RINGER

4.5.1 INTRODUCTION

Domestic animals are kept in many different climates, from the harsh cold environments of the north to the hot arid environment of the deserts and the hot, wet climates of the tropics. Stressful climatic conditions, either hot or cold, tend to aggravate the impact of xenobiotics on livestock production. The problem is further aggravated when modern intensive agricultural techniques, with increased stocking densities, place many animals in close proximity on a single site. Conversely, animals that are distributed over large rangelands are far less subject to the dangers of massive chemical exposures.

Chemical substances can enter the environment of domestic animals by direct application or in some instances by complex pathways. Application of pesticides to animals or their housing facilities, the use of fertilizers, including animal wastes, or herbicides or insecticides to cropland, and the use of sanitizing agents or veterinary drugs are examples of the direct exposure route, whereas indirect entrance may result from food-chain contamination, industrial wastes and accidents, or from combustion processes. The occurrence of chemical residues in domestic animal meat and by-products, including eggs and milk, reflects the increase in the use of agricultural chemicals as well as an increase in pollution of the environment in which the animals are reared and processed.

Tropical, arid, and subpolar regions may alter the utilization of these chemicals in animal agriculture. This is particularly true of veterinary drugs used in microbial infections, in the prevention of disease and infections, and in parasitic control or treatment. Phytotoxins are another case because of the impact that environmental conditions have on their diversity.

4.5.2 ANIMAL TOXICOSIS AND THE ENVIRONMENT

Toxicants, natural and synthetic, may impact on animal agriculture in different ways: (1) by directly or indirectly intoxicating animals (toxicoses), resulting in mortality or decreased production of edible food products; (2) by decreasing the availability or usability of nutritious feedstuffs due to the presence of naturally occurring toxins or added toxicants; and (3) by decreasing the wholesomeness of edible food products due to the presence of hazardous residues (Shull and Cheeke, 1983).

There is a wealth of information on domestic animal toxicology in the temperate zone (Ruckebusch *et al.*, 1983; Osweiler *et al.*, 1985). However, when we move to consider the arid, tropical or subpolar regions, we must also consider the effects that light, temperature (both cold and hot), and rainfall have on the toxicology of chemicals on these domestic animals. Domestic animals are homeotherms and as such can maintain a thermoneutral zone over a range of ambient temperatures. When the ambient temperature moves outside the thermoneutral zone the animal must alter its metabolic rate to maintain a constant core body temperature. Yousef (1985a) has presented an excellent review of domestic animal production and physiology under the stress of cold and heat in various regions of the world.

Animals reared outside the temperate region are subjected to large swings in temperatures beyond the thermoneutral zone. Animal metabolism and behaviour are altered, characterized by a shift in food and water consumption, passage rate of food through the digestive tract, hormone synthesis and release, panting and shivering, and reduced activity to name a few. With an altered metabolism there is a potential for xenobiotics to be handled differently by animals. The converse is also true in that thermoregulation can be affected by a variety of toxic substances, thus altering the ability of domestic animals to tolerate thermal shifts in ambient temperature. For example, it has been demonstrated that the organophosphates, parathion or chlorpyrifos, reduce the tolerance of animals to cold exposure (Ahdaya *et al.*, 1976; Rattner *et al.*, 1982; Maguire and Williams, 1987). Organophosphate and carbamate pesticides are known to be cholinesterase inhibitors and to cause hypothermia in animals. Acetylcholine is one of the hormones involved in maintenance of body temperature; thus, cholinesterase inhibitors have an effect on thermoregulation in mammals (Ahdaya *et al.*, 1976) and birds (Rattner *et al.*, 1982). The inability to thermoregulate may not be the cause of greater death losses, at least in birds; rather, the cause may be reduced insulation brought about by loss of subcutaneous fat, or depletion of carbohydrate and lipid reserves induced by decreased food intake (Rattner *et al.*, 1982).

Species that are indigenous to a region are better able to maintain a more constant core body temperature than non-adapted animals. This ability is brought about through anatomical and physiological adaptations that have occurred. The question has been asked, do arctic birds and mammals maintain body temperatures within the same range as species from temperate regions? Both birds and mammals subjected to environmental temperatures of −30°C and −50°C maintained their body temperatures within normal limits (Schmidt-Nielsen,

1979). Protected by insulation, animals in arctic climates do not need to eat more than animals in milder climates (Yousef, 1985b). The climate of the arctic region is not always cold, but rather is characterized by extreme variation from winter cold to summer heat and great differences in sunlight between winter and summer days (Irving, 1964). Non-adapted animals experience difficulty in adapting to these extremes.

In general, animals indigenous to tropical areas of the world are better adapted for heat exposure than those from temperate regions (Ingram and Mount, 1975; Yousef, 1982). In the tropical areas, animals such as cattle possess an increased ability to lose heat due to a greater surface area in the region of the dewlap and prepuce and increased numbers of sweat glands and the presence of short hair. Fat stores may be in humps or intermuscular rather than subcutaneous, which assists the conductance of heat from the core to the surface skin. Animals from the temperate region with high productivity have often failed to continue to yield as well when exposed to tropical climates with extremes in temperature.

In hot, arid climates adaptation mechanisms aid in the maintenance of normal body temperatures. Heat storage, insulation, panting, gular fluttering, and blood flow mechanisms to cool the brain are just some of the physiological adaptations used by animals in hot climates with limited water supplies (*see* Schmidt-Nielsen (1979) for review). Because domestic animals are homeotherms and indigenous animals are adapted to withstand climatic variations, the problems with xenobiotic exposure are not greatly different from those in the temperate region. However, several problems are amplified. At high ambient temperatures, there is an increase in water consumption and a concomitant decrease in feed consumption (National Research Council, 1981). If exposure to a chemical is via water the exposure will be increased and the animal may be at greater risk.

In cold climates, feed consumption increases during extremes of cold making exposure via the food chain an increased problem. It is known that food restriction and/or water deprivation may significantly alter the response of an animal to toxic chemicals (Baetjer, 1983). These changes in food and water consumption, which mark the principal metabolic shift in animals in response to environmental fluctuations, may contribute to toxicological differences in pesticides between the world's regions.

4.5.3 NATURAL TOXINS

Toxigenic fungi have ubiquitous geographical distribution influenced by climatic conditions, cultivation, and harvesting techniques, as well as storage procedures and the livestock production practices used. Mycotoxins occur in particular feeds and in particular regions (National Research Council, 1979; Smith, 1982). Aflatoxins are comparatively common in subtropical regions and depend on factors such as weak plants resulting from drought stress, insect or mechanical damage, climatic conditions before drying, and improper storage conditions

(Galtier and LeBars, 1983). In tropical areas of the world, mycotoxins in grain, protein concentrates, and other feedstuffs are a major problem because warm, humid environmental conditions favour fungal growth, and farming practices in many tropical areas are not sophisticated (Cheeke and Shull, 1985). Also, crop storage conditions are frequently inadequate in these areas. Thus, in hot humid regions the production of natural chemicals in feedstuffs may result in toxicosis, posing a problem for livestock production.

4.5.4 CHEMICAL ACCIDENTS WITH PESTICIDES

Chemical accidents that adversely affect animal agriculture have occurred in the past and will doubtless occur in the future. These problems are not unique to any area of the planet and do not hinge on climatic conditions. As an example, when organochlorine insecticides were being phased out because of their adverse effects on non-target organisms, together with their persistency in the environment and their carcinogenicity, the organophosphate insecticides were introduced as logical replacements. Some of these compounds were halogenated phenyl phosphonates and phosphonothionates that were lipid soluble, persistent, and of lower toxicity to mammals than the parathions and other widely used organophosphorus insecticides. However, a number of them were known to be delayed neurotoxins (Metcalf, 1982). Leptophos, one such chemical, was used in 1971 to control the cotton leafworm in Egypt. Some 1300 water buffalo died from paralysis and distal axonopathy characteristic of delayed neurotoxicity (Abou-Donia *et al.*, 1974). Human poisoning was also evident (Hassan *et al.*, 1978). Egypt was only one of about 50 countries into which leptophos was sold. This example of toxicosis in livestock is one where water from cropland collected in a river and water consumption in a hot, arid climate caused the death of many animals. This is not an isolated case of chemical toxicosis in livestock, but documentation in the literature is not common. In temperate regions there are numerous reports of poisoning in intensified animal production units (Shull and Cheeke, 1983). These accidents have included such chemicals as polychlorinated biphenyls, polybrominated biphenyls, tetrachlorodibenzo-p-dioxins, and organochlorine insecticides.

4.5.5 MINERALS

Lead is considered to be one of the major environmental pollutants and has been incriminated as a cause of accidental poisoning in domestic animals in more cases than any other substance (National Research Council, 1972). Lead that contaminates the environment is largely air-borne but is redeposited by dust into soil and water and is taken up by or exists on the surface of plants which are grazed by livestock. Cattle, sheep, and horses are good indicators of pollution on vegetation (Debackere, 1983). Lead toxicosis in cattle from the use of lead-based pigments in paint was common, as was poisoning of water fowl by spent lead shot. Restriction in the use of lead-based paints and, currently, in lead shot has reduced the problem in the United States. Lead from smelters may cause problems in horses grazing in adjacent areas. Dogs and cats give a very good indication of lead pollution in urban areas as the concentration of lead in their livers and kidneys increases with increased pollution (Debackere, 1983).

Animals can be exposed to mercury contamination from air, soil, water, and ingestion of contaminated feed. Mercury contamination results from fossil fuel combustion, agricultural fungicides, smelting of commercial ores, and through industrial discharge followed by water and, then, fish pollution. The fish are incorporated into animal feeds by way of fishmeals or protein concentrates. Toxicosis of domestic animals has also been due to the consumption of contaminated grain. In 1971, Iraqi authorities ordered 73 000 tons of wheat and 22 000 tons of barley from suppliers in Mexico and Canada, respectively, that were treated with mercury. This grain was used for planting but some was prepared into homemade bread. Oral ingestion may have included meat and other animal products obtained from the livestock given the treated grain (Bakir *et al.*, 1973). The latent period between dose and onset of symptoms may have given farmers a false sense of security since chickens given wheat for a period of a few days did not die (Bakir *et al.*, 1973). Hospital admission amounted to some 6530 cases (Clarkson *et al.*, 1976). Both man and animals are subject to mercury toxicosis through contaminated grains.

In Minamata Bay in Japan, methylmercury poisoning was observed in cats before human cases were recognized. The common thread between the disease in cats and that in humans was shown to be the consumption of mercury-contaminated fish (Hodges, 1976). Under modern feeding and management conditions of livestock production, cadmium toxicosis is relatively unimportant (Neathery and Miller, 1975) but does not preclude ingestion of recycled waste material, such as sewage sludge, in which cadmium may be concentrated. Cadmium is toxic and is an antagonist of zinc, iron, copper, and other elements. Some plants, such as clover, have the capacity to concentrate cadmium from soil.

Arsenic may result in contamination of livestock in areas surrounding smelters and where arsenicals are used for weed and insect control. Since fish are often high in arsenic, fishmeals may contribute sizeable quantities of arsenic to livestock. Non-ruminants are generally more susceptible to intoxication than are ruminants or horses. The degree of toxicity in ruminants is variable and may depend on the route of exposure, age, nutritional status, and exposure duration (Case, 1974; Selby *et al.*, 1974).

There are relatively few veterinary examples of acute copper toxicosis except in cases of accidental overdosing or the consumption of copper-containing compounds. Sheep are extremely sensitive to excess copper and therefore are good indicators of environmental pollution by copper. Concentrations two- to three-fold above normal grass copper concentrations of 8–15 ppm are toxic to sheep (Debackere, 1983).

Selenium is used in some areas as a supplement to animal diets; still, in other areas, there may be selenium toxicity due to high levels. Dietary selenium requirements are approximately 0.1 to 0.3 ppm, while toxic concentrations are about 10 to 50 times greater (National Research Council, 1980). When pasture is limited in dry weather, accumulator plants may be readily available and eaten by livestock, resulting in selenium poisoning. In the United States irrigation of arid land with water high in certain minerals, including selenium, has resulted in a large site where water fowl have shown evidence of what appears to be selenium toxicosis.

These are but some of the adverse effects minerals have produced on livestock production. In some cases climatic conditions alter their consumption but in most instances exposure is indirect.

4.5.6 ANIMALS AS POLLUTION INDICATORS

It is clear that domestic animals can serve as indicators for environmental pollution by chemicals. The role of domestic animals as indicators for environmental pollution through pesticides is negligible (Debackere, 1983). Based on present evidence, fish and birds appear more susceptible than mammals to pesticides, especially organophosphates, carbamates, and chlorinated hydrocarbons (Walker, 1983; Debackere, 1983). Physiological and anatomical differences are likely to affect susceptibility to a wide range of compounds. In domestic birds the excretory route via egg laying and the fact that blood from the gut goes to the kidney via the renal portal system prior to hepatic contact conveys an advantage to birds over mammals; however, the high body temperature, urinary release into the cloaca, and relatively small liver render birds more susceptible to pesticides (Walker, 1983). Physiological and biochemical evidence suggests that birds have less effective defence mechanisms than do mammals to xenobiotics.

Generally, indicator species are chosen for their toxicological susceptibility (Kenaga, 1978). Chickens are regularly used in the laboratory as predictive models for delayed neurotoxicity by organophosphorus chemicals. In addition to the chicken, the human, water buffalo, horse, cow, sheep, pig, dog and cat have been reported to be sensitive, while common laboratory animals, such as the rat, mouse, rabbit, guinea pig, hamster, and gerbil, are not. The adult chicken is utilized most frequently as the test animal; however, the cat can be used and may serve as an excellent model in extrapolation to man.

Wild birds are invaluable models for environmental toxicology due to their abundance, visibility, and diverse habitat associations (Hill and Hoffman, 1984), and are used to monitor pollution in urban as well as in aquatic environments.

In Guatemala, hogs fed seed wheat treated with organomercury as a fungicide developed blindness, lack of coordination, and posterior paralysis in 2 to 3 weeks (Ordonez *et al.*, 1966). Humans followed with similar signs of toxicosis from eating the same wheat.

When it comes to chemicals introduced into our environment by indirect pathways, as referred to earlier, domestic birds as well as wild ones, rabbits, cats, and cattle have been biological indicators for the presence of contamination.

4.5.7 REFERENCES

Abou-Donia, M. D., Othman, M. A., Tantawy, G., Khalil, A. Z., and Shawer, M. F. (1974). Neurotoxic effect of leptophos. *Experientia*, **30**, 63–64.

Ahdaya, S. M., Shah, P. V., and Guthrie, F. E. (1976). Thermoregulation in mice treated with parathion, carbaryl, or DDT. *Toxic Appl. Pharmac.*, **35**, 575–580.

Baetjer, A. M. (1983). Water deprivation and food restriction on toxicity of parathion and paraoxon. *Archs. Envir. Hlth.*, **38** (3), 168–171.

Bakir, F., Damluji, S. F., Amin-Zaki, L., Murtadha, M., Khalidi, A., Al-Rawi, N. Y., Tikriti, S., Dhahir, H. I., Clarkson, T. W., Smith, J. C., and Dohert, R. A. (1973). Methylmercury poisoning in Iraq. *Science*, **181**, 230–241.

Case, A. A. (1974). Toxicity of various chemical agents to sheep. *J. Am. Vet. Med. Ass.*, **164**, 277.

Cheeke, P. R., and Shull, L. R. (1985). *Natural Toxicants in Feeds and Poisonous Plants*. AVI Publishing Co., Westport, CT.

Clarkson, T. W., Amin-Zaki, L., and Al-Tikriti, S. K. (1976). An outbreak of methylmercury poisoning due to consumption of contaminated grain. *Fed. Proc.*, **35** (12), 2395–2399.

Debackere, M. (1983). Environmental pollution: the animal as source, indicator, and transmitter. In: *Veterinary Pharmacology and Toxicology*, pp. 595–608. AVI Publishing Co., Westport, CT.

Galtier, P., and LeBars, J. (1983). Mycotoxin residue problem and human health hazard. In: *Veterinary Pharmacology and Toxicology*, pp. 625–640. AVI Publishing Co., Westport, CT.

Hassan, A., Abdel-Hamid, F. B., Abou-Zeid, A., Moktar, D. A., Abdel-Pazek, A. A., and Ibrahain, M. S. (1978). Clinical observations and biochemical studies of humans exposed to leptophos. *Chemosphere*, **7**, 283–290.

Hill, E. F., and Hoffman, D. J. (1984). Avian models for toxicity testing. *J. Am. Coll. Toxicol.*, **3** (6), 357–376.

Hodges, L. (1976). *Environmental Pollution*, 2nd edn, pp. 230–236. Holt, Rinehart and Winston, New York.

Ingram D. L., and Mount, L. E. (1975). *Man and Animals in Hot Environments*. Springer-Verlag, New York.

Irving, L. (1964). Maintenance of warmth in arctic animals. *Symp. Zool. Soc., London*, 13 : 1. Cited in: Yousef, M. K., (1985) *Stress Physiology in Livestock*, vol. II (chap. 10). CRC Press Inc., Boca Raton, FL.

Kenaga, E. E. (1978). Test organisms and methods useful for early assessment of acute toxicity of chemicals. *Environ. Sci. Technol.* **12**, 1322–1329.

Maguire, C. C., and Williams, B. A. (1987). Cold stress and acute organophosphorus exposure: interaction effects on juvenile northern bobwhite *Arch. Environ. Contam. Toxicol.* **16**, 477–481.

Metcalf, R. L. (1982). Historical perspective of organophosphorus ester-induced delayed neurotoxicity. *Neurotoxicol.*, **3** (4), 269–284.

National Research Council (1972). *Lead: Airborne Lead in Perspective*. National Academy of Sciences, Washington, D.C.

National Research Council (1979). *Interactions of Mycotoxins in Animal Production*. National Academy of Sciences, Washington, D.C.

National Research Council (1980). *Mineral Tolerance of Domestic Animals*. National Academy of Sciences, Washington, D.C.

National Research Council (1981). *Effect of Environment on Nutrient Requirements of Domestic Animals*. National Academy of Sciences, Washington, D.C.

Neathery, M. W., and Miller, W. J. (1975). Metabolism and toxicity of cadmium, mercury and lead in animals: a review. *J. Dairy Sci.*, **58** (12), 1767–1781.

Ordonez, J. V., Carrillo, J. A., Miranda, C. M., and Yale, J. L. (1966). Organic mercury identified as the cause of poisoning in humans and hogs. *Science*, **172**, 65–67.

Osweiler, G. D., Carson, T. L., Buck, W. B., and Van Gelder, G. A. (1985). *Clinical and Diagnostic Veterinary Toxicology*, 3rd edn. Kendall/Hunt Publishing Co., Dubuque, IA.

Rattner, B. A., Sileo, L., and Scanes, C. G. (1982). Hormonal responses and tolerance to cold of female quail following parathion ingestion. *Pestic. Biochem. Physiol.*, **18**, 132–138.

Ruckebusch, Y., Toutain, P. L., and Koritz, G. D. (1983). *Veterinary Pharmacology and Toxicology*. AVI Publishing Co., Westport, CT.

Schmidt-Nielsen, K. (1979). *Animal Physiology: Adaptation and Environment*, 2nd edn. Cambridge University Press, New York.

Selby, L. A., Case, A. A., Dorn, C. R., and Wagstaff, D. J. (1974). Public health hazards associated with arsenic poisoning in cattle. *J. Am. Vet. Med. Ass.*, **165**, 1010.

Shull, L. R., and Cheeke, P. R. (1983). Effects of synthetic and natural toxicants on livestock. *J. Anim. Sci.*, **57** (2), 330–354.

Smith, J. E. (1982). Mycotoxins and poultry management. *World's Poult. Sci. J.*, **38** (3), 201–212.

Walker, C. H. (1983). Pesticides and birds—mechanisms of selective toxicity. *Agric. Ecosystems and Environ.*, **9**, 211–226.

Yousef, M. K. (1982). *Animal Production in the Tropics*. Praeger Publishers, New York.

Yousef, M. K. (1985a). *Stress Physiology in Livestock*, vol. I. CRC Press Inc., Boca Raton, FL.

Yousef, M. K. (1985b). *Stress Physiology in Livestock*, vol. II. CRC Press Inc., Boca Raton, FL.

Chapter 5

Case Studies

Ecotoxicology and Climate
Edited by P. Bourdeau, J. A. Haines, W. Klein and C. R. Krishna Murti

5.1 Environmental Pollution in Coastal Areas of India

R. SEN GUPTA, SUGANDHINI NAIK AND V. V. R. VARADACHARI

5.1.1 ADDITIONS FROM LAND

The chemistry and biology of coastal waters are very vulnerable to additions of biodegradable and stable compounds from land. It has been estimated (Qasim and Sen Gupta, 1983) that in 1984, 5 million tonnes of fertilizers, 55 000 tonnes of pesticides, and 125 000 tonnes of synthetic detergents were used in India (Table 5.1.1). Roughly about 25% of all these can be expected to ultimately end up in the sea every year. Some of these substances are biodegradable while others are persistent. Their cumulative effect over a long period could be quite harmful to the coastal marine environment. These effects are, as yet, not very perceptible over all the Indian coast. But near a few big cities and industrial conglomerates the effects are, indeed, becoming near disastrous. In the 15 years from 1959 to 1974 phosphate-phosphorus concentration in the nearshore waters of Bombay increased from 0.82 to 1.13 μmol/l, i.e. by about 40% (Sen Gupta and Sankaranarayanan, 1975). The present concentration (1984) is around 2 μmol/l (Zingde, 1985). The dissolved oxygen concentration decreased from 4.71 ml/l to near zero in 1983 (Parulekar *et al.*, 1986). High values of phosphate-phosphorus have also been observed at nearshore Madras (Sen Gupta, unpublished data).

Similar situations can be expected to have taken place with regard to some toxic heavy metals. Indian rivers annually add about 1600 million tonnes of sediment to the seas around (Holeman, 1968), a major portion of which can be expected to settle in the nearshore regions. Most of the heavy metals are transported to the sea this way. This can be illustrated by our observations in the estuarine regions of River Hooghly. These were carried out in September when the freshwater runoff, and consequently suspended solids, can be expected to be maximum. Several heavy metals were analysed in suspended sediments, collected by filtering a large volume of water, from two stations every three hours over two tidal cycles in the final 125 km stretch of the river. The mean concentration of metals came to 2.14% of suspended solids on a wet weight basis.

Table 5.1.1 Population, related data, and some estimates of pollutants entering the sea around India (as of 1984)

Population	720 million
Coastal population (25% of total)	180 million
Area of the country	3.276 million km^2
Agricultural area	1.65 million km^2
Exclusive economic zone	2.015 million km^2
River runoff (annual mean)	1645 km^3
Rainfall per year (on land)	3500 thousand million m^3
Rainfall per year (on Bay of Bengal)	6500 thousand million m^3
Rainfall per year (on Arabian Sea)	6100 thousand million m^3
Domestic sewage added to the sea by coastal population per year (at 60 l per head per day)	3900 million m^3
Industrial effluents added to the sea by coastal industries per year	390 million m^3
Sewage and effluents added by the rivers to the sea per year	50 million m^3
Solid waste and garbage generated by coastal population per year (at 0.8 kg per head per day)	53 million tonnes
Fertilizer used per year (at 30.5 kg/ha/year)	5 million tonnes*
Pesticides used per year (at 336 g/ha/year)	55 000 tonnes*
Synthetic detergents used per year	125 000 tonnes*
Oil transported across the Arabian Sea in 1983	513 million tonnes
Oil transported to Western Hemisphere in 1983	291 million tonnes
Oil transported to Far East and Japan in 1983	222 million tonnes
Tar deposition on beaches along the west coast of India per year	750–1000 tonnes

*Approximately 25% of this amount is expected to end up in the sea

Another set of similar observations was carried out at three stations, spread over a distance of about 80 km, in the region of the river mouth. The mean concentration of metals at the westernmost station, which is in the direction of the main outflow from the river, was 1.91% of suspended solids. However, if the mean value of observation at all the three stations is taken the concentration of metals comes to 1.45% of suspended solids. This would mean that approximately 11–32% of the metals in suspended form can be expected to settle to the bottom at the confluence of the River Hooghly and the Bay of Bengal.

5.1.2 SEWAGE AND EFFLUENTS

Domestic sewage and industrial effluents are discharged in the water courses in and around India in untreated or partially treated form. These, naturally, add a variety of pollutants which include, among others, certain toxic heavy metals and metalloids. The total volume of all discharges from the environs of Bombay is around 365 million tonnes (MT) per year (Sabnis, 1984). Similar discharges from the environs of Calcutta are around 350 MT every year (Ghose *et al.*, 1973). These figures would, perhaps, help towards ascertaining the total volume of domestic sewage and industrial effluents generated and released in India.

The expected composition of such discharges is presented in Table 5.1.2. The data are from the annual releases to the Mahim River bay and creek in the city of Bombay (Sabnis, 1984). The Bay occupies an area of 64 km^2 and is influenced by semi-diurnal tide with a maximum height of 3 m. It once had good fisheries and an oyster bed, and its fringing mangroves used to be visited by migratory birds. Today it is one of the most industrialized and densely populated areas of Bombay. Birds hardly flock there and the fisheries are dead, as no fauna can survive in the toxic environment. The Bay receives 64 MT domestic sewage and 0.9 MT industrial effluents every year. These discharges were initially untreated, but are now partially treated. The resulting concentration of H_2S–S in waters of the Bay ranges from 1.5 to 98.4 μmol/l, depending on the stages of the tide. The release of effluents of hydrocarbon origin has so heavily contaminated the creek that it has become common practice to recover the oil by soaking sorbents. Geochronology of sediments using ^{210}Pb gave the maximum period since deterioration started as 54 years, which is roughly about the time when pumping of untreated sewage into the river started.

5.1.3 TOXIC METALS IN MARINE BIOTA

It may be assumed, perhaps, that due to the discharge of such a large volume of contaminants, concentrations of heavy metals in the marine biota around India

Table 5.1.2 Pollutants discharged into Mahim Bay (Bombay) every year (in tonnes). From Sabnis (1984)

Dissolved solids	92 619
Chlorides	37 495
BOD	16 480
Suspended solids	15 649
Sulphates	4 791
Nitrogen	2 236
Phosphorus	383
Iron	162
Manganese	32
Zinc	16
Copper	7
Nickel	5
Cobalt	2
Lead	0.7

will be fairly high. That such is not yet the case can be seen from the values presented in Table 5.1.3, which presents the concentrations of heavy metals in zooplankton and in muscles of fishes of commercial importance, including sharks. It can be seen from the table that concentrations of almost all of the metals, particularly the toxic metals Pb, Cd, and Hg, are within the permissible limits for human consumption. It has also been observed that concentrations of all these metals in the livers of the dolphin, barracuda, shark, skipjack tuna, and yellowfin tuna are significantly higher than in their muscles. This indicates that most of these metals are assimilated by these fishes in a fat-soluble form. Acceptable correlation with their sex, size, and stage of maturity was observed in the different fish tissues.

The absence of mercury in zooplankton indicates that the metal is assimilated by fishes, probably by other pathways of the marine food chain. Mercury concentrations in the muscles of sharks and skipjack tuna are the highest of all the values (Table 5.1.3). Concentrations of the toxic heavy metals, Hg, Cd, and Pb, in different tissues of the fishes (Table 5.1.4) indicate that their highest occurrence is in the liver and the kidney. However, the highest concentrations of mercury, observed in muscles, are still much lower than the internationally permissible maximum of 0.5 ppm for human consumption. Analyses of mercury in the muscles of several commercially important fishes from the inshore regions and from a polluted creek in and around Bombay in 1975 gave values ranging from 0.04 to 0.57 ppm on a fresh weight basis (Tejam and Haldar, 1975). A high concentration of mercury, 0.2–7.3 ppm on a fresh weight basis, was observed in muscles of crab from the same area in 1980 (Ganeshan *et al.*, 1980).

5.1.4 PESTICIDES RESIDUES

India is predominantly an agricultural country. Large quantities of pesticides, herbicides, fungicides, etc. are used in agriculture and disease-vector control. Their residues will ultimately find their way into the sea. Organophosphorus pesticides will undergo biodegradation, but the organochlorines and polychlorinated biphenyls are persistent in the marine environment. Their main transport route to the sea is through the atmosphere, while a part may also be transported with sewage and effluents.

A few measurements of DDT and its isomers (expressed as t-DDT) have been conducted in zooplankton and water from the eastern part of the Arabian Sea and from the air over it. During a series of observations in 1978, concentrations of t-DDT were found to vary from 0.05 to 3.21 ppm wet weight in zooplankton samples from the eastern Arabian Sea (Kureishy *et al.*, 1978). A recent observation from the same area (Kannan and Sen Gupta, 1987) in a few zooplankton samples indicated values from 0.379 to 1.63 ppm wet weight. Both the sets of values, however, indicated a relatively decreasing trend from nearshore to offshore regions (Figure 5.1.1). Concentrations of t-DDT in the waters of

Table 5.1.3 Concentration ranges of a few essential and non-essential heavy metals (ppm wet weight) in zooplankton, crustaceans, bivalves and muscles of certain fishes from the northern Indian Ocean. ND = not detectable. Sources: Kureishy *et al.* (1979, 1981, 1983)

	Essential heavy metals						Non-essential heavy metals		
	Cu	Fe	Mn	Zn	Ni	Co	Pb	Cd	Hg
Zooplankton	2.0–5.0	35.0–94.0	3.0–7.0	8.0–31.0	0.2–3.0	ND–4.0	1.0 –12.6	0.02–5.99	ND
Prawns (6 spp.)	3.5–24.0	–	–	–	–	–	1.0	<0.2 –2.5	ND–0.17
Crabs	0.7–13.5	–	–	–	–	–	<1.0 –7.88	0.61–1.12	0.004–0.01
Clams	–	–	–	–	–	–	1.28	1.66	0.06
Oysters	45.0	–	–	–	–	–	<1.0	1.36	0.02
Mussels	–	–	–	–	–	–	1.31	1.38	0.09
Flying fish	0.1–0.7	4.0–62.0	ND–3.7	4.0–21.0	ND–0.9	0.2–1.3	1.8 –5.76	ND–0.65	ND–0.07
Silver bellies	1.0–1.6	–	–	–	–	–	<1–3.21	0.58–2.11	0.0001–0.01
Malabar anchovies	4.4	–	–	–	–	–	<1	0.7	0.01
Sardines (2 spp.)	0.03	8.0–10.0	0.2	4.5–6.3	–	0.7–1.1	<1	ND–0.62	ND–0.01
Mackerel (2 spp.)	1.0–1.3	12.0	0.01	6.0	ND	1.8	<1	0.22–1.62	0.01–0.02
Jew fish (2 spp.)	ND–0.8	6.0–8.0	0.3–10.0	4.0–4.8	ND	0.7–1.1	<1–1.14	0.19–0.42	0.006–0.01
Perch (3 spp.)	0.2–0.7	6.0–29.0	ND–0.1	3.4–6.1	0.3–0.5	ND	<1	ND–1.47	0.007–0.1
Pilot fish	0.1–4.9	–	–	–	–	–	<1–2.95	ND–0.83	ND–0.02
Scianid (2 spp.)	0.1–0.3	–	–	–	–	–	<1	0.86–1.36	ND–0.02
Sole	–	–	–	–	–	–	<1	0.35	0.01
Pomfret	–	–	–	–	–	–	<1	0.73	0.01
Catfish	–	–	–	–	–	–	<1.02	0.92	0.06
Trevally (2 spp.)	ND–0.7	5.0–11.0	0.1–9.0	2.0–5.0	ND–0.6	ND–1.2	<1	ND–0.62	0.018–0.08
Grunter	0.36	–	–	–	–	–	2.7	ND	0.24
Talang	0.4	–	–	–	–	–	<1	ND	0.36
Tuna (4 spp.)	0.3–3.0	7.0–164.0	0.1–7.5	4.0–12.0	ND–4.0	ND–3.2	<1–3.3	ND–2.00	0.004–0.22
Dolphin	0.2–1.7	13.0–39.0	ND–3.1	5.0–9.0	0.1–1.2	ND–1.9	<1–2.95	ND–0.95	0.01–0.14
Seer fish	0.4	–	–	–	–	–	<1–1.5	0.25–0.66	0.09–0.11
Barracuda	0.1–0.5	4.0–17.0	0.2–3.1	3.3–5.8	0.1–0.3	0.6–1.9	<1	ND–0.28	0.06–0.2
Sea pike	–	–	–	–	–	–	1.46	ND	0.11
Sharks (4 spp.)	0.14–1.1	10.0–57.0	ND–2.0	4.5–12.0	ND–0.3	ND–3.8	<1–6.02	ND–0.81	0.02–0.21

Table 5.1.4 Ranges and average concentrations of a few toxic heavy metals (ppm wet weight) in different body parts of fishes from the northern Indian Ocean. ND = not detectable. Source: Kureishy *et al.* (1979, 1981, 1983)

Body parts	Mercury		Cadmium		Lead	
	Range	Average	Range	Average	Range	Average
Muscle	ND–0.36	0.07	ND–3.24	0.59	1–3.43	1.11
Liver	ND–0.04	0.01	1.2–87.3	20.18	1–17.62	3.8
Gill	ND–0.03	0.016	ND–0.76	0.42	1–7.0	3.14
Heart	ND–0.08	0.026	ND–1.91	0.54	1–3.4	1.36
Kidney	ND–0.04	0.015	0.38–36.69	9.02	1–69.46	8.61
Gonads	ND–0.03	0.015	ND–8.06	1.25	1–4.76	1.36

the eastern Arabian Sea were from 0.06 to 0.16 ng/l while those in the air over the same area varied from 0.93 to 10.9 ng/m^3 (Tanabe and Tatsukawa, 1980). These observations would clearly indicate that residues of organochlorines in the marine environment of the Arabian Sea are cause for concern.

5.1.5 PETROLEUM HYDROCARBONS

In 1983 the global marine transport of oil was 1206 MT, of which 513 MT (42.5% of the total) was shipped from the Gulf countries (British Petroleum, 1984; Sen Gupta and Qasim, 1985). The main routes of marine transport of oil from the Gulf countries are across the Arabian Sea. One of these is through the Mozambique Channel around South Africa to the Western Hemisphere; while the other is along the west coast of India, around Sri Lanka, across the southern Bay of Bengal, through the Malacca Strait to the Far East, including Japan. In 1983, 291 MT of oil was shipped to the Western Hemisphere and 222 MT to the Far East, including Japan, from the Gulf countries. This, coupled with increasing emphasis on offshore oil exploration in many countries in the area, makes the northern Indian Ocean very vulnerable to oil pollution.

Oil pollution is caused both by accidents and by operational discharges, e.g. tanker disasters, ballast water, and bilge washings. The presence of about 75% of oil in the marine environment is due to operational discharges. Fortunately, only a few tanker disasters have occurred along these routes so far.

The effects of oil pollution on the sea are: presence of oil slicks; dissolved/dispersed petroleum hydrocarbons; floating petroleum residues (tar balls); and tar on beaches. A total of 6689 observations on oil slicks and other floating pollutants revealed positive results for oil slicks on 5582 or 83.5% of the total (Sen Gupta and Kureishy, 1981).

A summary of observations on floating petroleum residues (tar balls) from the surface of the Arabian Sea is presented in Table 5.1.5. The concentrations vary in time and space; they are occasionally very high along the tanker routes. In the Arabian Sea the concentrations range from 0 to 6.0 mg/m^2 with a mean

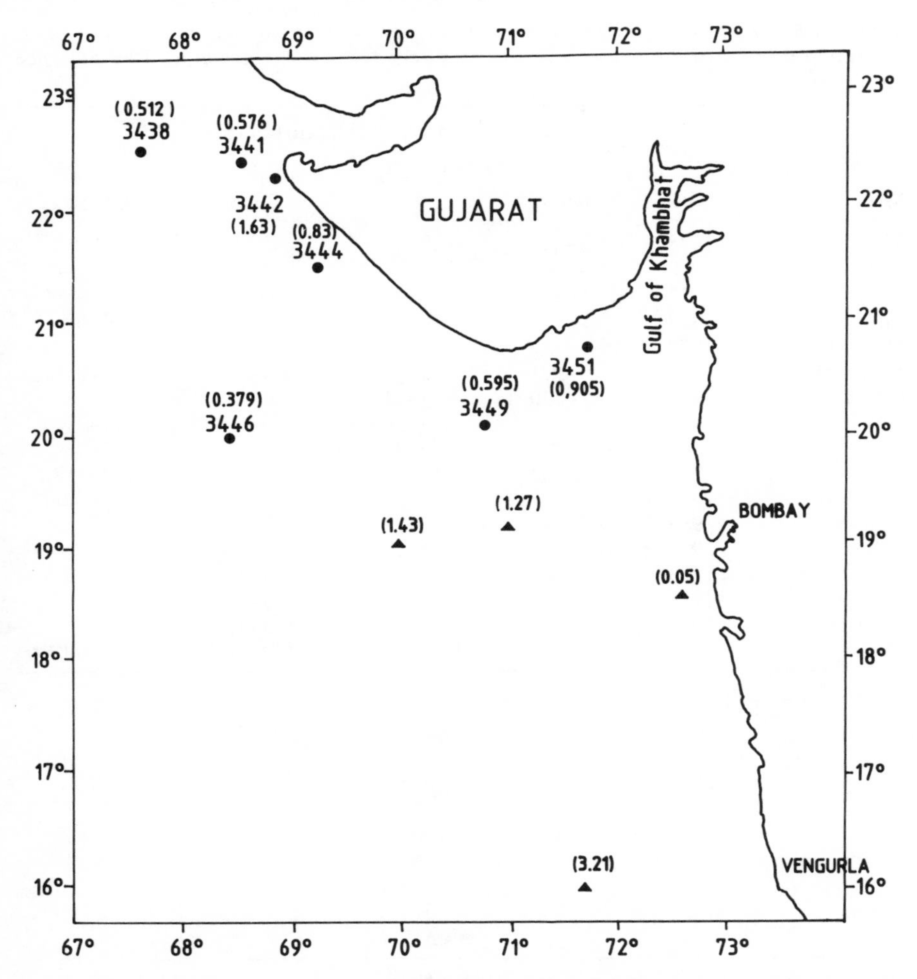

Figure 5.1.1 Points of observation for t-DDT in zooplankton. Numbers in parentheses are concentrations in ppm wet weight

of 0.59 mg/m^2. The range on the Bay of Bengal tanker route varies from 0 to 69.75 mg/m^2, with a mean of 1.52 mg/m^2. This would indicate that the tanker route across the southern Bay of Bengal is relatively more polluted than those across the Arabian Sea.

A number of observations were taken along the western-bound tanker route across the Arabian Sea during June–September 1983 (Table 5.1.5). Near absence of floating tar particles in this area during this period was expected, as the surface currents of the Arabian Sea normally flow towards the Indian west coast during the southwest monsoon months (June–September).

Table 5.1.5 Dissolved/dispersed hydrocarbons, particulate petroleum residues (tar balls) in, and transport of oil across the Arabian Sea. ND = not detectable; PI = presence indicated

Month/year		Dissolved/dispersed hydrocarbons (μg/l)			Tar balls (mg/m^2)
		0 m	10 m	20 m	
March	1978	0.6–22.9	0.9–18.1	25–28	0.02–0.32
June	1978	–	–	–	1.26–3.46
December	1978	21.5–42.8	16.8–42.5	28.6–37.5	0–0.54
May	1979	–	–	–	0.09–6.0
December	1979	18.6–41.6	16.9–34.15	10.4–24.9	–
February	1980	6.4–9.0	3.1–6.3	2.4–4.5	0–0.53
October	1980	47–230	77–210	–	–
November	1980	2.6–55.6	1.9–36.0	–	–
January	1981	4.0–9.5	ND–2.5	–	–
March	1981	200–305	130–277	–	0.30–112.2
February	1982	–	–	–	0–0.06
January	1983	9.7–24.2	6.7–13.2	2.2–7.2	0–0.06
July	1983*	3.0–8.9	ND–1.2	ND–1.0	0–PI
September	1983*	7.4–16.2	3.4–8.5	ND–1.2	0–PI
October	1983*	5.7–10.6	2.1–5.4	ND–1.1	0–PI
November	1983*	5.1–12.4	1.4–6.1	ND–1.6	
May	1984	22.9–114.6	18.6–62.2	18.9–72.0	0–0.06

Transport of Oil

Year	Volume (million tonnes)
1977	1024.4
1978	974.9
1979	1009.6
1980	869.3
1981	725.4
1982	578.7
1983	512.9

*Indicates data from the west-bound tanker route. The rest are from the east bound tanker route

The concentrations of dissolved and dispersed petroleum hydrocarbons in the upper 20 m of the tanker routes across the Arabian Sea are also presented in Table 5.1.5. The ranges are almost uniform, excepting some occasional high values. However, some seasonal variations can be observed which are due to varying intensities of tanker traffic and changes in meteorological factors. The quantum of oil pollution in the northern Indian Ocean due to this, as of end 1983, is presented in Figure 5.1.2.

One very interesting factor can be observed from the values in Table 5.1.5. The ranges did neither vary nor change appreciably from 1978 to 1984. Marine transport of oil and its products from the Middle East countries, however,

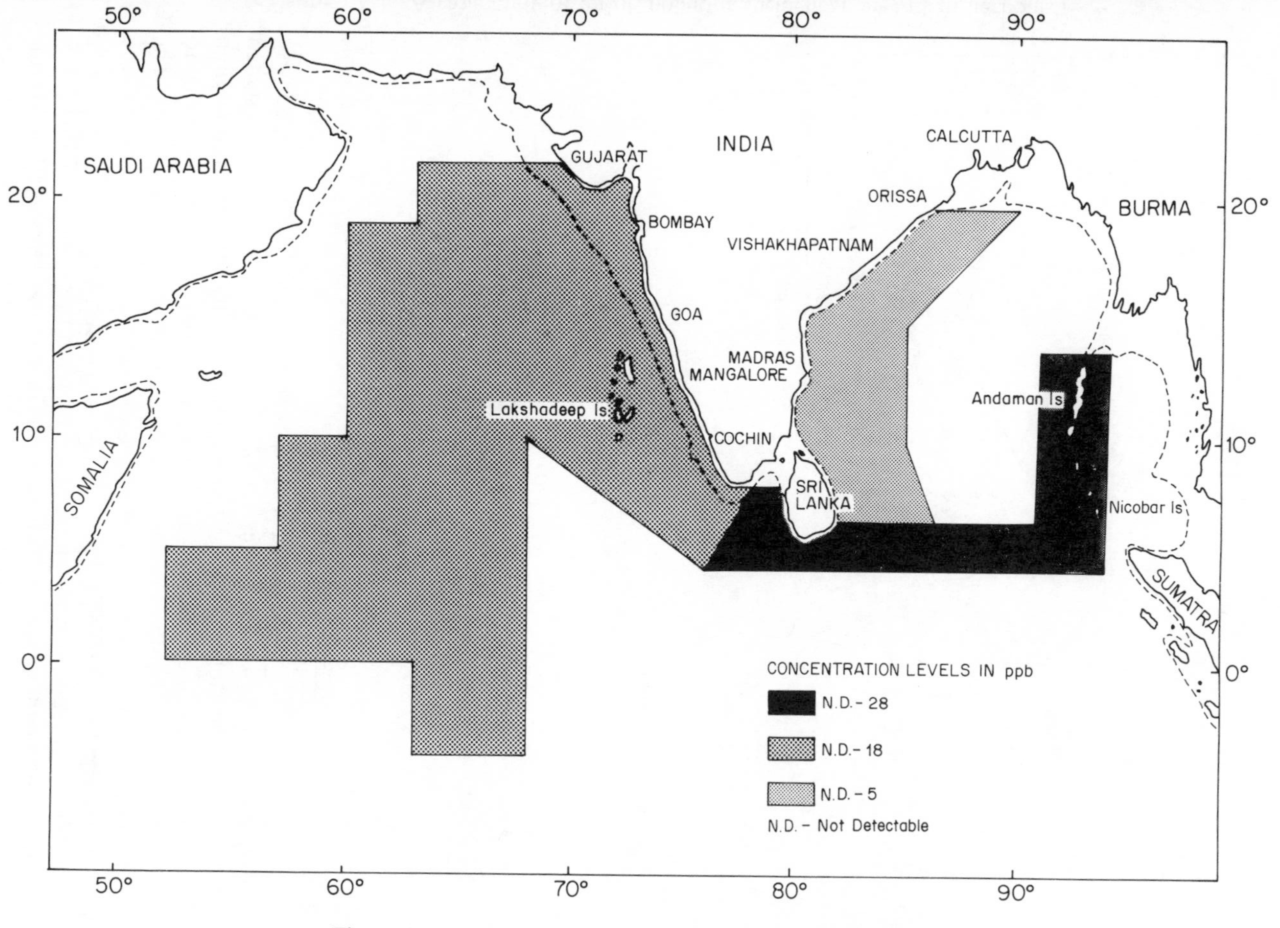

Figure 5.1.2 Dissolved petroleum residues in Indian Ocean

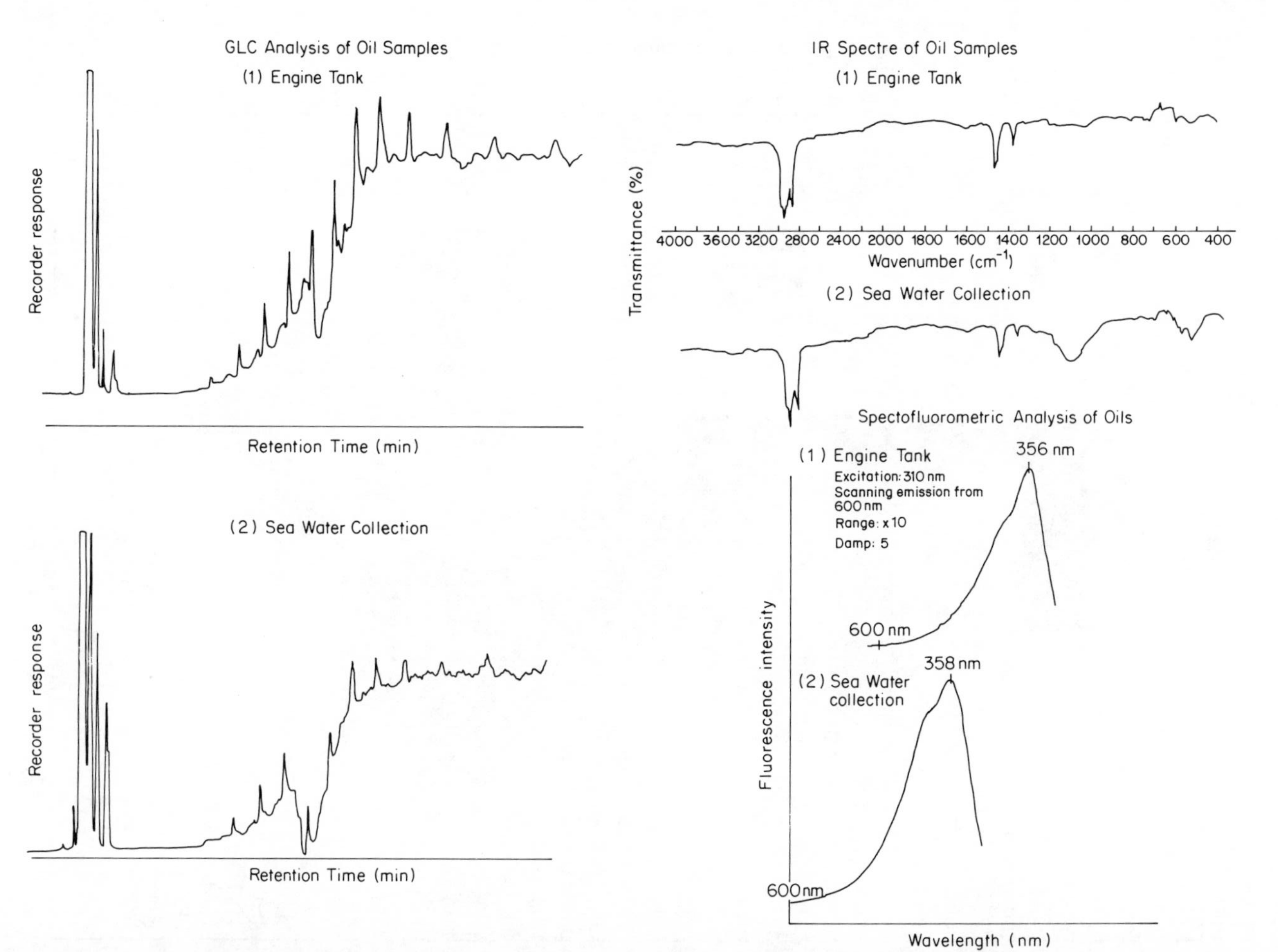

Figure 5.1.3 A case study of an oil pollution incident in an Indian harbour

decreased significantly over the same period (Table 5.1.5). This leads us to two conclusions. Oil pollution in the surface waters of the Arabian Sea did not undergo much quantitative alteration over the years and is not directly related to the volume transported. The pollution is mainly due to operational discharges of bilge and bunker washings of the tankers on their way to the Gulf ports for loading the cargo of oil.

Deposition of tar-like residues on the beaches is a chronic problem along the west coast of India. However, this is a seasonal phenomenon depending on the pattern of coastal circulation, largely regulated by the monsoons. Records from the west coast of India during the years 1975 and 1976 indicate a range from 22 to 448 mg/m^2, with a peak value on one occasion of 1386 g/m^2. The computed total deposits on beaches along the west coast of India are 1000 and 750 tonnes for 1975 and 1976 respectively (Table 5.1.1).

In the nearshore area of Bombay high concentrations of dissolved/dispersed hydrocarbons ranged from 2 to 46 $\mu g/l$ in water and from 4 to 32 $\mu g/g$ dry weight in sediments. In the Bombay harbour region the concentrations, after a tanker fire accident, ranged from 27 to 105 $\mu g/l$ at the surface and from 36 to 59 $\mu g/l$ at 5 m. The concentrations in sediments increased from 1–26 $\mu g/g$ to 40–512 $\mu g/g$ after the accident. In a polluted creek in Bombay concentrations in sediments were in the high range of 142–393 $\mu g/g$.

Cargo ships and oil tankers, visiting Indian ports, may cause oil pollution in harbour waters by operational or accidental discharges of oil and oily ballasts. According to MARPOL 1973/1978 (Convention of the International Maritime Organization) such spills are highly irregular and the owners of the ships are liable to pay compensation to the harbour authorities for clean-up operations. It is difficult, however, to identify the source of pollution and to match it to oil in harbour waters. The analytical techniques for such matching are: UV/spectrofluorometry, IR-spectrometry, and gas chromatography. A case study of one such incident in an Indian harbour (Bhosle *et al.*, 1987) is presented in Figure 5.1.3.

5.1.6 REFERENCES

Bhosle, N. B., Fondekar, S. P. and Sen Gupta, R. (1987). Sampling and analysis of petroleum hydrocarbons in the marine environment. *Indian J. of Mar. Sci.*, **16**, 192–195.

British Petroleum (1984). *British Petroleum Statistical Review of World Energy*. British Petroleum, London.

Ganeshan, R., Shah, P. K., Turel, Z. R., and Haldar, B. C. (1980). Study of heavy elements (Cd, Hg and Se) in the environment around Bombay by radiochemical neutron activation analysis. *Trans. 4th. Internat. Conf. on Nuclear Methods in Environ. and Environmental Res.* Univ. Missouri, USA.

Ghose, B. B., Ray, P., and Gopalakrishnan, V. (1973). Survey and characterization of waste waters discharged into the Hooghly estuary. *J. Inland Fisheries Soc. India*, **5**, 82–101.

Holeman, J. N. (1978). The sediment yield of major rivers of the world. *Wat. Resour. Res.*, **4**, 787–797.

Kannan, S. T., and Sen Gupta, R. (1987). Organochlorine residues in zooplankton off the Saurashtra Coast, India. *Mar. Pollut. Bull.*, **18** (2), 92–94.

Kureishy, T. W., George, M. D., and Sen Gupta, R. (1978). DDT concentration in zooplankton from the Arabian Sea. *Indian J. of Mar. Sci.*, **7**, 54–55.

Kureishy, T. W., George, M. D., and Sen Gupta, R. (1979). Total mercury content in some marine fish from the Indian Ocean. *Mar. Pollut. Bull.*, **10**, 357–360.

Kureishy. T. W., Sanzgiri, S., and Bragania, A. M. (1981). Some heavy metals in fishes from the Andaman Sea. *Indian J. of Mar. Sci.*, **10**, 303–307.

Kureishy, T. W., Sanzgiri, S., George, M. D., and Braganca, A. M. (1983). Mercury, cadmium and lead in different tissues of fishes from the Andaman Sea. *Indian J. of Mar. Sci.*, **12**, 60–63.

Parulekar, A. H., Ansari, Z. A., Harkantra, J., and Rodrigues, A. C. (1986). Long-term variations in benthic macro-invertrebrate assemblage of Bombay waters. In: Thompsen, M. F., Sarojini, R., and Nagabhushanam, R. (eds), *Biology of Marine Benthic Organisms*, pp. 485–494. Oxford and IBH Publ. Co., New Delhi.

Qasim, S. Z., and Sen Gupta, R. (1983). Environmental characteristics of the ocean. In: Pfafflin, J. R., and Zeigler, E. N. (eds), *Encyclopedia of Environmental Science and Engineering*, vol. I, pp. 294–309. Gordon and Breach Science Publishers, New York.

Sabnis, M. M. (1984). *Studies on some Major and Minor Elements in the Polluted Mahim River Estuary*. Doctoral dissertation, University of Bombay.

Sen Gupta, R., and Sankaranarayanan, V. N. (1975). Pollution studies off Bombay. *Mahasagar—Bull. Nat. Inst. of Oceanography*, **7**, 73–78.

Sen Gupta, R., and Kureishy, T. W. (1981). Present state of oil pollution in the northern Indian Ocean. *Mar. Pollut. Bull.*, **12**, 295–301.

Sen Gupta, R., and Qasim, S. Z. (1985). Petroleum—its present and future. *Mahasagar—Bull. Nat. Inst. of Oceanography*, **18**, 79–90.

Tanabe, S., and Tatsukawa, R. (1980). Chlorinated hydrocarbons in the North Pacific and the Indian Ocean. *J. Oceanogr. Soc. Japan.*, **36**, 217–226.

Tejam, B., and Haldar, B. C. (1975). A preliminary survey of mercury in fish from Bombay and Thana environment. *J. Environ,. Hlth.*, **17**, 9–16.

Zingde, M. D. (1985). Waste water effluents and coastal marine environment of Bombay. Proceedings, Sea water quality demands, pp. 20.1–20.23. Naval Chemical and Metallurgical Laboratory, Bombay.

Ecotoxicology and Climate
Edited by P. Bourdeau, J. A. Haines, W. Klein and C. R. Krishna Murti

5.2 *Biodegradation of Pesticides in Tropical Rice Ecosystems*

N. SETHUNATHAN

5.2.1 INTRODUCTION

Tropical and/or subtropical conditions exist in most of the developing countries. Hot and humid environments in the tropics and subtropics favour the build-up of a myriad of pathogens and insects harmful to the agricultural and plantation crops of economic importance. Yet, use of pesticides in the developing countries of the tropics has been negligible. For instance, pesticide use in India amounted to only 338 g/ha as compared to 1490 g/ha in Japan (Krishna Murti and Dikshith, 1982). This does not necessarily imply that problems of pesticide contamination do not exist in developing countries. In recent years, with the advent of modern agricultural technology in developing countries, there has been a steady increase in the use of pesticides on some economically important crops. There is concern over environmental problems arising from the massive use of pesticides in irrigated rice and other crops (cotton in particular), even in developing countries of the tropics.

In India and probably many other countries in the tropics, insecticides outstrip other groups of pesticides in consumption, while herbicides top the list in developed countries (Table 5.2.1). In recent years, the use of organochlorine pesticides has been steadily declining in developed countries due to concern over their long-term persistence and consequent ecological disturbances in the environment. But, organochlorine insecticides account for 70% of all pesticides used in India during 1983–84 (Anonymous, 1984) and probably in many developing countries, because of their low price, indigenous availability, broad-spectrum action, and long-term efficacy.

Recently, however, there has been concern in several developing countries over the environmental hazard from the heavy and regular use of organochlorine pesticides in public health and agriculture. This concern, based on the persistence data generated from studies in temperate environments of the developed countries, is unwarranted at least for some organochlorines, such as hexachlorocyclohexane (HCH). According to reports, hot and humid conditions

Table 5.2.1 Consumption of pesticides per category in India and developed countries. From Mrinalini (1983). Reproduced with permission from Pesticides, Bombay

Pesticide group	India	Developed countries
Insecticides	75%	35%
Herbicides	7%	45%
Fungicides	15%	15%
Others	3%	5%

such as exist in most of the developing countries of the tropics and subtropics effect more rapid breakdown of certain biodegradable pesticides than cooler conditions in temperate environments (Talekar *et al.*, 1977; 1983a, 1983b). Thus accumulation problems from the intensive use of pesticides may be less pronounced under tropical conditions than under temperate conditions. Again, persistence of pesticides in the environment is greatly influenced by the cultivation practices used for specific crops. For instance, the common practice of flooding in rice culture and of organic amendments for most agricultural crops hastens the degradation of several pesticides. The recent trend to ban or restrict the use of certain organochlorine pesticides in some developing countries based on research conducted elsewhere, mostly under temperate conditions not relevant to local tropical situations, merits caution. Admittedly, information on the fate and behaviour of pesticides under tropical conditions is fragmentary. Available literature (Sethunathan, 1973; Sethunathan *et al.*, 1977; Sethunathan and Siddaramappa, 1978; Sethunathan *et al.*, 1982, 1983; Rajagopal *et al.*, 1984a) indicates that biodegradation is an important means of detoxification of several commonly used pesticides, insecticides in particular, in tropical soils. In this section, an attempt is made to review the progress made in research on the degradation of certain pesticides in tropical soils and, wherever possible, to compare their relative persistence in tropical and temperate soils.

5.2.2 ORGANOCHLORINES

Among the organochlorines, hexachlorocyclohexane (HCH) is probably the most widely used insecticide in many developing countries. Hexachlorocyclohexane accounts for about 56% of all pesticides used in India. Organochlorine pesticides, HCH, DDT, methoxychlor, heptachlor, aldrin, dieldrin, and endrin, were considered as recalcitrant to biodegradation, because of their extreme stability and consequent long-term persistence (for several years) in aerobic soil and water environments. But, not all these insecticides are stable in anaerobic environments. The first convincing evidence of anaerobic instability of an organochlorine was provided when HCH disappeared fairly rapidly (within 90 days) from a tropical Philippine soil upon flooding, a common practice in rice culture (Raghu and MacRae, 1966).

Interestingly, gamma-HCH was not accumulated in the soil even after repeated application of HCH granules to flooded rice paddies in the Philippines (IRRI, 1967). Flooded rice fields were treated with HCH granules at the rate of 3 kg/ha, with two or four applications per season for two seasons in a year and for two years. Irrespective of the total amount of gamma-HCH applied in two years (a total of 48 kg/ha at four applications/season and 24 kg/ha at two applications/season), its concentration declined to low levels in soil (0.014–0.018 ppm) and water (0.1 ppm), and was identical in both plots within 20 days after every application.

Most recently, Yoshida and Yamaya (1984) studied the relative persistence of gamma-HCH in 11 soil samples collected from temperate, subtropical, and tropical regions under submerged conditions. Most of the gamma-HCH was degraded in eight soil samples from all regions in 70 days, and the rate of degradation increased with increased incubation temperature in the range of 20°C to 40°C. Interestingly, a tropical soil (Seputih soil from Indonesia) effected most rapid degradation of gamma-HCH, especially at higher temperatures.

That the alpha, beta, and delta isomers of HCH, like gamma-HCH, show a greater persistence in temperate soil (Tsukano, 1973) than in tropical soil (Castro and Yoshida (1974) under flooded conditions is best illustrated by data derived from two independent studies and brought together in Figure 5.2.1. In tropical soil (Philippines), all the four isomers of HCH reached low levels (3.4–17.1% of the original level) in 21 days. But, in temperate soil (Japan), only gamma-HCH declined to low levels (about 7% of the original level), while other isomers persisted with recoveries of 50–75% of the original level even after 28 days. As for the HCH isomers, other organochlorines, such as DDT, methoxychlor, heptachlor, and endrin, but not aldrin and dieldrin, were more rapidly decomposed in flooded soils than in non-flooded soils (Yoshida and Castro, 1970; Castro and Yoshida, 1974). Degradation occurred only in non-sterile soil samples (Castro and Yoshida, 1971).

A flooded soil differs from a non-flooded soil in physical, chemical, and microbial characteristics (Ponnamperuma, 1972; Gambrell and Patrick, 1978). Within a few days after flooding a soil becomes predominantly anaerobic due to diminished oxygen supply. Following flooding, the pH of acid or alkaline soils stabilizes at near neutrality while the redox potential drops rapidly. Soil constituents (nitrate, ferric, manganese, sulphate ions) are almost sequentially reduced under submergence.

More importantly, under such reducing conditions the anaerobes (facultative and obligate) become dominant. It is these anaerobic microorganisms that effected the rapid degradation of HCH/DDT and other anaerobically unstable organochlorines. Specifically, in flooded soils an obligate anaerobe, *Clostridium sphaenodes*, isolated from HCH-enriched flooded soil was capable of degrading alpha- and gamma-HCH, but not beta- and delta-HCH, under anaerobic conditions (MacRae *et al.*, 1969; Sethunathan *et al.*, 1969; Heritage and MacRae,

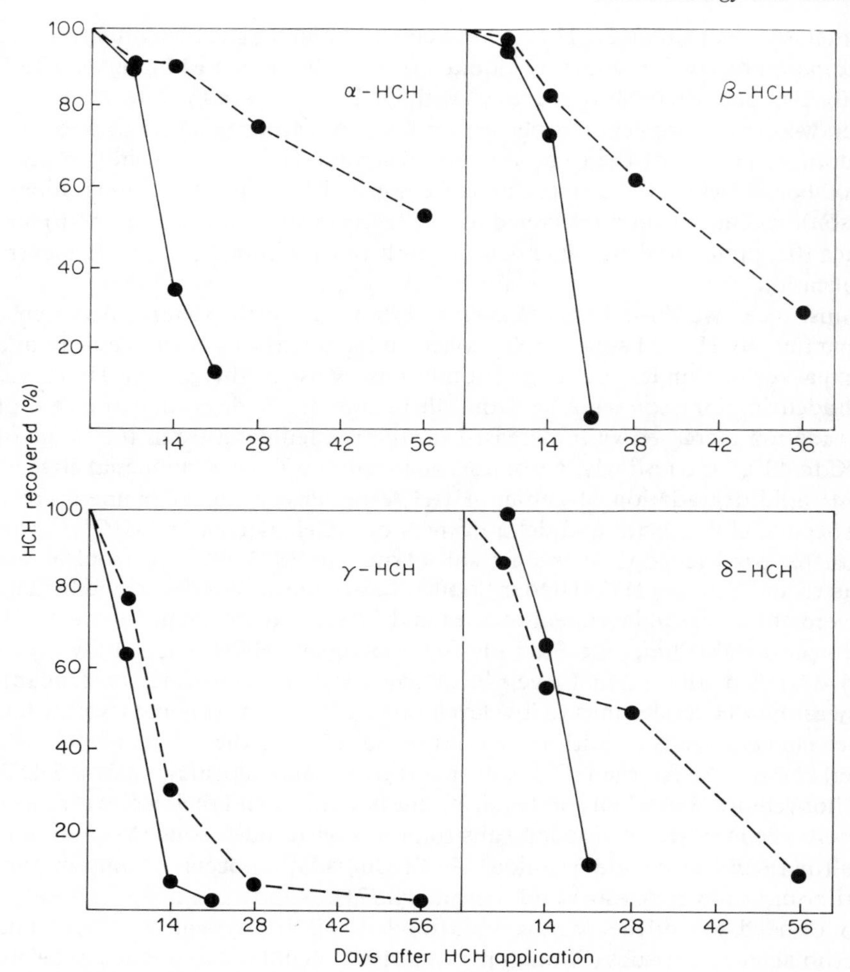

Figure 5.2.1 Relative persistence of isomers of hexachlorocyclohexane (HCH) in tropical (●———●) and temperate (●— — —●) soils under flooded conditions. Adapted from Castro and Yoshida (1974) (tropical soil) and from Tsukano (1973) (temperate soil)

1979). The same bacterium also degraded related organochlorines, such as DDT, methoxychlor, and heptachlor (Sethunathan and Yoshida, 1973a) under anaerobic conditions. In a more recent study (Yoshida and Yamaya, 1984) with soil samples from temperate, subtropical, and tropical regions, a gamma-HCH degrading bacterium isolated from each of eight soils was identified as *Clostridium* sp. Moreover, the gamma-HCH degrading activity of the isolates was more pronounced at 40°C than at 20°C and 30°C. Under actual field conditions in the tropics, high soil temperatures may accelerate the degradation of HCH isomers, especially under flooded soil conditions. Indeed, a rise in soil temperature from 20°C to 40°C accelerated the reduction of the soil under submergence (Cho and Ponnamperuma, 1971), and such intense reducing conditions would catalyze the degradation of HCH isomers (Siddaramappa and Sethunathan, 1975), mediated essentially by the dominant anaerobic microorganisms.

Evidence suggests that the microorganisms effect the degradation of HCH by a process known as co-metabolism, i.e. without utilizing it as an energy source for proliferation. In such co-metabolic reactions, an adequate supply of organic sources would accelerate the proliferation of these anaerobic microorganisms and thereby the degradation of pesticides in soil environments. In fact, the addition of organic sources, such as rice straw or green manure, stimulated the degradation of HCH and DDT in a soil under flooded, but not under non-flooded conditions (Guenzi and Beard, 1968; Yoshida and Castro, 1970; Siddaramappa and Sethunathan, 1975; Ferreira and Raghu, 1981).

Worldwide research has shown the ubiquitous occurrence, in anaerobic ecosystems, of a wide range of microorganisms, bacteria in particular, that readily degrade DDT (Sethunathan *et al.*, 1983). Thus, available evidence indicates that gamma-HCH and DDT undergo fairly rapid degradation in microbially active soils capable of attaining redox potential of -50 to -100 mV upon flooding (Sethunathan and Siddaramappa, 1978).

A flooded soil planted to rice differs from a flooded unplanted soil in physico-chemical and microbiological characteristics. A flooded planted soil is more dynamic than a flooded unplanted soil, because of the intense microbial activity and complex reactions in the nutrient-rich rhizosphere of the rice plant. Besides, a flooded unplanted soil is predominantly anaerobic, while significant transport of oxygen from the foliage of the rice plant to the rhizosphere occurs in a flooded soil planted to rice. However, in a study on the stability of HCH isomers in a flooded soil with and without rice plants, alpha- and gamma-HCH decreased to less than 5% of the original level in 30 days in both planted and unplanted soils (Table 5.2.2; Brahmaprakash *et al.*, 1985). Beta-HCH appeared to be more stable than alpha- and gamma-HCH, with a recovery of 30.9% from the planted soil and 50.6% from unplanted soil. Measurements of redox potential showed that flooded soil planted with rice registered a potential of $+200$ mV as compared to a potential of -300 mV in flooded unplanted soil. Evidently, degradation of aerobically stable isomers of HCH in a flooded soil was not retarded in the presence of rice plants despite the ability to transport oxygen to the root region and to maintain the flooded soil in an aerobic state. Possibly, degradation of HCH isomers occurred at anaerobic microsites, which abound in a flooded soil planted with rice.

The use of biodegradable organochlorine insecticides in flooded rice paddies, despite their instability in anaerobic environments, still poses problems of great concern:

Table 5.2.2 Mineralization of different isomers of HCH in a flooded soil with and without rice plants (Brahmaprakash *et al.*, 1985). Reproduced with permission from Springer-Verlag & Heidelberg

	Alpha-HCH*		Beta-HCH*		Gamma-HCH*	
Days/ Fraction	Unplanted	Planted	Unplanted	Planted	Unplanted	Planted
	(percent radioactivity recovered)					
O day						
HCH	87.9	78.7	80.6	76.3	96.8	82.3
Soil-bound	0.5	0.8	0.5	0.6	0.4	0.5
15 days						
HCH	18.9	22.1	50.7	44.3	5.2	8.22
Soil-bound	1.1	1.2	1.4	1.3	1.1	0.95
30 days						
HCH	5.5	4.3	50.6	30.9	2.2	2.4
Soil-bound	1.9	1.5	2.1	2.3	1.5	1.3

*^{14}C-alpha-HCH, ^{14}C-beta-HCH, and ^{14}C-gamma-HCH were added at 1.66×10^5 dpm/100 g of soil

(i) Commercial formulations of HCH contain several isomers, of which the gamma isomer is readily biodegradable. Beta-HCH, albeit a minor constituent in HCH formulations, is more resistant to anaerobic degradations than gamma-HCH and, therefore, accumulates in the environment, raising concern over its entry into the food chain. Although beta-HCH undergoes degradation, but at a slow rate, in flooded soil, attempts to isolate a microorganism capable of degrading beta-HCH in pure culture have been unsuccessful.

(ii) Gamma-HCH, despite its instability in anaerobic flooded rice field soils, accumulates in rice straw, a major cattle feed in several countries, and eventually enters the food chain.

(iii) DDT is readily degraded in anaerobic environments, but its degradation product, DDD, which is as toxic as DDT, resists further degradation. Likewise, endrin is unstable in anaerobic systems (Gowda and Sethunathan, 1976, 1977) but its degradation products are highly stable. Thus, there is a need to understand the toxicological significance of such stable metabolites formed from parent pesticides.

5.2.3 ORGANOPHOSPHATES

Parathion, until recently the most widely used organophosphorus insecticide in agriculture and public health, has now been banned from commercial use because of its high mammalian toxicity. However, parathion still serves as a model insecticide in research on the metabolism of organophosphorus pesticides in soil and water environments. Degradation of parathion in a tropical

(Philippines) rice soil occurred more rapidly under flooded conditions than under non-flooded conditions (Sethunathan and Yoshida, 1973b; Sudhakar Barik *et al.*, 1979) and temperate (Katan *et al.*, 1976) soils proceeded essentially by nitro group reduction to aminoparathion after the first application. The rate of degradation of parathion in both soil samples was accelerated after two or three applications of parathion or its hydrolysis product *p*-nitrophenol (Sudhakar Barik *et al.*, 1979; Ferris and Lichtenstein, 1980). Pretreatment of the temperate soil (cranberry) with parathion or *p*-nitrophenol led to the accelerated mineralization of subsequently added parathion to CO_2 (Ferris and Lichtenstein, 1980).

In studies with tropical soil, CO_2 was not quantified, but the pathway of parathion degradation shifted from nitro group reduction after the first application of parathion, essentially to hydrolysis after two or three applications of parathion or *p*-nitrophenol (Sudhakar Barik *et al.* 1979). The shift in the pathway of degradation was a direct consequence of the proliferation of parathion-hydrolysing microorganisms that utilized *p*-nitrophenol as the energy source.

A *Pseudomonas* sp. ATCC 29353, isolated from a flooded tropical soil after three repeated additions of parathion, readily hydrolysed parathion and then metabolized *p*-nitrophenol to nitrite (Siddaramappa *et al.*, 1973) and CO_2 (Sudhakar Barik *et al.*, 1976). Such accelerated biological hydrolysis of parathion has been demonstrated also in a temperate soil after repeated applications of the insecticide (Munnecke *et al.*, 1982). According to these observations, accelerated degradation of parathion through biological hydrolysis can be of common occurrence in both tropical and temperate regions. Parathion is short-lived in soil environments, especially under flooded conditions. This necessitates the frequent application of this insecticide at short intervals for efficient control of harmful pests. But, such frequent applications of parathion may, in turn, promote the build-up of parathion-hydrolysing microorganisms that undermine its efficacy.

Methyl parathion and fenitrothion, structurally related to parathion, are used increasingly in agriculture because of their relatively low mammalian toxicity. Adhya *et al.* (1981a) reported fairly rapid decomposition of parathion, methyl parathion, and fenitrothion to their respective amino analogues in a tropical (Indian) soil under flooded conditions. There are some independent reports on the degradation of fenitrothion in tropical (Indian) and temperate (Japanese) soils. Takimoto *et al.* (1976) found that fenitrothion was readily converted to aminofenitrothion in four different soils from Japan under flooded conditions.

Miyamoto (1977) studied the relative persistence of fenitrothion in four Japanese soils under flooded and non-flooded conditions. In two soils, the rate of degradation of fenitrothion was almost similar under both water regimes. But in the other two soils (Moriyama and Katano), soil submergence accelerated the degradation of fenitrothion. Thus, the half-life of fenitrothion was about

4 days in Moriyama and Katano soils under flooded conditions, as compared to 28 days in Moriyama soil and 22 days in Katano soil under non-flooded conditions. Furthermore, the accumulation of metabolites differed with moisture regime. In non-flooded soils, 3-methyl-4-nitrophenol and carbon dioxide were the major products of decomposition of fenitrothion, while aminofenitrothion was formed as the major metabolite in all flooded soils.

In some flooded soils 50–66% of fenitrothion applied was converted to aminofenitrothion within 7 days. According to a recent study in our laboratory, degradation of fenitrothion in five tropical soils (from India) proceeded more rapidly under flooded conditions than under non-flooded conditions (Adhya *et al.*, 1987). The half-life of fenitrothion in five soils was 3.9, 6.6, 5.6, 5.4, and 10.9 days under flooded conditions, as compared to the corresponding values of 4.4, 10.5, 15.4, 20, and 97 days under non-flooded conditions. As in temperate soils, aminofenitrothion accumulated in three of the five tropical soils under flooded conditions, while 3-methyl-4-nitrophenol was the major metabolite in all non-flooded soils. These incubation studies under well-defined laboratory conditions would suggest identical rates and pathways of degradation of fenitrothion in tropical and temperate soils; but under actual field conditions, its degradation may be faster in the hot environment of the tropics than in cooler climates of the temperate region.

Diazinon is yet another organophosphorus insecticide widely used in rice culture. This insecticide lost its efficacy after regular and intensive use of three or more years in tropical rice paddies of the Philippines (IRRI, 1970) and Bangladesh. Likewise, in a recent bioassay study from India (Rao and Rao, 1984), brown planthoppers were released on potted rice plants at 2 days after one, two, three, four, and five applications of granular diazinon (at 1.5 kg a.i./ha at 30-day intervals) to the soil. The efficacy of diazinon in controlling brown planthoppers decreased progressively with successive applications, the mortality being 100% after the first application, 33% after the second application, and 0% after the third application.

In a study on the mechanism of this phenomenon, diazinon was found to disappear within 3–6 days after its incubation with paddy water from rice fields (three locations in the Philippines) that had been previously treated with diazinon (Sethunathan and Pathak, 1972). But the loss of this insecticide from water from fields with no history of diazinon use was negligible even after 20 days. The diazinon-degrading factor, found in paddy water from diazinon-treated rice fields, was inactivated following autoclaving. Moreover, a *Flavobacterium* sp. ATCC 27551, isolated from paddy water samples that had been previously treated with diazinon, showed an exceptional capacity to hydrolyse diazinon and then metabolize its hydrolysis product, 2-isopropyl-4-methyl-6-hydroxypyrimidine to CO_2 (Sethunathan and Yoshida, 1973c).

Undoubtedly, intensive use of diazinon in Philippine rice fields has eventually led to a build-up of diazinon-hydrolysing soil microbes under actual tropical field conditions. Also, repeated field application to a temperate soil from England led to accelerated degradation of diazinon (Forest *et al.*, 1981). A *Flavobacterium* sp. isolated from this conditioned soil hydrolysed diazinon. This would, at least in part, explain the decreased efficacy of diazinon in rice fields after its repeated applications.

Growing cells (Adhya *et al.*, 1981b) and phosphotriesterases (Brown, 1980) of *Flavobacterium* sp. 27551, isolated from diazinon-treated rice field in Philippines, hydrolysed both diethyl (diazinon, parathion) and dimethyl (methyl parathion and fenitrothion) phosphorothioates. The *Flavobacterium* sp., isolated from temperate soil, hydrolysed diazinon and parathion (Forest *et al.*, 1981). The *Pseudomonas* sp. 29353, isolated from an Indian soil enriched with parathion, hydrolysed both diethyl phosphorothioates (parathion and diazinon), despite the differences in ring moiety, but not the dimethoxy compounds, methyl parathion and fenitrothion. The ability of *Pseudomonas* sp. to hydrolyse parathion and diazinon, but not methyl parathion, suggests that the alkyl constituent attached to phosphorus is more important than the aromatic portion in determining the susceptibility of these phosphorothioates to bacterial hydrolysis. Thus, these versatile bacteria, *Flavobacterium* sp. in particular, isolated from tropical agricultural soils are equipped with powerful hydrolases for decontamination of systems polluted with parathion and related organophosphorus insecticides.

5.2.4 CARBAMATES

Until the mid-1960s, carbamate insecticides were seldom used in India, and probably not in any other tropical countries. In recent years, however, carbamate pesticides are used increasingly in the tropics for controlling specific pests not controlled by organochlorines and organophosphates. For instance, the N-methylcarbamates, carbaryl and carbofuran, are used increasingly as the most effective insecticides against brown planthoppers (*Nilaparvata lugens* Stål.), a major pest of rice in the tropics and subtropics. In India, carbaryl is probably second only to hexachlorocyclohexane in terms of the estimate for pesticide use during the period 1977–82 (Krishna Murti and Dikshith, 1982). Since carbamate use in tropical countries is more recent, the metabolism of carbamate pesticides under tropical conditions is less well understood than that of organochlorines and organophosphates.

Reports on the persistence of carbaryl in flooded soil are contradictory. Carbaryl disappeared completely from the non-sterilized suspension of a Philippine soil within 8 days, as based on UV-absorption spectral analysis of the supernatant of the soil suspension (IRRI, 1966). However, in a flooded alluvial soil from Malaysia, Gill and Yeoh (1980) showed a half-life of about 7 weeks. Carbaryl was more persistent in an acid sulphate soil than in the alluvial soil. The accumulation of the hydrolysis product, 1-naphthol, correlated with the disappearance of carbaryl. There are reports of more rapid (Venkateswarlu *et al.*, 1980; Rajagopal, 1984) degradation of carbaryl in tropical soils under flooded conditions than under non-flooded conditions. In another study (Brahmaprakash, 1984), soil degradation of carbaryl occurred at almost identical rates under both water regimes. Degradation occurred both in sterile and non-sterile soils, but more rapidly in the latter.

There is considerable literature on the metabolism of carbaryl and/or 1-naphthol in cultures of microorganisms isolated from temperate environments and in microbial enzyme systems (Bollag and Liu, 1971; Bollag, 1981). A few instances of degradation of carbaryl in pure and mixed cultures of micro-organisms isolated from tropical soils have been reported. A strain of *Achromobacter* sp. converted carbaryl in a mineral salts medium to 1-naphthol, hydroquinone, catechol, and pyruvate (Sud *et al.*, 1972). Also, this bacterium grew well with all the four metabolities as sole source of carbon.

In a more recent study (Rajagopal *et al.*, 1983; Rajagopal, 1984), suspensions from three (an alluvial, a laterite, and a sodic) tropical flooded soils, that had been previously treated three times with carbaryl or 1-naphthol, effected a more rapid degradation of carbaryl in a mineral salts medium than did the suspensions from the same soils never before exposed to carbaryl. This indicated enrichment of soils with carbaryl-degrading microorganisms following application of not only carbaryl but also its hydrolysis product, 1-naphthol. Also, bacteria isolated from these soil enrichments showed exceptional capacity to degrade carbaryl in a mineral salts medium (Rajagopal *et al.*, 1984c). Interestingly, degradation of carbaryl by soil enrichments and bacterial cultures was more pronounced in the absence than in the presence of ammonium nitrogen, possibly due to the preferential utilization of the inorganic nitrogen by the microorganisms. Evidently, an N-free environment is especially suitable for the proliferation of microorganisms that degrade nitrogenous pesticides, such as carbaryl.

In yet another study (Hirata *et al.*, 1984), repeated applications of carbaryl to grey humic and red yellow latosol increased the rate of degradation of carbaryl, possibly due to a rapid increase in the number of microorganisms using the pesticide as a substrate. Such microbial enrichment after carbaryl application suggests that microbial degradation of carbaryl is of great significance in neutral soils. Evidence suggests some microbial degradation of carbaryl even in an alkaline soil (sodic) where chemical hydrolysis is considerable (Rajagopal *et al.*, 1983).

Enrichment cultures and pure cultures of bacteria from carbaryl-amended soils degraded not only carbaryl but also carbofuran with almost equal ability. The degradation of carbaryl by all soil enrichments and bacterial cultures proceeded essentially by hydrolysis to 1-naphthol. Within 5 days of incubation with soil enrichment cultures, 1-naphthol and an unidentified metabolite were formed from carbaryl (Rajagopal, 1984). 1-Naphthol was metabolized further, but slowly, to 1,4-naphthoquinone by a *Bacillus* sp. Conversion of ring ^{14}C in carbaryl to $^{14}CO_2$ or other volatiles by enrichment cultures and pure cultures of bacteria (Rajagopal *et al.*, 1984c) was negligible, even after 40 days of incubation. Consequently, most of the ring ^{14}C in carbaryl accumulated in the medium.

In recent years, there has been considerable concern over the environmental hazard from the intensive use of carbofuran in flooded rice paddies in the tropics and subtropics. Carbofuran degradation in some tropical soils was faster under flooded conditions than under non-flooded conditions (Venkateswarlu *et al.*, 1977; Li and Wong, 1980; Rajagopal, 1984). Furthermore, the degradation of carbofuran in flooded soils was more rapid under undisturbed conditions (predominantly anaerobic) than under aerobic conditions provided by shaking (Venkateswarlu and Sethunathan, 1978). Under continued anaerobiosis of undisturbed flooded soils, the hydrolysis product, carbofuran phenol, accumulated, but when the undisturbed soil was returned to an aerobic condition, carbofuran phenol decreased rapidly. The addition of rice straw also accelerated the hydrolysis of carbofuran to carbofuran phenol in predominantly anaerobic flooded soils (Venkateswarlu and Sethunathan, 1979).

Fluctuations in aerobic status and organic amendments alter the microbial activities in the soil and thereby influence the persistence of soil-applied carbaryl, carbofuran and their metabolites. The degradation of these carbamates in all soils proceeded by hydrolysis to 1-naphthol and carbofuran phenol, respectively, which showed recalcitrance to further degradation. Also, evolution of $^{14}CO_2$ from the ring ^{14}C in carbaryl and carbofuran (Venkateswarlu and Sethunathan, 1979; Rajagopal and Sethunathan, 1984) was negligible (less than 0.3% of the applied ring ^{14}C) in 40 days; during this period, 27% of the carbonyl ^{14}C was evolved as $^{14}CO_2$. Most of the ring ^{14}C in carbaryl and carbofuran accumulated in the soil as the hydrolysis products plus soil-bound residues (Rajagopal and Sethunathan, 1984).

In recent years, there has been increasing concern over the decreased efficacy of carbofuran in controlling insect pests in cornfields of temperate countries, viz. USA and Canada (Fox, 1983) and in flooded rice paddies of the Philippines in the tropics (IRRI, 1977). Two main groups of pesticides were affected by decreased efficacy in cornfields of the USA: carbofuran insecticides and thiocarbamate herbicides. This problem was aggravated by repeated use of the same pesticide in successive years. Research in temperate countries provided convincing evidence of accelerated degradation of carbofuran on its incubation with soil samples from fields with a 2–4 year history of continuous carbofuran use (Kaufman, 1983). Interestingly, accelerated degradation of carbofuran occurred in some, but not all, re-treated soil samples. Read (1983) found that carbofuran was degraded 600–1000 times faster in soil samples from Canadian cornfields previously treated with this insecticide than in soil samples from fields with no history of carbofuran use. Likewise, Felsot *et al.* (1981) reported enhanced degradation of carbofuran in soil samples from cornfields in Illinois with a three-year history of continuous carbofuran use and reported a problem of a decreased efficacy of carbofuran.

In another study with a humic Mesisol from Canada, the decomposition of carbofuran was faster in re-treated soil than in soil with no history of carbofuran use (Greenhalgh and Belanger, 1981). No appreciable degradation of carbofuran occurred in 90 days in soil samples that had been sterilized before its addition; but where these sterile soil samples were inoculated with soils with a history of carbofuran use and then fortified with carbofuran, the insecticide disappeared almost completely in 30 days (Felsot *et al.*, 1981). Also, a *Pseudomonas* sp., isolated from re-treated soil, effectively degraded the insecticide in 30 days. Likewise, Read (1983) found that soil from re-treated cornfields lost its ability to degrade carbofuran when sterilized before the addition of the insecticide. According to these observations, accelerated degradation of carbofuran in re-treated soils was caused by soil microorganisms.

Evidence for a microbial role in the decreased efficacy of carbofuran in tropical rice fields is not conclusive. The persistence curves of carbofuran in a soil with and without a history of continuous carbofuran use were almost identical (Siddaramappa *et al.*, 1978; Venkateswarlu and Sethunathan, 1978). In another study (Rajagopal, 1984), carbofuran, applied to a soil, decreased to 14, 8, and 5% of the original level 40 days after one, two, and three applications, respectively. This slight increase in the rate of degradation of carbofuran after successive applications to the flooded soil does not suggest a distinct build-up of carbofuran-degrading microorganisms in the re-treated soil. However, when suspensions from the re-treated (viz. after three applications of carbofuran) and untreated soils were added to a mineral solution supplemented with carbofuran, the suspension from re-treated soil effected more rapid degradation of the insecticide than the suspension from the untreated soil (Rajagopal *et al.*, 1984b). Also, microorganisms isolated from these soil enrichment cultures were capable of degrading the insecticide (Rajagopal *et al.*, 1984b). This would suggest enrichment of microorganisms capable of degrading carbofuran after repeated applications of carbofuran or carbofuran phenol. It is not then clear why the degradation of carbofuran was not distinctly accelerated in the tropical soil after its repeated applications despite enrichment of carbofuran-degrading microorganisms.

The decreased efficacy of carbofuran insecticides in cornfields of the USA is analogous to the decreased efficacy of diazinon in controlling brown plant-hoppers in tropical rice fields of the Philippines, attributed, at least in part, to the build-up of diazinon-hydrolysing bacterium, *Flavobacterium* sp. ATCC 27551, in particular after repeated applications of the insecticide for seven consecutive seasons (Sethunathan, 1972; Sethunathan and Pathak, 1972; Sethunathan and Yoshida, 1973c). Carbamate and organophosphorus pesticides are short-lived and their dissipation is further hastened in a tropical environment.

This would necessitate more frequent application of these insecticides for effective pest control. However, such repeated applications, would, in turn, hasten the build-up of soil microbes that undermine the efficacy of these pesticides. Thus, the problem of reduced efficacy of pesticides associated with their rapid destruction by microorganisms may be more serious and widespread with short-lived carbamates and organophosphates than with more persistent pesticides.

5.2.5 RING CLEAVAGE REACTIONS

There is considerable evidence to indicate that ring cleavage reactions are retarded in predominantly anaerobic flooded soil not planted to rice. Less than 1–2% of the originally applied ring ^{14}C in HCH, parathion, diazinon, carbaryl, and carbofuran were evolved as $^{14}CO_2$ in 30–60 days from a flooded soil not planted to rice despite almost complete disappearance of the parent molecules (Sethunathan and Siddaramappa, 1978). Ring cleavage is an aerobic process mediated essentially by microorganisms. Increased soil aeration and intense microbial activity in the rhizosphere may accelerate ring cleavage reactions in a flooded soil planted to rice. Thus, in unplanted soil, less than 5.5% of the ring ^{14}C in parathion was evolved in 15 days under both flooded and non-flooded conditions (Reddy and Sethunathan, 1983a, 1983b). In soil planted with rice, 9.2% of the radio-carbon was evolved under non-flooded conditions and 22.6% under flooded conditions. Evidently, flooded soil planted with rice permits significant ring cleavage.

But not all pesticides undergo significant ring cleavage in flooded soil planted to rice. Evolution of $^{14}CO_2$ from ring ^{14}C in HCH, carbaryl, and carbofuran was negligible (less than 2% of original level) both in planted (to rice) and unplanted soils under flooded conditions (Brahmaprakash *et al.*, 1985; Figure 5.2.2). Despite slow ring cleavage, accumulation of soil-bound residues from gamma-HCH was negligible (less than 2.5% of original level), possibly because of the immediate volatilization of its degradation products, gamma-PCCH and gamma-TCCH, which are more volatile than gamma-HCH. Slow ring cleavage of carbaryl and carbofuran led to significant accumulation of ^{14}C as their hydrolysis products (1-naphthol or carbofuran phenol) and soil-bound residues both in planted and unplanted soils under flooded conditions (Brahmaprakash and Sethunathan, 1985).

5.2.6 CONCLUSIONS

We do not have comparative data for many pesticides on their persistence in temperate versus tropical soil environments. Even in limited instances, as for example HCH, where data have been generated by independent studies with tropical and temperate soils, a meaningful comparison is seldom possible,

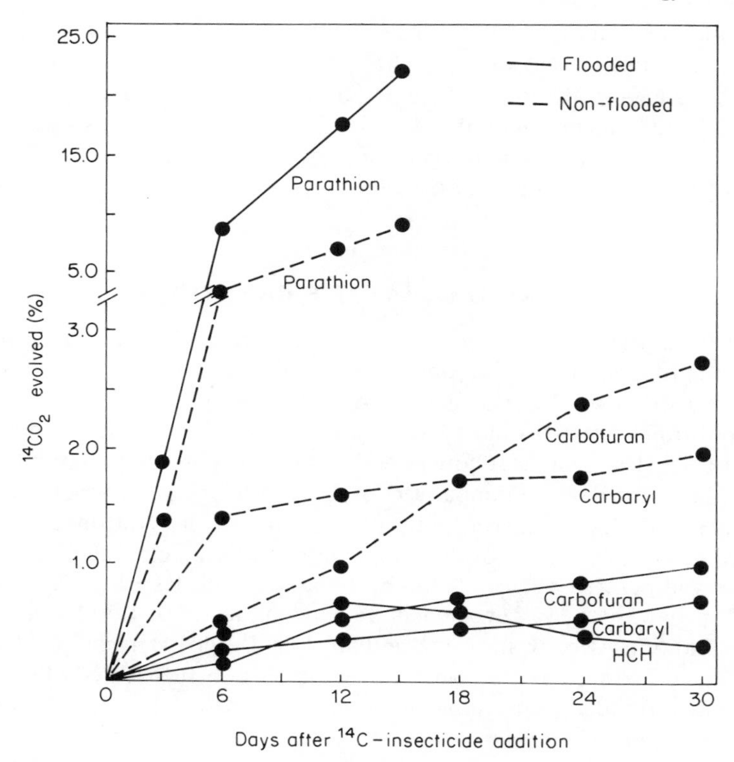

Figure 5.2.2 Cumulative $^{14}CO_2$ (percentage of the added ring ^{14}C) evolved from ring-labelled parathion (Reddy and Sethunathan, 1983a), hexachlorocyclohexane (HCH), carbofuran, and carbaryl (Brahmaprakash and Sethunathan, 1985; Brahmaprakash *et al.*, 1985) in a soil planted to rice under flooded and non-flooded conditions (Reproduced by permission of Elsevier Scientific Publishers B.V.)

because of several variables, such as soil type and experimental conditions. However, on the basis of the circumstantial evidence from soil persistence studies at different temperatures under laboratory conditions, most pesticides would undergo fairly rapid degradation under the high temperatures encountered in the tropics. There are also several reports of the rapid disappearance of several pesticides applied to tropical and subtropical agricultural soils under actual field conditions (Talekar *et al.*, 1977, 1983a, 1983b); but in most of these field studies only parent pesticide molecules were monitored.

Conclusions based on the monitoring of only parent pesticide molecules and not their degradation products can be misleading, since especially under hot and humid conditions of the tropics, substantial loss of pesticides can occur through phenomena like volatilization, lateral or vertical transport, and photodecomposition, besides microbial and chemical degradation. Available evidence indicates that degradation pathways of pesticides may be identical in tropical and temperate conditions, but the rate of these degradation processes may be faster in the tropics.

For a meaningful generalization of pesticide behaviour in tropical versus temperate conditions, there is an urgent need to generate data on the fate of more pesticides in the tropical soil environment.

5.2.7 REFERENCES

Adhya, T. K., Sudhakar Barik, and Sethunathan, N. (1981a). Stability of commercial formulation of fenitrothion, methyl parathion, and parathion in anaerobic soils. *J. Agric. Fd. Chem.*, **29**, 90–93.

Adhya, T. K., Sudhakar Barik, and Sethunathan, N. (1981b). Hydrolysis of selected organophosphorus insecticides by two bacteria isolated from flooded soil. *J. Appl. Bact.*, **50**, 167–172.

Adhya, T. K., Wahid, P. A., and Sethunathan, N. (1987). Persistence and bidegradation of selected organophosphorus insecticides in flooded versus non-flooded soils. *Biol. Fert. Soils*, **5**, 36–40.

Anonymous (1984). *Pesticides Info.*, **10**, 32.

Bollag, J. M. (1981). Oxidative coupling of aromatic compounds by enzymes from soil micro-organisms. In: Paul, E. A., and Ladd, J. N. (eds), *Soil Biochemistry*, pp. 113–152, vol. 5. Marcel Dekker Inc. New York.

Bollag, J. M., and Liu, S. Y. (1971). Degradation of Sevin by soil micro-organisms. *Soil Biol. Biochem.*, **3**, 337–345.

Brahmaprakash, G. P. (1984). *Relative Persistence of Pesticides in Rice Soils*. Ph.D. Thesis, Utkal University, Bhubaneswar.

Brahmaprakash, G. P., and Sethunathan, N. (1985). Metabolism of carbaryl and carbofuran in soil planted to rice. *Agric. Ecosystems Environ.*, **13**, 33–42.

Brahmaprakash, G. P., Reddy, B. R., and Sethunathan, N. (1985). Persistence of hexachlorocyclohexane isomers in soils planted to rice and by rhizosphere soil suspensions. *Biol. Fert. Soils*, **1**, 103–109.

Brown, K. A. (1980). Phosphotriesterases of *Flavobacterium* sp. *Soil Biol. Biochem.*, **12**, 105–112.

Castro, T. F., and Yoshida, T. (1971). Degradation of organochlorine insecticides in flooded soils in the Philippines. *J. Agric. Fd. Chem.*, **19**, 1168–1170.

Castro, T. F., and Yoshida, T. (1974). Effect of organic matter on the biodegradation of some organochlorine insecticides in submerged soils. *Soil Sci. Plant Nutr.*, **20**, 363–370.

Cho, D. Y., and Ponnamperuma, F. N. (1971). Influence of soil temperature on the chemical kinetics of flooded soils and the growth of rice. *Soil Sci.*, **112**, 184–194.

Felsot, A., Maddox, J. V., and Bruce, W. (1981). Enhanced microbial degradation of carbofuran in soils with histories of Furadan® use. *Bull. Environ. Contam. Toxicol.*, **26**, 781–788.

Ferreira, J., and Raghu, K. (1981). Decontamination of hexachlorocyclohexane isomers in soil by green manure application. *Environ. Tech. Lett.*, **2**, 357.

Ferris, I. G., and Lichtenstein, E. P. (1980). Interactions between agricultural chemicals and soil microflora and their effects on the degradation of ^{14}C parathion in a cranberry soil. *J. Agric. Fd. Chem.*, **28**, 1011–1019.

Forest, M., Lord, K. A., Walker, N., and Woodville, H. C. (1981). The influence of soil treatment on the bacterial degradation of diazinon and other organophosphorus insecticides. *Environ. Pollut. Ser. A*, **24**, 93–104.

Fox, J. L. (1983). Soil microbes pose problems for pesticides. *Science*, **221**, 1029–1031.

Gambrell, R. P., and Patrick, W. H. Jr. (1978). Chemical and microbiological properties of anaerobic soils and sediments. In: Hook, D. D., and Crawford, R. N. (eds), *Plant Life in Anaerobic Environments*, pp. 375–423. Ann Arbor, Michigan, U.S.A.

Gill, S. S., and Yeoh, C. L. (1980). Degradation of carbaryl in three components of paddy-field ecosystem of Malaysia. In: *Agrochemical Residue-Biota Interactions in Soil and Aquatic Ecosystems*, pp. 229–243. International Atomic Energy Agency, Vienna.

Gowda, T. K. S., and Sethunathan, N. (1976). Persistence of endrin in Indian rice soils under flooded conditions. *J. Agric. Fd. Chem.*, **24**, 750–753.

Gowda, T. K. S., and Sethunathan, N. (1977). Endrin decomposition in soils as influenced by aerobic and anaerobic conditions. *Soil Sci.*, **124**, 5–9.

Greenhalgh, R., and Belanger, A. (1981). Persistence and uptake of carbofuran in a humic Mesisol and the effects of drying and storing soil samples on residue levels. *J. Agric. Fd. Chem.*, **29**, 231–235.

Guenzi, W. D., and Beard, W. E. (1968). Anaerobic conversion of DDT to DDD and aerobic stability of DDT in soil. *Soil Sci. Soc. Am. Proc.*, **32**, 522–524.

Heritage, A. D., and MacRae, I. C. (1979). Degradation of hexachlorocyclohexane and structurally related substrates by *Clostridium sphenoides*. *Austr. J. Biol. Sci.*, **32**, 493–496.

Hirata, R., Luchini, L. C., Mesquita, T. B., and Ruegg, E. F. (1984). Effect of repeated applications of ^{14}C-carbaryl and of addition of glucose and cellulose to soil samples. *Pesquista Agropecuaria Brasileira*, **19C**, 79–84.

IRRI (1966). Annual Report for 1965, pp. 126–127. International Rice Research Institute, Los Banos, Philippines.

IRRI (1967). Annual Report for 1966, pp. 199–201. International Rice Research Institute, Los Banos, Philippines.

IRRI (1970). Annual Report for 1969, pp. 241–242. International Rice Research Institute, Los Banos, Philippines.

IRRI (1977). Annual Report 1976, pp. 204–207. International Rice Research Institute, Los Banos, Philippines.

Katan, J., Fuhremann, T. W., and Lichtenstein, E. P. (1976). Binding of ^{14}C-parathion in soil: A reassessment of pesticide persistence. *Science*, **193**, 891–894.

Kaufman, D. D. (1983). In: *Workshop on Problems and Progress in Enhanced Biodegradation of Agricultural Chemicals*. Dept. Agric., Beltsville, Maryland, U.S.A. (Cited by Fox, 1983.)

Krishna Murti, C. R., and Dikshith, T. S. S. (1982). Application of biodegradable pesticides in India. In: Matsumura F., and Krishna Murti, C.R. (eds), *Biodegradation of Pesticides*, pp. 257–305. Plenum Press, New York.

Li, G. C., and Wong, S. S. (1980). Distribution and accumulation of carbofuran in a rice-paddy model ecosystem. *Chih. Wu Pao Hu Hsueh Hui Hui K'an*, **22**, 337–345.

MacRae, I. C., Raghu, K., and Bautista, E. M. (1969). Anaerobic degradation of the insecticide lindane by *Clostridium* sp. *Nature*, **221**, 859.

Miyamoto, J. (1977). Degradation of fenitrothion in terrestrial and aquatic environments including photolytic and microbial reactions. In: Roberts, J. R., Greenhalgh, R., and Marshall, W. K. (eds), *Fenitrothion: The Long-term Effects of its Use in Forest Ecosystems*, pp. 105–134. NRC Assoc. Cttee on Scientific Criteria for Environ. Quality, Ottawa, Canada.

Mrinalini, N. (1983). Structure and growth of the pesticides industry in India. *Pesticides*, March, pp. 3–9.

Munnecke, D. M., Johnson, L. M., Talbat, H. W., and Barik, S. (1982). Microbial metabolism of enzymology of selected pesticides. In: Chakrabarty, A.M. (ed.), *Biodegradation and Detoxification of Environmental Pollutants*, pp. 1–32. CRC Press, USA.

Ponnamperuma, F. N. (1972). The chemistry of submerged soils. *Adv. Agron.*, **24**, 29–66.

Raghu, K., and MacRae, I. C. (1966). Biodegradation of the gamma-isomer of benzenehexachloride in submerged soils. *Science*, **154**, 262–264.

Rajagopal, B. S. (1984). *Microbial Metabolism of Carbamate Insecticides in Rice Soils.* Ph.D. Thesis, Utkal University, Bhubaneswar.

Rajagopal, B. S., and Sethunathan, N. (1984). Influence of nitrogen fertilizers on the persistence of carbaryl and carbofuran in flooded soils. *Pestic. Sci.*, **15**, 591–599.

Rajagopal, B. S., Chendrayan, K., Reddy, B. R., and Sethunathan, N. (1983). Persistence of carbaryl in flooded soils and its degradation by soil enrichment cultures. *Plant Soil*, **73**, 35–45.

Rajagopal, B. S., Brahmaprakash, G. P., Reddy, B. R., Singh, U. D., and Sethunathan, N. (1984a). Effect of persistence of selected carbamate pesticides in soil. *Residue Rev.*, **93**, 1–200.

Rajagopal, B. S., Brahmaprakash, G. P., and Sethunathan, N. (1984b). Degradation of carbofuran by enrichment cultures and pure cultures of bacteria from flooded soils. *Environ. Pollut.*, **36A**, 61–64.

Rajagopal, B. S., Rao, V. R., Nagendrappa, G., and Sethunathan, N. (1984c). Metabolism of carbaryl and carbofuran by soil enrichment and bacterial cultures. *Can. J. Microbiol.*, **31**, 1458–1466.

Rao, P. R. M., and Rao, P. S. P. (1984). Note on the effects of repeated soil applications of diazinon and carbofuran. *Pesticides*, **28** (12), 57–58.

Read, D. C. (1983). Enhanced microbial degradation of carbofuran and fensulfothion after repeated applications to acid mineral soil. *Agric. Ecosystems Environ.*, **10**, 37–46.

Reddy, B. R., and Sethunathan, N. (1983a). Mineralization of parathion in the rice rhizosphere. *Appl. Environ. Microbiol.*, **45**, 826–829.

Reddy, B. R., and Sethunathan, N. (1983b). Mineralization of parathion in the rhizosphere of rice and pearl millet. *J. Agric. Fd. Chem.*, **31**, 1379–1381.

Sethunathan, N. (1972). Diazinon degradation in submerged soil and rice-paddy water. In: *Fate of Organic Pesticides in Aquatic Environments*, pp. 244–255. Adv. Chem. Series, **111**.

Sethunathan, N. (1973). Microbial degradation of insecticides in flooded soil and in anaerobic cultures. *Residue Rev.*, **47**, 143–165.

Sethunathan, N., and Pathak, M. D. (1972). Increased biological hydrolysis of diazinon after repeated application in rice paddies. *J. Agric. Fd. Chem.*, **20**, 586–589.

Sethunathan, N., and Yoshida, T. (1973a). Degradation of chlorinated hydrocarbons by *Clostridium* sp. isolated from lindane-amended flooded soil. *Plant Soil*, **38**, 663–666.

Sethunathan, N., and Yoshida, T. (1973b). Parathion degradation in submerged rice soils in the Philippines. *J. Agric. Fd Chem.*, **21**, 504–506.

Sethunathan, N., and Yoshida, T. (1973c). A *Flavobacterium* sp. that degrades diazinon and parathion as sole carbon sources. *Can. J. Microbiol.*, **19**, 873–875.

Sethunathan, N., and Siddaramappa, R. (1978). Microbial degradation of pesticides in rice soils. In: *Rice and Soils*, pp. 479–497. International Rice Research Institute, Los Banos, Philippines.

Sethunathan, N., Bautista, E. M., and Yoshida, T. (1969). Degradation of benzene hexachloride by a soil bacterium. *Can. J. Microbiol.*, **15**, 1349–1354.

Sethunathan, N., Siddaramappa, R., Rajaram, K. P., Barik, S., and Wahid, P. A. (1977). Parathion: residues in soil and water. *Residue Rev.*, **68**, 91–122.

Sethunathan, N., Adhya, T. K., and Raghu, K. (1982). Microbial degradation of pesticides in tropical soils. In: Matsumura F., and Krishna Murti, C. R. (eds), *Biodegradation of Pesticides*, pp. 91–115. Plenum Press, New York.

Sethunathan, N., Rao, V. R., Raghu, K., and Adhya, T. K. (1983). Microbiology of rice soils. *CRC Critical Rev. Microbiol.*, **10**, 125–172.

Siddaramappa, R., and Sethunathan, N. (1975). Persistence of gamma-BHC and beta-BHC in Indian rice soils under flooded conditions. *Pestic. Sci.*, **6**, 395–403.

Siddaramappa, R., Rajaram, K. P., and Sethunathan, N. (1973). Degradation of parathion by bacteria isolated from flooded soil. *Appl. Microbiol.*, **26**, 846–849.

Siddaramappa, R., Tirol, A., Seiber, J. N., Heinrichs, E. A., and Watanabe, I. (1978). The degradation of carbofuran in paddy water and flooded soil of untreated and retreated rice fields. *J. Environ. Sci. Health.*, **13B**, 369–380.

Sud, R. K., Sud, A. K., and Gupta, K. G. (1972). Degradation of sevin (1-naphthyl N-methylcarbamate) by *Achromobacter* sp. *Arch. Mikrobiol.*, **87**, 353–358.

Sudhakar Barik, Siddaramappa, R., and Sethunathan, N. (1976). Metabolism of nitrophenols by bacteria isolated from parathion-amended flooded soil. *Antonie van Leeuwenhoek*, **42**, 461–470.

Sudhakar Barik, Wahid, P. A., Ramakrishna, C., and Sethunathan, N. (1979). A change in the degradation pathway of parathion after repeated applications to flooded soil. *J. Agric. Fd. Chem.*, **27**, 1391–1392.

Takimoto, Y., Hirota, M., Inui, H., and Miyamoto, J. (1976). Decomposition and leaching of radioactive Sumithion in 4 different soils under laboratory conditions. *J. Pestic. Sci.*, **1**, 131–143.

Talekar, N. S., Sun, L. T., Lee, E. M., and Chen, J. S. (1977). Persistence of some insecticides in subtropical soil. *J. Agric. Fd. Chem.*, **25**, 348–352.

Talekar, N. S., Chen, J. S., and Kao, H. T. (1983a). Persistence of fenvalerate in subtropical soil. *J. Econ. Entomol.*, **76**, 223–226.

Talekar, N. S., Kao, H. T., and Chen, J. S. (1983b). Persistence of selected insecticides in subtropical soil after repeated biweekly applications over two years. *J. Econ. Entomol.*, **76**, 711–716.

Tsukano, Y. (1973). Factors affecting disappearance of BHC isomers from rice field soil. *JARC*, **7**, 93–97.

Venkateswarlu, K., and Sethunathan, N. (1978). Degradation of carbofuran in rice soils as influenced by repeated application and exposure to aerobic conditions following anaerobiosis. *J. Agric. Fd. Chem.*, **26**, 1148–1151.

Venkateswarlu, K., and Sethunathan, N. (1979). Metabolism of carbofuran in rice straw-amended and unamended rice soils. *J. Environ. Qual.*, **8**, 365–368.

Venkateswarlu, K., Gowda, T. K. S., and Sethunathan, N. (1977). Persistence and biodegradation of carbofuran in flooded soils. *J. Agric. Fd. Chem.*, **25**, 553–556.

Venkateswarlu, K., Chendrayan, K., and Sethunathan, N. (1980). Persistence and biodegradation of carbaryl in soils. *J. Environ. Sci. Health*, **15B**, 421–429.

Yoshida, T., and Castro, T. F. (1970). Degradation of gamma-BHC in rice soils. *Soil Sci. Soc. Am. Proc.*, **34**, 440–442.

Yoshida, T., and Yamaya, Y. (1984). Microbial degradation of insecticide gamma-hexachlorocyclohexane (gamma-HCH) in soils. *Jap. J. Soil Sci. Plant Nutr.*, **55**, 97–102.

Ecotoxicology and Climate
Edited by P. Bourdeau, J. A. Haines, W. Klein and C. R. Krishna Murti

5.3 Effects of Insecticides in Rice Ecosystems in Southeast Asia

E. D. MAGALLONA

5.3.1 INTRODUCTION

Rice is an important crop worldwide, with about 144.1 million hectares being devoted to its production and about 50% of the global population being rice eaters. Of this area, more than 90% is in Asia (Medrana, 1983). Southeast Asia, which is composed of Burma, Indonesia, Cambodia, Laos, Malaysia, Philippines, Thailand, and Vietnam, produces 20% of the world's total, with a production of 77 122 000 tonnes.

The predominant production system for lowland rice is the paddy or flooded system. In the Philippines, for example, out of the total 3.5 million hectares used in production, the upland or dry system, which relies on rain, is used on only 0.4 million hectares (Mabbayad *et al.*, 1983). The rest is either irrigated lowland (65%) or rainfed lowland (35%).

Unfortunately, just like any other crop, the rice production system is also infested by a host of pests—insects, weeds, diseases, rodents, etc. They exact a heavy toll on crop production efforts. Sanchez (1983) estimates that this amounts to 8.4 million tonnes for the Philippines in 1983, while Reddy (1978) mentioned 15–30% of production potential in the Asian and Pacific region for crops in general. In an area generally regarded as having an agricultural economy in deficit, these losses become all the more significant.

To solve these pest problems, pesticides have been resorted to and it appears that their use will continue, at least in the near future, notwithstanding increased interest in integrated pest management; furthermore, integrated pest management does not preclude pesticide usage. Thus, in the Philippines, Magallona and Mercado (1978) pointed out that 30% of the area devoted to rice was treated with insecticides. Subsequently, Antazo and Magallona (1982) reported that in a government survey of five major rice-producing areas, there was 100% use of pesticides in farming. Most farmers apply pesticides twice during the cropping season while others use them three times or more. In 1984, 39% of the pesticide volume went to rice production. Staring (1984) also mentioned that in India, rice accounts for 29.5% of the value of pesticides while in Indonesia the rice farmers are the main consumers of subsidized insecticides, with about 80–90% of the total being used in this crop. Malaysia also used 28.5% of the 1980 insecticide market in rice, and Thailand reported that rice accounted for 35% of all insecticides consumed in 1980.

Given this widespread insecticide utilization in rice production and considering that pesticides are largely misunderstood, the inevitable question is, 'What are the biological effects of these compounds on components of the ecosystem?'. This is a legitimate concern which we try to address here. However, the coverage is mainly on insecticides as (1) they are the most widely used, (2) they are generally conceded to have the most dramatic environmental impact, and (3) there are more published reports on the environmental impact of this particular group among the pesticides.

5.3.2 INSECTICIDE USAGE

5.3.2.1 Evolution of Insecticide Usage

Insecticide use in rice production essentially followed the general trend for other crops, which was from organochlorines to organophosphates to carbamates to pyrethroids. Currently, the insect growth regulators (IGRs), which exert insecticidal activity through inhibition of chitin synthesis, are a very promising group of compounds. Some interest is also generated in pesticides derived from plants, especially those from all seeds of neem and related plants.

The organochlorines, with the possible exception of endosulfan and lindane, have made their exit from the overall pesticide picture. This situation, however, is not a consequence of the adverse effects on components of the paddy system, but rather because of higher biological efficacy of the newer compounds. In a sense too, the sanctions against these organochlorides meted out by developed countries were adopted by tropical developing countries without studies on their relevance to the different agroecosystem setting. However, in the case of BHC, preference is placed on the higher purity materials, preferably lindane, the 99.9% gamma-isomer, instead of the approximately 16% isomer previously available. The main reasons for this shift have been the problems, recognized in temperate countries, which are associated with the impurities.

On the other hand, technical BHC, the term used for the lower purity material, is still much cheaper than the purified materials for the paddy rice farmer and is thus more affordable. For this reason and in a display of its independence from the line of thinking that what is good for the developed countries should be automatically adopted by developing countries, India has continued to manufacture and use the technical material (Rajak, 1982). This action is also based on the finding that in the tropics, especially with flooded lowland rice

Table 5.3.1 Major insecticides recommended against stemborers and hoppers in selected Asian countries*

PEST/INSECTICIDE	Bangladesh	India	Indonesia	Malaysia	Philippines	Sri Lanka	Thailand
STEMBORER							
Carbofuran	–	+	+	+	+	+	+
Diazinon	+	–	+	+	+	+	–
Gamma-BHC	–	+	–	+	+	–	–
Endosulfan	–	–	+	+	+	+	–
GREEN LEAFHOPPER (*Nephotettix* spp.)							
BPMC	–	–	–	+	+	–	+
Carbaryl	–	–	+	+	+	–	+
Carbofuran	–	+	+	+	+	+	+
Isoprocarb (MIPC)	–	–	–	+	+	–	+
BROWN PLANTHOPPER (*Nilaparvata lugens*)							
BPMC	–	–	+	+	+	–	+
Carbaryl	+	+	+	+	+	+	–
Carbofuran	+	+	+	–	+	–	–
Diazinon	–	+	+	–	+	–	–
Isoprocarb (MIPC)	–	–	+	–	+	–	+
Monocrotophos	+	–	+	–	+	–	+

*Compiled from Antazo and Magallona (1982); Amin *et al.* (1982); Malik and Khan (1982); Partoatmodjo and Alimoeso (1982); Peries (1982); Rajak (1982); Rumakon *et al.* (1982); URARTIP (1985).

production, lindane does not persist; *see* Sethunathan (1973) for an excellent review.

Organophosphates and carbamates feature prominently in the insecticide recommendations in Asian countries as seen in Table 5.3.1. The carbamates are especially useful against the hoppers, while the organophosphates' role could be traced to their being more numerous so that at least a few compounds are bound to be effective against some major insect pest. Of the carbamates, carbofuran appears to be the most widely recommended as a granule, while diazinon and monocrotophos are the organophosphates' counterparts.

The pyrethroids did not catch on in paddy rice crop protection mainly because of the resurgence problem associated with the brownplanthopper (Chelliah and Heinrichs, 1984). However, in the Philippines, cypermethrin has been registered for rice. Furthermore, Stephenson *et al.* (1984) showed that the environmental hazard from the use of this compound in rice is low. The ICRs and the botanical pesticides appear to be more promising at the moment but it remains to be seen if they will have a significant position in the future.

5.3.2.2 Rice Production and Use of Insecticides

To better understand the biological effects of insecticides, it is very important to look at when and how these compounds are applied in rice production. This pertains not only to effects on components of a rice ecosystem or organisms that by chance or design enter such an ecosystem, but also to aspects such as pollution which affect other organisms in the environment. In particular, this refers to paddy water that is intentionally drained as the rice grains start to ripen.

The growth of rice can be divided into four stages: (a) seedling stage; (b) vegetative stage; (c) reproductive stage; and (d) ripening stage. These growth stages and the insect pests usually present are shown in Figure 5.3.1. Throughout most of the growing season, water is conserved in the paddy. At the ripening stage the paddy may be drained, slightly during the milky dough stage, and then fully at the yellow ripe to full ripe stage; the latter stage is usually two weeks before harvest. This is because non-draining prolongs ripening of the grains and makes harvesting difficult.

Thus insecticide application against specific pests is carried out at the following general intervals before harvest:

seedling pests—100 days
whorl maggot—75 days
stemborers—15–100 days
hoppers—15–100 days
rice bug—15–100 days

That insecticide application is normally directed against the major pests, e.g. the seedling pests, whorl maggots, stemborers, and hoppers, means that there

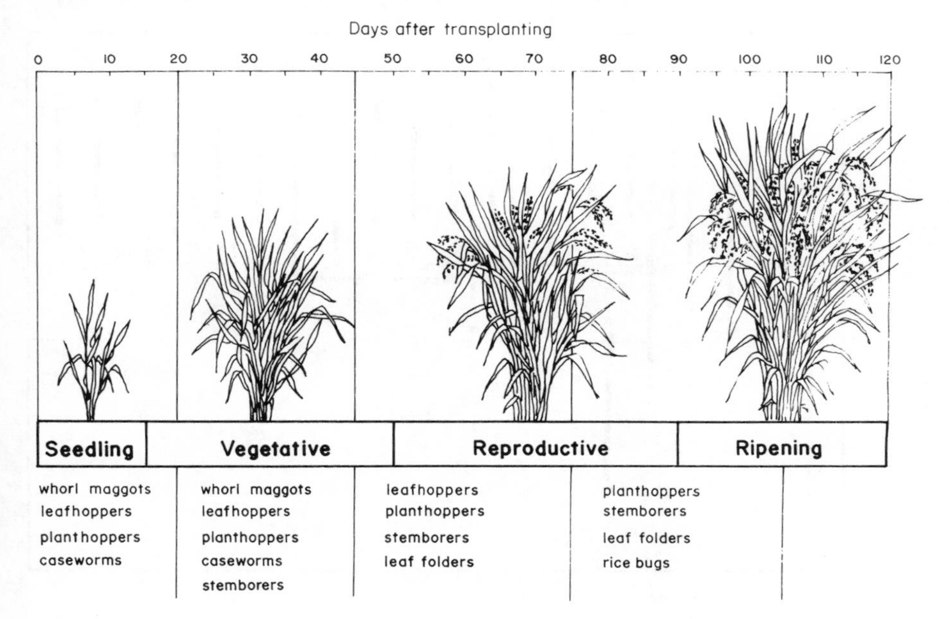

Figure 5.3.1 Representative growth stages of the rice plant showing the associated major insect pests

is an interval of at least 20–30 days between last application and harvest, or 10–20 days from last application to drainage of pond water. This is because insecticides to control the hoppers and the borers are not necessary late in the growing period, even in the presence of the pests, short of infestation levels.

During the interval between application and drainage (or harvest), the insecticide is subject to dilution, degradation, transfer, and other factors. The small amounts that may be left at the time of draining the paddy could be transported to other bodies of water (rivers, lakes) where they could be taken up by flora and/or fauna or could settle in muds. Any effect of the pesticide depends primarily on its concentration in the substrate and the inherent susceptibility of the flora or fauna concerned.

In view of this, it is important to understand the effect of the above factors on the applied insecticide, especially under tropical conditions.

5.3.3 FATE OF PESTICIDES

5.3.3.1 The Tropical Ecosystem

On a comparative basis, the tropical ecosystem is generally conceded to result in lower pesticide persistence than the cooler temperate system. Sunlight is more

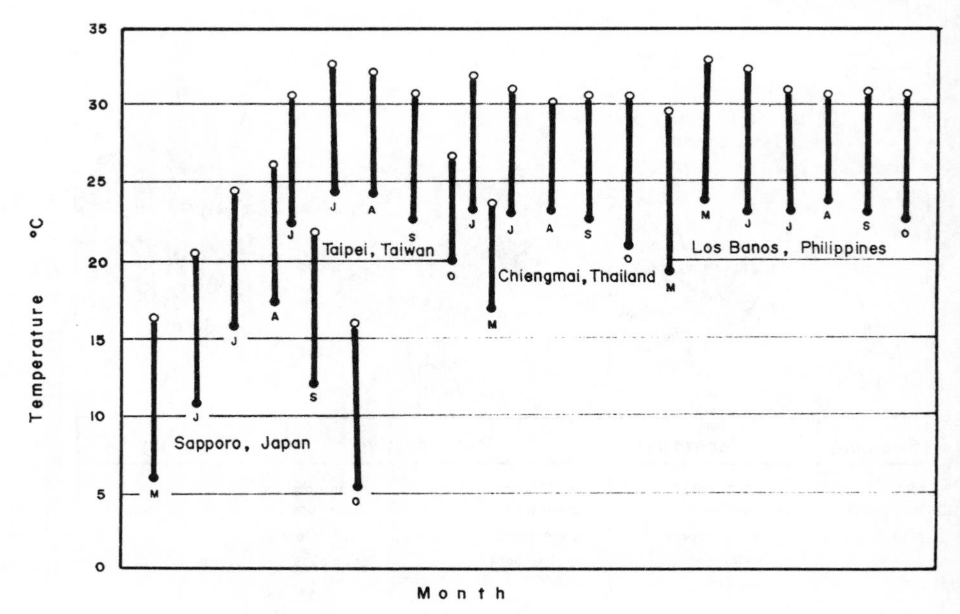

Figure 5.3.2 Mean temperatures and temperature ranges during the main rice-growing months at four locations in Asia (after Rice Production Manual, Philippines, 1969)

intense, causing not only direct photolytic effects but also such related conditions as warmer days and nights, more rapid reactions because of elevated temperature, greater volatilization and codistillation. As seen in Figure 5.3.2 for example, Los Banos, Philippines, would have a mean temperature fluctuating between about 26.5 to 28.5°C, whereas Sapporo, Japan, would have a maximum of about 21.5° and lows of about 11.5°C. Consequently, insecticide degradation in the strictly chemical sense is expected to be faster in Los Banos than in Sapporo. Support of this contention can be seen in the work of Mikami *et al.* (1980). They showed that with fenvalerate, photodegradation ranged from about 4 days in summer to 13–15 days in winter. They further calculated that at 40°N latitude, the half-life for disappearance was 4.1 and 12.4 days in summer and winter, respectively. The role of sunlight was further demonstrated by the half-life of 2, 3, and 18 days in Kudaira light clay, Azuchi sandy clay loam, and Katano sandy loam soils on exposure to sunlight, versus 55–83% remaining in these soils after 20 days without sunlight.

Rainfall is responsible primarily for the washing off of pesticides from their treatment sites, transport through erosion and solution, dilution of pesticides in aquatic environments, and for leaching and hydrolytic reactions. Rainfall is quite heavy in the Southeast Asian region, which is visited periodically by typhoons and monsoons. For selected parts of the Philippines, the pattern is

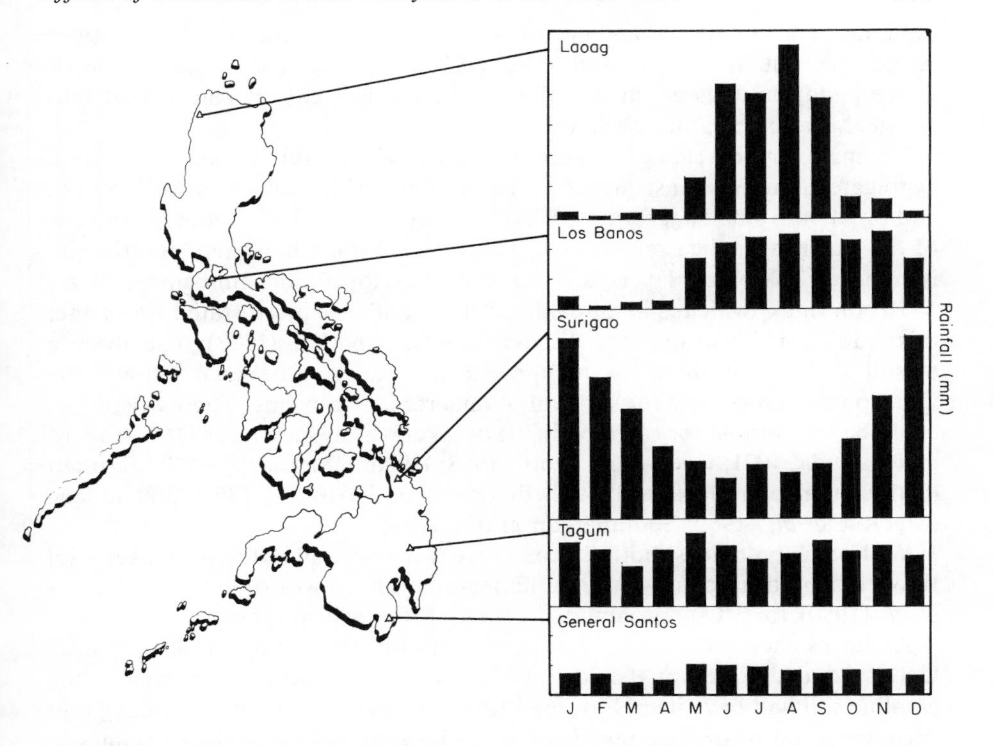

Figure 5.3.3 Rainfall pattern in some parts of the Philippines (Rice Production Manual, Philippines, 1983)

shown in Figure 5.3.3. There is a wide variation of rainfall pattern even within the tropical system, but in most cases rainfall is considered to be high. Again, rainfall is higher than in Japan or Korea so that it is expected to have a more pronounced effect on insecticides.

The warmer climate also gives rise to faster breakdown of pesticide molecules in accordance with basic chemical reaction principles. This is further reinforced by a richer, more diverse microorganism population, which may use the pesticide molecule for its metabolism or co-metabolism. Hirose *et al.* (1979; in Hashimoto, 1982) however, found that the toxicity of diazinon, fenitrothion, and phentnoate to carp and daphnids increased at higher temperature.

5.3.3.2 The Paddy Rice System

Within the tropical ecosystem, we have to consider the paddy rice system as different and unique for several reasons, among them being: (1) its essentially flooded nature; (2) dry, sun-exposed condition after flooding; and (3) distribution of inputs in different ways and means as a consequence of the puddled situation. As pointed out by Villegas and Feuer (1970), a puddled soil such as that of a rice paddy undergoes three major changes which can be categorized into physical, biological, and chemical.

The main physical change that occurs consists of break-up of the soil structural aggregates. This is a consequence of the flooding, which causes the soil to swell and break into small aggregates. Water loss by percolation is reduced. In view of the water layer the exchange of gases between the atmosphere and the soil is impeded. Two soil layers are therefore developed: the thin upper layer, 1–10 mm thick, which is in an oxidized state; and a reduced state lower zone; this gradient is illustrated and discussed by Sethunathan (1973). The absence of soil air brought about by the impeded gas exchange between soil and the atmosphere allows only for survival of anaerobic organisms. These organisms could be responsible for rapid pesticide breakdown, as shown for DDT (Castro and Yoshida, 1971; Guenzi and Beard, 1968; Hill and McCarty, 1967), diazinon (Sethunathan and MacRae, 1969; Sethunathan and Yoshida, 1969), and lindane (MacRae *et al.*, 1969; Sethunathan *et al.*, 1969).

With some pesticides, both the oxidative (aerobic) and reductive (anaerobic) flooded conditions could cause rapid degradation, as was found by Oyamada *et al.* (1980) for ^{14}C-naproanilide in three different soil types.

As far as chemical change is concerned, soil pH is an important indicator. Under flooded conditions, acid soil will tend to have its pH increased while alkaline soil will have its pH lowered towards neutral. Liming, fertilizing and decomposition of organic matter are likewise expected to produce changes in the soil chemistry. Considering that the paddy soil is the ultimate sink of applied pesticides (Bajet and Magallona, 1980; Takase and Nakamura, 1974; Tejada, 1983; Varca and Magallona, 1982) these changes assume significance because it is here that pesticides could be degraded/transformed to other molecular entities. The effect of pH may be direct (hydrolysis) or indirect (mediated by specific types of microorganisms).

Another factor to consider is the occurrence of bound residues. Bound residues cannot be detected using conventional residue analysis techniques so that by effectively reducing detected residue levels, the pesticide may be considered to have a shorter persistence than is actually the case. Ogawa *et al.* (1976) found that 1/6 to 1/3 of the applied radioactivity of ^{14}C-BPMC remained as bound residues 30 days after treatment. With isoprocarb, this amounted to 33% after 6 months (Magallona *et al.* 1985). Furthermore, our unpublished data showed 38% bound residues for carbofuran.

These bound residues could also be available to plants if they are in soil or to animals if they are in feed. Magallona *et al.* (1985) showed that rice and watermelon seedlings can take up bound isoprocarb residues, which are formed as a consequence of use in rice crop protection. Raghu and Drego (1985) likewise showed that bound lindane residues from flooded soils were bioavailable, as

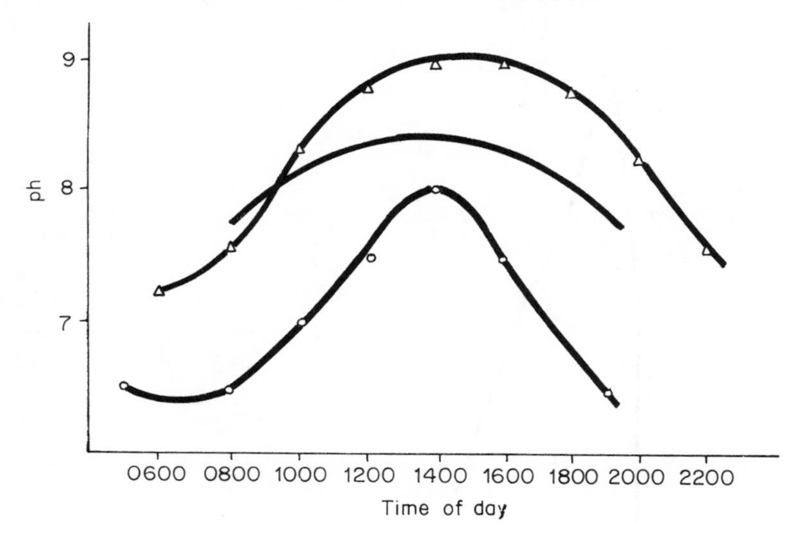

Figure 5.3.4 Changes in field paddy water pH with time of day as observed by (a) Catahan (1977); (b) Obcemea *et al.* (1977) and Bajet and Magallona (1982). (Reproduced by permission of the Philippine Association of Entomologists)

evidenced by $^{14}CO_2$ evolution. Formation of bound residues of lindane was favoured by neutral or alkaline conditions in a flooded soil, and was less in acidic soil.

The paddy water itself undergoes changes in pH depending on the time of day as a consequence of gas flows in the aqueous carbonate system, represented by the reaction below (Obcemea *et al.*, 1977).

$$CO_2(\text{atm.}) + H_2O \rightleftharpoons H_2CO_3 \rightleftharpoons H^+ + HCO_3^- \rightleftharpoons H^+ + CO_3 \rightleftharpoons CaCO_3$$

These pH changes are seen in Figure 5.3.4 as reported by Obcemea *et al.* (1977), Catahan (1977), and Bajet and Magallona (1982). In almost all cases, there is a pH peak ranging from 8 to 9.5 at about 1300 to 1600 hours, with lows in the morning and at dusk. Considering that the organophosphates and carbamates are easily hydrolysed at high pH, their degradation is expected to be rapid. This is documented in Figure 5.3.5 for several compounds we have studied in our laboratory. On the other hand, diazinon degradation seems to be favoured by high pH (Sethunathan and MacRae, 1969).

Some pesticides, however, appear to be persistent in the rice paddy. Thus, Takase and Nakamura (1974) as cited by Masuda (1979), observed a t1/2 of 50 days in paddy soil at 28°C for disulfoton and its five oxidative metabolites. These chemicals persisted beyond 12 weeks. On the other hand, Takase *et al.*

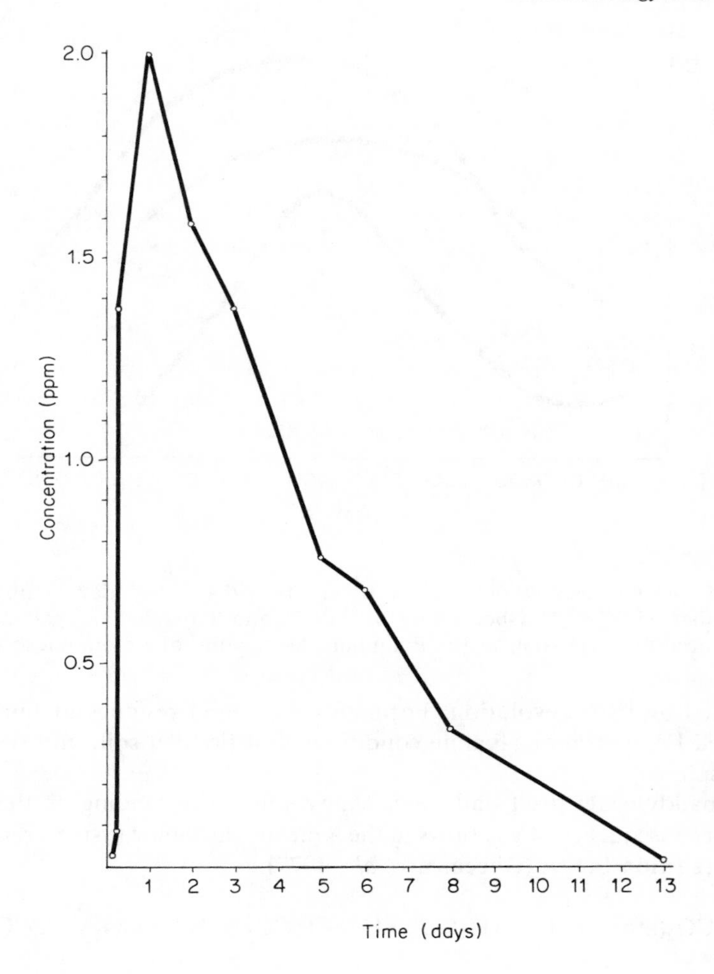

Figure 5.3.5 Concentration of carbofuran in paddy water after granular insecticide application (from Argente *et al.*, 1977). (Reproduced by permission of the Philippine Association of Entomologists)

(1971) recovered only 36% of the applied material in soil 10 days after application. Propaphos showed a slow decrease in paddy soil, remaining at about 10% of the initial concentration at 45 days after application (Asaka *et al.*, 1978). Masuda (1979) found that disulfoton and its oxidative metabolites were at constant levels around the rice plant for 25 days, although there was a rapid initial decline from 490 mg/kg at transplanting to 17 mg/kg after 7 days. The same trend was observed by Tanaka *et al.* (1976).

5.3.3.3 Effects on Insecticides

The net effect of the above factors is generally a rapid decline in pesticide levels in paddy rice (Sethunathan, 1973; Sethunathan *et al.*, 1975; Magallona, 1983). The loss mechanisms will not be discussed further here as they are adequately covered by Sethunathan in Section 5.2. of this book. However, for purposes of completeness in coverage, some discussions are included when we deal with the specific compounds, gamma-BHC, endosulfan, diazinon, carbofuran, isoprocarb, and BPMC.

Two additional aspects about insecticides appear to deserve mention here. They are (1) the conversion of pesticides to biologically active products, and (2) a differentiation between persistence and degradation. Some pesticides are known to be converted to biologically active products in the environment. The more familiar cases are conversion of aldrin to dieldrin by epoxidation, parathion to paraoxon by S to O substitution, heptachlor to heptachlor epoxide by epoxidation, and aldicarb to its sulphoxide and sulphone by oxidation. Carbofuran is also converted to the 3-keto and the 3-hydroxy compounds, which show some biological activity.

As far as residues in food are concerned, these biologically active transformation products are of interest, but even in this case they are considered only insofar as they are judged to be 'toxicologically significant', a term which loosely combines the magnitude of their formation and their adverse toxicological effects. Much less concern is expressed on the impact of these moieties on the environment, especially in tropical developing countries where research is not well-supported. Thus, some observed effects may not be easily reconciled with the levels or persistence of the parent compound but could well be due to these products.

Theoretically, also, a distinction should be made between 'persistence' and 'degradation'. The first term could include loss from the substrate of interest through a variety of mechanisms whereas the second term generally leads to non-toxic products. Thus we should always be concerned with the material balance in a system and this should include pesticides introduced through, say, irrigation waters. Furthermore, we have to be concerned if our ecotoxicant of interest is simply transferred from one substrate to another, and the more so if the latter is 'vulnerable'.

From the ecotoxicological standpoint, it does not necessarily follow that a degradation product innocuous against mammalian species would be harmless to other organisms as well. For example, 1-naphthol, a degradation product of carbaryl, is more toxic to goldfish (*Carassius auratus*) and killifish (*Fundulus heteroclitus*) (Shea and Berry, 1983). In fact, it is thought that 1-naphthol may be responsible for many effects attributed to carbaryl in view of this general trend with molluscs, fish, and protozoa. A BHC isomer, delta-BHC, is also virtually harmless to mammals but a very powerful molluscicide.

Of course, some compounds are transformed to other insecticidally-active compounds, as with aldrin, which is converted to dieldrin in guppy although scarcely found in *Daphnia* (Tsuge *et al.*, 1980).

5.3.4 EFFECTS OF INSECTICIDES ON SOME PADDY COMPONENTS

Biological effect is a function of the temporal concentration of the insecticide in a substrate. Immediately upon application, the concentration is still high so that chances of eliciting a response/effect from an organism are higher. As the concentration goes down, this effect may be somewhat mitigated.

The above is generally true in the case of acute effects. A more complicated phenomenon is bioaccumulation, wherein a pesticide is accumulated in the organism through continuous low-level uptake for a substrate, in most cases, water. The significance of this concentration on the organisms is generally unknown but it could cause a build-up to hazardous levels for the consumer of this organism. Some mortalities on the part of the primary organism are also possible. In most cases, once the organism is removed from contact with the pesticide, the residue level starts to decrease, indicating that the whole process is one of equilibrium between two substrates. Of course, in the case of some organisms, bioaccumulation may be viewed as a decontamination process whereby the pesticide is removed from the effective environment.

If from Figure 5.3.1, we consider that an applied pesticide is essentially confined to the paddy except for the quantity which escapes through volatilization and unintentional flooding, then its effects are likewise confined to organisms in the paddy system. These organisms include the pests themselves, the parasitoids and predators, plants, disease organisms (especially insect pathogens), and other animals. Man could be affected if he comes in and performs some tasks while pesticide levels are high or during the process of pesticide application. Fish, organisms which are not normally found in significant quantities in rice paddies, have recently become important because of the newly developed rice-fish culture. This system seeks to take advantage of water in the flooded field to raise fish in order to help fill the protein gap in most Southeast Asian countries. Furthermore, some of the fish in nearby aquatic systems may be useful for other purposes, such as predating on mosquito larvae.

Formulation type is also important in assessing ecosystem effects. Thus, the sprayables may not affect soil microorganisms to an extent comparable to the granular compounds. Conversely, non-herbivorous terrestrial organisms would not be directly affected by granulars to a significant degree but would be significantly affected by sprayables. The adverse effects of granulars may also be mitigated by the phased release, which occurs together with rapid breakdown for most compounds (Kanauchi *et al.*, 1982; Magallona, 1983; Seiber *et al.*, 1978). This may affect the availability of the compound for control so that more economical application systems, such as root-zone application, are being tried (Siddaramappa *et al.*, 1979).

5.3.4.1 Effects of Pests

Toxicity is an expected effect of pesticides on the target pests, so it will not be discussed further here. Of interest are only (a) development of resistance, and (b) resurgence in the case of the brown planthopper, both of which may be viewed as undesirable effects on pests. In addition to the coverage here on these two topics, the papers of Nagata and Mochida (1984) on resistance development and Chelliah and Heinrichs (1984) for resurgence are recommended reading.

There have been numerous reports of insects becoming resistant to insecticides, and paddy insects are no exception. The Annual Reports of the International Rice Research Institute at Los Banos, Philippines, continually refer to development of resistance in the most important rice pests, either through comparison with greenhouse strains which have not been exposed to pesticides or to loss of effectiveness by compounds previously known to be effective. Let us summarize just a few of the studies.

(1) In a test of a green leafhopper (GLH) collected from four sites in the Philippines vs. IRRI greenhouse cultures, low-level resistance to acephate and parathion methyl in the Palayan collections and acephate in the Sta. Maria cultures was shown (IRRI, 1981). Resistance was not detected with isoprocarb and monocrotophos.
(2) Brown planthopper (BPH) collected at the IRRI farm, an area receiving heavy and varied insecticide treatment, was later found to be resistant to chlorpyrifos + BPMC, BPMC, and acephate (IRRI, 1983b).
(3) Field population levels of BPH (IRRI field and farmer's field at Calauan, Laguna) had low levels of resistance to BPMC, carbaryl, carbofuran, diazinon, malathion, and MTMC when applied topically (IRRI, 1984). By contact toxicity using a Potter spray tower, six commercially available insecticides showed significantly higher BPH mortality in the greenhouse than in the field populations, indicating resistance development; there was no significant difference in the three populations for BPMC + chlorpyrifos. When these insecticides were applied as a foliar spray, there was significantly lower BPH mortality in the field than in the greenhouse population (Table 5.3.2). In the case of GLH, only with carbofuran and isoprocarb was significantly lower mortality observed in the field population.

Resurgence is the other problem associated with pesticides, notably with the BPH. It is the phenomenon whereby the use of a pesticide results in a population increase over that of untreated controls. Destruction of biological control agents has been discounted as the cause of this phenomenon (IRRI, 1981). As pointed out by Chelliah and Heinrichs (1984), this could be due to a variety of causes.

Table 5.3.2 Toxicity* of some insecticides to three populations of BPH by different methods of application (mg/kg bodyweight) (after IRRI, 1984)

	IRRI greenhouse			IRRI field			Calauan farmer's field
	Topical[†]	Contact[‡]	Foliar	Topical	Contact	Foliar	Topical
BPMC	1.85	–	–	10.38 (5.61)	–	–	2.21 (1.19)
Carbaryl	2.64	25.0	35.0	2.36 (0.89)	5.0 (5.0)	7.5 (4.7)	1.81 (0.69)
Carbofuran	0.29	97.5	80.0	0.85 (2.93)	12.5 (7.8)	2.5 (32)	0.80 (2.75)
Diazinon	6.34	–	–	14.21	–	–	–
Malathion	44.58	–	–	60.78	–	–	45 (1.02)
MIMC	3.32	–	–	2.79	–	–	46.57 (1.04)
BPMC	–	82.5	–	–	20.0	–	2.51 (0.76)
Monocrotophos	–	97.5	82.5	–	45.0 (2.2)	2.5 (33)	–
BPMC + chlorpyrifos	–	65.0	60.0	–	47.5 (1.4)	7.5 (8)	–
Acephate	–	52.5	75.0	–	15.0 (3.5)	7.5 (10)	–
Isoprocarb	–	50.0	52.5	–	7.5 (6.7)	2.5 (27)	–

*Values in parenthesis are the Estimated Resistance Ratio (ERR), obtained as follows:

$$ERR = \frac{\text{Toxicity to field population}}{\text{Toxicity to IRRI greenhouse population}}$$

[†]Using purified materials
[‡]Using commercial formulations

However, some insecticides can result in resurgence, probably by providing for a favourable environment for the BPH.

At the International Rice Research Institute (IRRI, 1981) 16 insecticides used for rice control in Asia were found to cause resurgence. On the other hand, carbofuran flowable, ethylan (Perthane), technical BHC (20EC), BPMC (50EC), carbophenothion (48EC), MTMC (30EC), and chlorpyrifos (40EC) significantly reduced BPH population. Application rates seem to influence occurrence of resurgence. Thus, TN1 plants treated with decamethrin, a resurgence-inducing insecticide, had the highest level of free-amino nitrogen, but the plants sprayed with ethylan, a non-resurgence-inducing insecticide, had significantly lower levels. Insecticide application did not affect levels of starch, sugar, and total nitrogen in the plants.

5.3.4.2 Effects on Parasitoids and Predators

This aspect of pesticide use has been studied quite extensively with the BPH, a more recent major pest of rice. Among the more important of these beneficial organisms are the spiders, *Lycosa pseudoannulata* (Boes, et Str.) and *Collitrichia formosana* (Oi); the mirid bug, *Cyrtorhinus lividipennis* (Reuter), and the small water strider, *Microbelia doughlasi atrolineata*. To some extent, one may also consider the fungus, *Metarrhizium anisopliae*.

Toxicity of some insecticides to these beneficial organisms is to be expected because of their biological affinity to the target pest as well as their inevitable contact with the applied chemical. This is particularly true with chemicals that are applied by spraying. With granulars having systemic activity, these beneficial organisms should be exposed to less pesticides and because they do not feed on dead pests, toxicity to these organisms should theoretically be low.

When 12 commercially available insecticides were tested against the BPH and three of its predators, it was found that decamethrin was most toxic (IRRI, 1983a, 1984). All insecticides were toxic to *C. lividipennis*, but many were non-toxic to the spiders, *L. pseudoannulata* and *M. d. atrolineata*. This could be attributed to the greater physiological similarity of *C. lividipennis* to the BPH. On the other hand, decamethrin was selectively more toxic to the spider than the other insecticides. Carbosulfan and carbofuran were selectively toxic to the BPH compared to the first two predators. Gavarra and Raros (1973) also reported that parathion methyl is more toxic than carbaryl to *Lycosa*.

In Thailand, Rumakon *et al.* (1982) reported that one day after application of the insecticides BPMC, carbofuran, and isoprocarb, the populations of *C. lividipennis*, *Oligosita* sp., and *Tetragnatha* spp. were significantly reduced compared to controls, but these natural enemies were able to recover so that 14–15 days later their populations did not vary from controls. Amin *et al.* (1982) also reported that fenvalerate, BPMC, propuxur, endosulfan, and diazinon reduced the insect populations in a rice plot. Toxicity to *C. lividipennis* was

Table 5.3.3 Selective toxicity* of five insecticides to some organisms in a paddy rice relative to the brown planthopper (BPH) (after Amin *et al.*, 1982)

Arthropod	Fenvalerate	BPMC	Propuxur	Endosulfan	Diazinon
Nilaparvata lugens	1.0(28.8)[†]	1.0(41.4)	1.0(51.9)	1.0(56.1)	1.0(46.7)
Sogatella fucifera	1.1	1.2	1.3	1.3	1.7
Cyrtorhinus lividipennis	1.1	0.9	1.0	1.4	1.0
Spiders *Lycosa* sp., *Tetrognatha*, etc.)	1.4	3.9	2.4	3.1	1.9
Paedeaerus sp.	0.8	6.5	1.3	1.01	1.2
Casnoides sp.	0.4	0.8	0.75	3.9	1.2
Staphilinid	0.5	0.6	0.7	0.9	0.8
Carabids	1.1	0.9	1.2	0.8	0.73

*Based on the formula: selective toxicity = $\frac{\text{\% reduction in BPH population compared to controls}}{\text{\% reduction in organisms concerned compared to controls}}$

[†]Numbers in parenthesis refer to % reduction in BPH population.

essentially the same as that to the BPH, but the insecticides, and in particular BPMC, were less toxic to the spiders (Table 5.3.3).

Populations of the damselflies, *Agriocnemis pygmaea*, *A. femina femina*, and *Ishimura senegalensis*, which were predators of some rice pests, were reduced by insecticide treatment of paddy rice (JICA, 1981). It was suspected that in addition to killing the nymphs of damselflies which live in the soil, insecticides also affect the flying adults.

With *Anagrus* sp., a parasite of stemborer eggs, exposure to foliar spray of five insecticides resulted in more than 97% mortality. Some insecticides also reduced hatchability, while buprofezin, endosulfan, BPMC, and diazinon were not significantly different from the untreated checks (IRRI, 1981). Buprofezin was also found to be non-toxic to the three BPH predators (IRRI, 1983a, 1984).

The botanical insecticides appear to have a better selective toxicity to the predators than the conventional pesticides. Neem and Chinaberry oils were slightly toxic to the mirid bug at LD_{50} of 50 μgs/insect while Custard apple oil was moderately to highly toxic at 10–50 μg/insect. All three oils were essentially non-toxic to *L. pseudoannulata* (IRRI, 1983b).

5.3.4.3 Effects on Insect Pathogens

Some of the insecticides have also been shown to inhibit insect pathogens. For example, when mixed with culture media of *Metarrhizium anisopliae* and *Beauveria bassiana*, monocrotophos, BPMC, carbosulfan, and azinphos ethyl + BPMC reduced spore germination considerably (IRRI, 1984). However, when these insecticides were applied to the media surface, spore germination was high. Azinphos ethyl + BPMC was most toxic to *M. anisopliae*, while BPMC was the most toxic to *B. bassiana*. The least toxic of the insecticides to both fungi was monocrotophos.

Cadatal (1969) also found a wide variability in the effects of nine insecticides and three fungicides on the development of *B. bassiana*, and *Entomopthora* sp. Supracide, carbaryl, endosulfan, and endrin exhibited partial to complete inhibition of growth and sporulation at concentrations equivalent to the field recommendation. Fenitrothion, chlorfenvinphos, lindane, diazinon, and DDT were innocuous. The fungicides Panogen and Granosan L were toxic, while Kazumin allowed complete development.

5.3.4.4 Effects on Larger Animals

Of interest here are fishes, edible snails, toads, frogs, and ostracods. The importance of fishes as food sources as well as components of the ecosystem is well recognized so that fish toxicity data is a requirement for registration. However, the organisms used are mainly temperate fishes, and while the data thereby obtained could be a useful guide, for our purposes they cannot take the place of toxicity data obtained with tropical fishes that could be integrated with rice production. Rainbow trout or zebra fish are the recommended fishes on which to determine 96-hour LC_{50} for registration purposes (FAO, 1981), while *Tilapia* sp. is the fish of interest to us, as it has been found suited for rice–fish culture, being edible and a fast weight gainer. As Nebeker *et al.* (1983) pointed out with endosulfan, sensitivity may vary among species over as much as three orders of magnitude. Of course, it is recognized that pesticide manufacturers in general have easier access to these two temperate fishes than *Tilapia*, and the data generated can already be good indicators of the toxicity potential of a compound.

In this regard, the data gathered by the Freshwater Aquaculture Center of the Central Luzon State University in the Philippines is of considerable importance (Table 5.3.4). In this series of tests, the Median Tolerance Limits (TL_{50} or TLM) were obtained separately using a static bioassay system. Formulated products were used in conditions simulating those in the actual paddy. This should explain some observed inconsistencies, which could be due to: (a) variations in the active ingredient content of the product; (b) differences in release rates of pesticides from granules, as in the case of diazinon; (c) differing ages of fishes; and (d) differing micro test conditions.

Azinphos ethyl is one of the most toxic materials, followed by carbofuran, BPMC + chlorpyrifos, MTMC, BPMC, isoprocarb, endosulfan, and diazinon as spray, monocrotophos, and MTMC. There is considerable inconsistency in the results with isoprocarb sold by two major suppliers and there seems to be no plausible explanation at this time. Data from our laboratory with acetone solutions of the purified carbamates show methomyl to be the most toxic followed by carbofuran, carbosulfan, and BPMC, in that order (Table 5.3.5).

Table 5.3.4 24-hours TL50* of some pesticides to fishes associated with rice fish culture (consolidated from CLSU-FAC, 1978b, 1979a, 1979b, 1981b)†

Insecticide	*Tilapia nilotica* F.P.	*Tilapia nilotica* A.I.	*T. mosambica* F.P.	*T. mosambica* A.I.	*Carassius carassius* F.P.	*Carassius carassius* A.I.
Methyl parathion						
(Parapest 50EC)	26.5	13.25				
BPMC						
a. Baycarb I (50%)	5.9	2.95				
II	2.8	1.4	3.15	1.58	30.5	15.25
b. Shellcarb (50%)	6.75	3.85				
Diazinon						
a. Basudin 20EC	79.5	15.9			69.5	13.9
b. Diagran 5G	77.0	3.85			45.4	2.27
Endosulfan						
(Thiodan 35EC)	8.2	2.87			7.3	2.56
Isoprocarb						
a. Mipcin (50%)	61.7	30.85			43.6	21.8
b. Hytox (50%)	6.05	3.02	6.0	3.0	34.75	17.38
MTMC (Hopcin 50EC)	1.95	0.98	3.12	1.56	29.5	14.75
Azinphos-ethyl						
(Gusathion A, 40%)	0.0275	0.011	0.23	0.009	0.086	0.03
Monocrotophos						
(Azodrin 202R, 30%)	70 (48 hrs)	21	47.6 (48 hrs)	14.28	–	–
BPMC + Chlorpyrifos						
(Brodan EC; 31.5% total a.i.)	2.35	0.74	1.62	0.51	–	–
Carbofuran						
(Furadan F, 12%)	2.27 (48 hrs)	0.27	2.24 (48 hrs)	0.27	–	–
MTMC						
a. Tsumacide 50WP	87	43.5	85	52.5	–	–
b. Hopcin 50EC	1.95	0.98	3.12	1.56	29.5	14.75

*Expressed in ppm; formulated product (F.P.) was used in test, and data for active ingredient (A.I.) were computed based on declared active ingredient content of the product
†For purposes of rough comparison, the active ingredients are assumed to be equally toxic to fishes.

To reconcile pesticide use for paddy rice crop protection with the presence of fish, trenches are constructed in the middle of the paddy. One day before spray application, the paddy water is drained and the fishes are confined to the trench, which still contains some water. After several days, water is reintroduced to a depth of 7–11 cm, hopefully avoiding fish toxicity by giving adequate time for insecticide degradation. With this practice, *Bacillus thuringiensis*, BPMC, BPMC + chlorpyrifos, carbaryl, cypermethrin, isoprocarb, monocrotophos, and parathion methyl have acceptable fish recoveries, whereas azinphos ethyl is still very toxic (CLSU-FAC, 1978a, 1981a, 1981b). In persistence tests, the fish can be introduced safely by reflooding the

Table 5.3.5 Toxicity of some carbamate insecticides to *Tilapia nilotica* (Santiago and Magallona, 1982, reproduced with permission of the National Crop Protection Center, UPLB)

Insecticide	Material tested	LC 50 (mg/l)	Slope
Aldicarb	Purified	0.0013	0.31
Methomyl	Standard	0.04	0.64
Carbofuran	Purified	0.09	0.69
Carbosulfan	Standard	0.18	0.02
BPMC	Standard	0.96	0.17
Isoprocarb	Purified	1.6	0.79
Carbaryl	Purified	2.3	0.73
Landrin	Standard	5.1	0.48
MTMC	Purified	9.5	0.15

paddy within one week after application of monocrotophos, BPMC + chlorpyrifos (as spray), isoprocarb, permethrin, and carbaryl (CISU-FAC, 1980, 1981b).

Basha *et al.* (1983) found that the order of toxicity of three insecticides to *Tilapia mossambica* was: malathion > HCH > carbaryl, with 48-hour LC50 of 0.37, 3.2, and 5.5 mg/l, respectively. They later showed that sublethal doses of these compounds cause respiratory distress in this organism (Basha *et al.*, 1984).

It should be pointed out, however, that considering the use of these fishes for food, one should be concerned not only with acute toxicity but also with bioaccumulation. Unfortunately, there is a paucity of data in this regard. However, with BPMC, carbaryl, carbofuran, chlorpyrifos, endosulfan, gamma-BHC, isoprocarb, isothioate, and XMC this does not appear to be a problem (Bajet and Magallona, 1982; Medina-Lucero, 1980; Tejada, 1983; Varca and Magallona, 1982; Gorbach *et al.*, 1971a, 1971b; Zulkifli *et al.*, 1983; Argente *et al.*, 1977; Tsuge *et al.*, 1980). On the other hand, it should be pointed out that with lindane, significant amounts accumulate in fertilized fish eggs (Ramamoorthy, 1985). Apparently, bioconcentration potential is related to water solubility and partitioning coefficient (Kanazawa, 1981).

Much less is known about the effect of pesticides on other organisms, probably because of the difficulties involved in the assay as well as in the interpretation of its importance. *Pila luzonica* Reeve, an edible freshwater snail that is considered a delicacy in some parts of the Philippines, is one organism that has been studied in some detail. Guerrero and Guerrero (1980) found gamma-BHC at 2.8 and 4.5 mg/kg in this organism and *Vivipara angularis* Muller (another edible snail), respectively, when these were cultured in rice paddies, indicating their capacity for insecticide uptake and possible bioconcentration. Low-level uptake of residue by *P. luzonica* was likewise observed with isoprocarb, BPMC, and carbofuran/carbosulfan (Bajet and Magallona, 1982; Varca and

Magallona, 1982; Tejada, 1983). This is expected for these compounds, which have high polarities, but the earlier lipophilic compounds, e.g. DDT and BHC, could have resulted in bioaccumulation by this organism. Bajet and Magallona (1982) also reported an LC50 of 25.3 mg/l for this snail.

Perez (1981) showed that rice-field ostracods are adversely affected by malathion (most toxic), carbaryl, and methyl parathion (least toxic). These ostracods are predators of the blue-green algae.

With endosulfan, Gorbach *et al.* (1971a, 1971b) observed killing of all Brachyura as well as the majority of Coleoptera and larvae of *Tipulidae* on the first day of application in a rice field in Indonesia, although these organisms reappeared after 5 days. Tubificidae, Hydrocorisidae, Cyclopidae, and Gastropoda showed no signs of mortality.

5.3.4.5 Uptake by and Effects on Plants

Of the many plant species in a paddy system, uptake of pesticides is of interest only in the rice plant and edible plants like *Ipomoea aquatica*. With rice our interest is on pesticide residues, and it has been shown that normal application of carbofuran, lindane, endosulfan, BPMC, and carbosulfan did not result in residues in grains. This is primarily due to the long interval normally observed between insecticide application and harvest, and the rapid decline of most of these pesticides in the rice paddy, despite the presence of some residues in the rice plant, e.g. as in the case of isoprothiolane which has a t 1/2 of about a month in the rice plant (Kanauchi *et al.*, 1982).

With *I. aquatica* a different picture is presented because it can be harvested and consumed as a vegetable at any time so that residues taken up by the plant could be a health hazard. Uptake of radiolabelled isoprocarb (Bajet and Magallona, 1982), BPMC (Varca and Magallona, 1982) and carbofuran (Tejada, 1983) from paddy water was shown to occur in this plant. However, it is not known what levels could be considered safe or hazardous to health, and the setting of a maximum residue limit does not appear practicable at this time.

Muralikrishna and Venkateswarlu (1984) showed that at 5–10 ppm in soil, carbaryl, endosulfan, and parathion are not harmful to the soil algal population in both flooded and unflooded soils. Carbaryl and parathion had more pronounced effects on unflooded soil, but endosulfan had little effect up to 25 ppm in both water regimes.

5.3.4.6 Model Ecosystems Studies

Some attempts have been made to evaluate the fate and effects of pesticides using model ecosystems where a food chain is constituted and the pesticide introduced. The distribution and fate of pesticides in these components are followed. This exercise presupposes that the pesticide will behave in the ecosystem in the same manner as in the actual environment. Radiolabelled materials are often used, enabling possible quantification as well as identification of some transformation products.

One of the earlier models is that by Sastrodihardjo *et al.* (1978). Their model consisted of a series of containers wherein each paddy constituent (rice, carp, guppy, mudworms, water snails) was placed separately. Water was continuously passed through the system at the rate of 10 l/hour. The insecticides phosphamidon and endrin were applied at the recommended rate.

Our laboratory uses two systems, one for field level studies and another for the laboratory. These models are used primarily to assess the fate of pesticides (degradation and distribution), but some adverse effects such as fish mortality could also be observed. The laboratory model using radiolabelled compounds has the following components (Tejada, 1983):

compartment:	fully enclosed glass aquarium 40.6 cm width × 74.9 cm length × 40.1 cm depth
air circulation:	through vacuum pump with polyurethane foam trap
light:	fluorescent lamp
plant:	rice
animals:	*Tilapia nilotica* *Pila luzonica*

For field level studies, a concrete 289.5 × 23.1 cm compartment is used for growing rice; a trench can be dug in this compartment if rice–fish culture is of interest, in which case *Tilapia* fingerlings are introduced. At the appropriate time, the paddy water is drained into a pond containing fish (*Tilapia*), snails (*Pila luzonica*), toads (*Bufo marinus*), and the aquatic plant (*Ipomoea aquatica*) (Bajet and Magallona, 1982; Bautista *et al.*, 1985; Tejada, 1983; Varca and Magallona, 1982; Zulkifli *et al.*, 1983).

The emphasis in the Sastrodihardjo *et al.* (1978) model was assessment of the effect of phosphamidon. The insecticide adversely affected carps but not guppy and the water snail, *Limnaea* sp. With endrin treatment, mortality to carp, guppy, and water snail was about 20%. Neither insecticide affected the mudworms, *Tubifex* sp. Worms that were placed for 72 hours in 1 mg/l phosphamidon solution resulted in a 40% mortality in the total fish population but not in guppy. Endrin-treated worms caused greater lethal effects in guppies and rice carps.

5.3.5 FATE AND EFFECTS OF SELECTED INSECTICIDES

From Table 5.3.1 it can be seen that carbofuran, gamma-BHC, diazinon, endosulfan, and carbaryl are still widely used in rice production. The new compounds isoprocarb, BPMC, and monocrotophos have likewise gained in popularity. The discussions on persistence will be limited to only a few compounds for which there are adequate data. It is recognized that persistence and biological effects could be mitigated by residence time of a compound in a substrate. On the other hand, a compound which may not be highly toxic may have a more significant impact if it persists in the environment for a long time.

5.3.5.1 Gamma-BHC

This compound is still useful against stemborers. Interest is in the gamma isomer, the highly purified material known as lindane. The dislike for the other isomers, especially the alpha isomer, stems from their possible carcinogenicity, longer persistence, and lack of activity against insects. Furthermore, the delta isomer is a powerful molluscicide and this could be responsible for adverse effects on edible snails.

Breakdown of this compound and three of its isomers in the flooded soil is more rapid than in unflooded soil, being mediated by anaerobic organisms such as *Clostridium* sp. (Sethunathan, 1973; Raghu and MacRae, 1966; MacRae *et al.*, 1967; Sethunathan *et al.*, 1969; MacRae *et al.*, 1969). Matsumura *et al.* (1976) also showed that gamma-BHC was metabolized by 71 of 354 microorganisms isolated from the environment.

This loss was quite rapid even at three times the recommended application rate. In the case of the gamma isomer, there was a decline from 15 μg/g of soil initial concentration to about 0.25 μg/g in 70 days, whereas in unflooded soil this remained at about 11 μg/g. There were no significant differences in the decline of the different isomers. The degradation of BHC and other such persistent compounds as DDT, methoxychlor, and heptachlor, is aided by decomposition of organic matter, like rice stubble.

Medina-Lucero (1980) obtained essentially similar results with t 1/2 of 21.6, 5.3, 17.8, and 10.8 days in paddy soil, plants, suspended soil particles, and paddy water, respectively. Repeated application did not result in accumulation of residues. However, she observed that the alpha isomer was much more persistent, still being present after one month, whereas lindane was no longer detected. Toxic effects on fish were observed immediately after application.

In a survey of pesticides in rice paddies in Malaysia, Meier *et al.* (1983) found that in all samples of fish, sediment, and water, beta-HCH was higher than gamma-HCH. The average biomagnification ration in fish was comparable in both isomers—13.4 for the beta isomer and 11.3 for gamma-HCH.

Yamato *et al.* (1983) studied the comparative bioaccumulation and elimination of HCh isomers in the short-necked clam (*Venerupis japonica*) and guppy (*Poecilia reticulata*). Guppy rapidly bioaccumulated HCH isomers with the following bioaccumulation ratio:

alpha: 706; beta: 1043; gamma: 697; delta: 648.

Decline was constant when the organism was transferred to HCH-free water. With clams, absorption was rapid, reaching a plateau on day 3. The following bioaccumulation ratios were obtained.

alpha: 161; beta: 127; gamma: 121; delta: 272.

5.3.5.2 Endosulfan

Just like diazinon and gamma-BHC, this is an 'old' compound that is still recommended against some rice pests. It is generally applied as a spray from an emulsifiable concentrate formulation.

In the paddy system, endosulfan is not persistent and t 1/2 of 2.3 and 3.8 days were observed in suspended soil particles for endosulfan I and endosulfan II respectively (Medina-Lucero, 1980). In plants, these were 1.3 and 1.8 days respectively for the two metabolites. Being a non-persistent compound, endosulfan did not accumulate with repeated applications.

Gorbach *et al.* (1971b) also found that endosulfan residues in rice paddy fields declined rapidly both in water and in paddy mud. Immediately after treatment with 1.4 l of Thiodan 35 EC per hectare, the initial concentration in the water was 0.2–0.55 mg/l. This fell to less than 0.00087 mg/l within five days. Fish mortality using *Puntius javanicus*, an endosulfan-sensitive tropical fish, was observed only on the day of application. In the case of paddy mud, the low initial level of 0.41–1.55 mg/kg was reduced by 75% after 15 days.

Massive endosulfan application in Indonesia (Gorbach *et al.*, 1971a) resulted in only 0.00046 mg/l in river and canal waters, which is about 1/3 of LC100 of 0.00125 mg/l to *P. javanicus*. A rapid decrease in residues was noted. In general, endosulfan residues were low in the preceding substrates, fish ponds, and the sea, because of degradation and dilution effects.

When endosulfan was applied to rice plants at 15, 30, 40, 60, and 90 days after transplanting and then the paddy was drained into a pond containing

Table 5.3.6 Residues of endosulfan in a paddy rice system (after Bautista *et al.*, 1985)

Substrate	Days after last spraying			
	0	5	10	20
Paddy rice	0.08	0.08	0.002	nil
Pond water	0.003	0.002	0.002	nil
Paddy soil	0.50	0.19	0.09	0.13
Rice straw	0.11	0.46	0.05	0.04
Rice grain	0.12	0.30	0.03	0.02
Ipomoea equatica	0.25	0.04	0.03	0.003
Pila luzonica	0.03	nil	0.3	nil
Pila luzonica eggs	0.01	nil	nil	nil
Toad	nil	0.07	0.01	nil

T. nilotica and snails, the residues were very low in snails, snail eggs, toads, and pond water (Table 5.3.6) (Bautista *et al.* 1985).

The toxicity of endosulfan and its metabolites to several aquatic organisms was investigated by Knauf and Schulze (1973) who obtained very variable results. In the group of the more sensitive organisms, with 48 hr LC50 of 0.001 to 0.01 mg/l, were, in order of decreasing sensitivity, the fishes *Idus melanotus*, *Lebistes reticulatus*, and *Carassius auratus*. Another group of organisms had LC50 in the 0.08 to 1 mg/l range. These were *Daphnia magna*, the insects *Aedes aegypti* and *Chironomus thummi*, and the molluscs *Planorbis corneus*, *Limnaea stagnalis*, and *Physa fontinalis*. *Tubifex tubifex* and *Artemisia solina* belonged to a third group with LC50 in the 8 to 30 ppm (mg/litre) range.

In the green alga, *Chlorella vulgaris*, the concentration of 0.001 mg/l endosulfan which was found in Indonesia by Gorbach *et al.* (1971a, 1971b) did not have any effect on cell division, photosynthetic activity, or biomass production. Photosynthetic activity remained unimpaired even with concentrations as high as 50 mg/l, but biomass and cell division were adversely affected. Measured by these parameters, a concentration of 2 mg/l would have no adverse effects.

5.3.5.3 Diazinon

This organophosphate continues to be effective against the stemborers by granular application and against the hoppers as a spray. Hydrolysis in flooded soil is aided by anaerobic organisms like *Arthrobacter*, *Flavobacterium*, and *Streptomyces*, as evidenced by: (1) faster degradation in non-sterilized than in sterilized paddy soil; and (2) isolation of these organisms which cause the hydrolysis in laboratory cultures (Sethunathan and MacRae, 1969; Sethunathan and Yoshida, 1969; Sethunathan and Pathak, 1972). Repeated application of diazinon in paddy fields enhanced the activity of these microorganisms to the extent that the efficacy of the compound against the stemborers was reduced as a consequence of rapid breakdown. In non-sterilized submerged soil, t 1/2 of 8.8 days was obtained, whereas in submerged sterilized soil it was 33.8 days.

Hirano and Yushima (1969) found that the release of granular material into paddy water reached a maximum in 3 days after application. A rapid decrease followed. Volatilization of diazinon was observed from water (Takashi and Masui, 1974); not only is this important in the control of certain pests but also in loss of residues.

5.3.5.4 Carbofuran

This is the most widely used insecticide in rice crop protection at present. It is applied as a granule because of its high inhalation toxicity. Advantage is taken of its systemic activity and relative stability in the environment (Magallona,

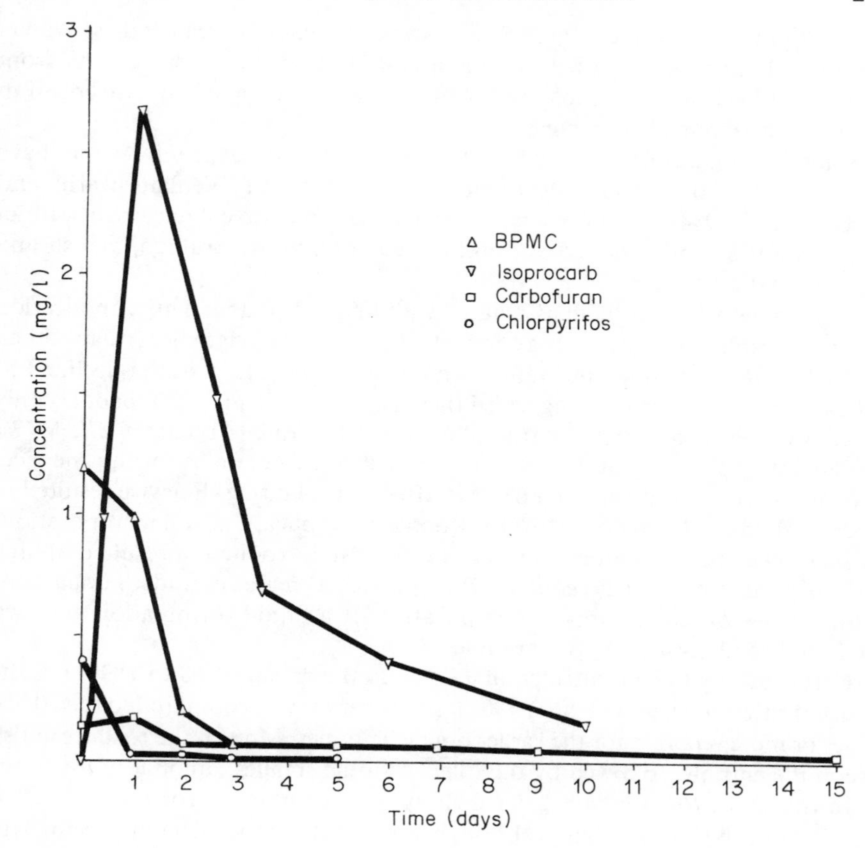

Figure 5.3.6 Decline curves for some insecticides in paddy water based on the work of Bajet and Magallona (1982) for isoprocarb, Zulkifli *et al.* (1983) for BPMC+ chlorpyrifos, and Tejada (1983) for carbofuran

1980). Around 1975–79 a flowable formulation was introduced but this did not catch on and health problems were also observed.

Carbofuran decline in water has been shown by Catahan (1977) to be pH dependent: at pH 7, t 1/2 is 10 days, decreasing to 0.58 days at pH 8.7, and 0.052 days at pH 10. In the field, photodecomposition contributes only a small part to the overall mechanisms. If we consider this in relation to Figure 5.3.4, which shows wide pH fluctation in paddy water, then carbofuran degradation can be expected to be rapid. This is borne out by field data. Thus, Argente *et al.* (1977) found a maximum level of 2.0 mg/l about 1 day after granular application followed by a steady decline (Figure 5.3.5). Half-life was about 4 days. In fact, Siddaramappa *et al.* (1977) reported that complete hydrolysis to the 7-phenol occurred in just 5 days after application because these waters have a pH range of 8.5 to 9.5. Tejada (1983) obtained similar results from carbofuran released in the rice paddy as a consequence of carbosulfan application (Figure 5.3.6).

Soil microorganisms like *Azospirillum lipoferum* and *Streptomyces* spp. have also been found to cause degradation of carbofuran (Venkateswarlu and Sethunathan, 1984). The presence of rice plants in soil also enhanced carbofuran degradation in both flooded and unflooded conditions (Brahmaprakash and Sethunathan, 1985).

High toxicity is a problem associated with this compound. Thus, application of the flowable formulation as a spray at 2 kg a.i. (active ingredient)/ha resulted in 100% *Tilapia nilotica* mortality, which persisted until residues declined to 0.03–0.6 mg/l. It was thus suggested that fish should be placed in paddies only after two weeks from broadcast application at this rate (Argente *et al.*, 1977). Broadcast application at 1 kg a.i./ha carried out four times during the rice-growing season (3, 23, 43, and 63 days after transplanting) likewise resulted in 100% fish kill (CLSU-FAC, 1979b). Root-zone application, soil incorporation, or broadcasting of granules one week before fish introduction resulted in high fish survival although Cho *et al.* (1978) reported decreases in spider populations with the use of single root-zone application of a liquid formulation and two or four broadcastings of the granule.

Carbofuran was concentrated in *Tilapia* to the extent of 68 to 189 times the concentration in water (Tejada, 1983); this is very low if compared to the 10 000 times or more reported for the earlier organochlorines. Most of the residues in fish are in the entrails, followed by the fillet, with the smallest amount in the head.

In the snail, *Pila luzonica*, the shell appears to have a protective effect so that there are slightly higher although not significantly different residues in the shell than in the flesh. A bioconcentration factor of 38.3 was obtained after 7 days.

5.3.5.5 Isoprocarb

This carbamate insecticide has become widely used against the hoppers, which are now considered the major insect pests of lowland rice. Its main market is rice so that it is registered only in a few countries, which unfortunately do not require extensive environmental impact assessment for registration.

Bajet and Magallona (1982) showed that like carbofuran, the decline of this compound in paddy water is pH dependent. At pH 8, the concentration in paddy water decreased from 5 to 0.025 mg/l after 3 hours. Half-life at pH 6 is 15 days, while at pH 7 two decline curves were observed, the first one of which lasted for about 2 days with t 1/2 of 1.4 days and the second of 88.8 days. Half-life in unflooded and flooded soils was 6.4 and 38.5 days, respectively, which is in agreement with the results of Ogawa *et al.* (1977).

Among the major metabolites in the soil, the following were tentatively identified based on the metabolic pathway proposed by Ogawa *et al.* (1977) for the compound:

(1) 2-isopropyl phenol
(2) 2-(1-hydroxy, 1-methylethyl) phenyl N-methylcarbamate
(3) 2-(1-hydroxy, 1-methylethyl) phenol

Radiolabelled isoprocarb could be detected in rice plants 6 hours after exposure, the radioactivity being concentrated on the stem and older leaves. After 2 days, this radioactivity was concentrated on the tip of the leaves. In *Ipomoea aquatica* the radioactivity was concentrated in the veins and veinlets of the old leaves, distribution being difussed in the young leaves. This uptake was demonstrated within 3 days of exposure.

In *Tilapia nilotica* and *Pila luzonica*, isoprocarb was found one day after exposure. Radioactivity was concentrated in the head area of the fish. In snails, an LC50 of 26.3 mg/l was obtained.

At the recommended rate of 0.75 kg a.i./ha isoprocarb did not decrease parasitism of *Tryporyza incertulas* (Walker) eggs (Than, 1976). However, spray application in the rice-paddy system resulted in only about 40% fish recovery (CLSU-FAC, 1978a, 1981b). *Tilapia* fingerlings may be introduced safely about 5 days after spray application (CLSU-FAC, 1980).

Quite recently, it was shown that bound residues of the compound are formed in paddy soil and this may contribute to the reported loss figures (Magallona *et al.*, 1985). These residues could be desorbed from the soil upon treatment with 0.01M $CaCl_2$. Low-level uptake of the bound residues has been demonstrated in rice and watermelon seedlings.

5.3.5.6 BPMC

This is another carbamate insecticide finding widespread use against the brown planthoppers and the green leafhoppers. In the Philippines, a popular emulsifiable concentrate contains 21% chlorpyrifos and 10.5% BPMC (Antazo and Magallona, 1982; URARTIP, 1985). In Indonesia, mixtures of BPMC with diazinon, fenitrothion, and phenthoate are recommended against the stemborers. (Partoatmodjo and Alimoeso, 1982).

Spray application of BPMC results in concentration of residues in rice leaves, stems, paddy water, and paddy soil, in decreasing order (NCPC, 1982; Varca and Magallona, 1982). Decline in paddy water followed pseudo first-order kinetics, with t 1/2 of 13.9 hours. In paddy soil, there was a rapid decline from 0.06 mg/kg 3 hours after spray application to 0.01 mg/kg after 3 days. In rice leaves, residues were 38.8 mg/kg 3 hours after application, but after one day only about 10% of this amount could be detected.

In a related study, Zulkifli *et al.* (1983) obtained a t 1/2 of 0.9 days in paddy water for BPMC applied using a BPMC + chlorpyrifos formulation. Residues were taken up by *Tilapia*, with highest levels found in entrails. As a consequence of uptake, residues were found in the fillet, peaking at about 2 days after exposure to pesticide-containing water. A 5.8 times magnification factor was observed in fish versus water levels; this may be considered low.

BPMC and BPMC + chlorpyrifos are toxic to *Tilapia* (CLSU-FAC, 1978b, 1980). With the latter, fish toxicity was observed up to 15 to 20 days after application. This does not correlate very well with our findings on the t 1/2 of 0.9 days (Zulkifli *et al.*, 1983) and 13.9 hours (Varca and Magallona, 1982), but inconsistencies could be attributed to widely differing experimental conditions or possibly the formation of toxic metabolites.

5.3.6 CONCLUSION

The application of insecticides in rice production, while low on a per unit area basis in Southeast Asia compared to the more developed parts of the world, is nonetheless still a cause for concern. This is particularly so in the paddy system of rice production, where it has been averred that pesticide use has caused such organisms as the snails to be wiped out in some localities.

Limited data so far available point to some adverse effects in certain organisms within the paddy system, while with other organisms the information is not adequate to draw valid conclusions. Of course, there is a wide variability of observed effects among the organisms. One mitigating factor to consider is the rather rapid degradation of pesticides in general under flooded conditions both in the soil and water phases. These are brought about by action of anaerobic microorganisms as well as hydrolysis due to fluctuating pH, more so with the carbamates and organophosphates. Thus, with many rice pesticides currently in use and especially the carbamates, their rapid decomposition mitigates against long-term ecological effects.

The occurrence of resistance in the pests, resurgence in pest population after application of certain chemicals, and toxicity to beneficial organisms, including fishes recommended in the rice–fish culture, are among the major adverse effects discussed. In addition to these, there are other observations which cannot be considered adverse at this time, e.g. uptake by plants and limited uptake by aquatic organisms, but which nevertheless require attention.

All these problems associated with the current pesticides which, it may be pointed out, are different from the earlier group of compounds, point to the need for chemicals with more favourable environmental impact. This aspect is already being considered in the design of new chemicals but it will be a few years before these compounds become commercialized. Furthermore, research using tropical agro-ecosystems needs to be done locally. Indeed, the paucity of information shown here points to the need for greater research support if we are to understand the impact of these inputs on the tropical environment.

5.3.7 REFERENCES

Amin, S. M., Lim, B. K., and Yong, Y. C. (1982). *Pesticide (Insecticide) Use and Their Specificity on Rice in Peninsular Malaysia*. Country paper presented during the Working Group Meeting on Pesticide Use Specificity, FAO, Bangkok, Thailand, Nov. 23–26.

Antazo, T. A., and Magallona, E. D. (1982). *Pesticide Use and Specificity on Rice: the Philippine Experience*. Country paper presented during the FAO Working Group Meeting on Pesticide Use and Specificity, Bangkok, Thailand, Nov. 23–26.

Argente, A. M., Seiber, J. N., and Magallona, E. D. (1977). *Residues of Carbofuran in Paddy-Reared Fish (Tilapia mossambica) Resulting from Treatment of Rice Paddies With Furadan Insecticide*. Paper presented at the 8th Annual Convention of the Pest Control Council of the Philippines, Bacolod City, Philippines.

Asaka, S., Kawauchi, N., Koyoma, S., and Emura, K. (1978). *Pestic. Sci.*, **3**, 305–310 (from Masuda, T. 1979).

Bajet, C. M., and Magallona, E. D. (1982). Chemodynamics of isoprocarb in the rice paddy environment. *Philipp. Ent.*, **5** (4), 355–371.

Basha, S. M., Prasada Rao, K. S., Sambasna Rao, K. R. S., and Ramana Rao, K. V. (1983). Differential toxicity of malathion, BHC and carbaryl to the freshwater fish, *Tilapia mossambica* (Peters). *Bull. Environment. Contam. Toxicol.*, **31** (5), 543–546.

Basha, S. M., Prasada Rao, K. S., Sambasna Rao, K. R. S., and Ramana Rao, K. V. (1984). Respiratory potentials of the fish (*Tilapia mossambica*) under malathion, carbaryl and lindane intoxication. *Bull. Environment. Contam. Toxicol.*, **32** (5), 570–574.

Bautista, E. R. B., Siac, L. P., Nuguid, Z. F., Abad, L. V., de la Cruz, M. D., and Magallona E. D. (1985). *Fate of Persistent Insecticides in a Paddy Rice Ecosystem*. Report submitted to International Atomic Energy Agency for IAEA Research Contract #3400/RB.

Brahmaprakash, G. P., and Sethunathan, N. (1985). Metabolism of carbaryl and carbosulfan in soil planted to rice. *Agric., Ecosystems and Env.*, **13**, 33–42.

Cadatal, T. D. (1969). *Effect of Chemical Pesticides on the Development of Fungi Pathogenic to some Rice Insects*. M.S. Thesis (unpublished), Dept. of Entomology, College of Agriculture, Univ. of the Phil., College, Laguna, Philippines.

Castro, T. F., and Yoshida, T. (1971). Degradation of organochlorine insecticides in flooded soils in the Philippines. *J. Agr. Fd. Chem.*, **19**, 375.

Catahan, M. P. (1977). *The Persistence of Carbofuran in the Rice Paddy Water*. M.S. Thesis in Chemistry, Univ. of the Phil. at Los Banos, College, Laguna, Philippines.

Chelliah, S., and Heinrichs, E. A. (1984). Factors contributing to brown planthopper resurgence. *Proc. FAO/IRRI Workshop on judicious and efficient use of insecticides on rice*, pp. 107–115. International Rice Research Institute, Los Banos, Laguna, Philippines.

Cho, S. Y., Lee, S. R., and Ryu, J. K. (1978). Effects of carbofuran root-zone placement on the spider populations in the paddy fields. *Korean J. of Plt. Prot.*, **17** (2), 99–104.

CLSU-FAC (1978a). *Technical Report No. 13*. Freshwater Aquaculture Center, Central Luzon State University, Munoz, Nueva Ecija, Philippines.

CLSU-FAC (1978b). *Technical Report No. 14*. Freshwater Aquaculture Center, Central Luzon State University, Munoz, Nueva Ecija, Philippines.

CLSU-FAC (1979a). *Technical Report No. 15*. Freshwater Aquaculture Center, Central Luzon State University, Munoz, Nueva Ecija, Philippines.

CLSU-FAC (1979b). *Technical Report No. 16*. Freshwater Aquaculture Center, Central Luzon State University, Munoz, Nueva Ecija, Philippines.

CLSU-FAC (1980). *Technical Report No. 17*. Freshwater Aquaculture Center, Central Luzon State University, Munoz, Nueva Ecija, Philippines.

CLSU-FAC (1981a). *Technical Report No. 19*. Freshwater Aquaculture Center, Central Luzon State University, Munoz, Nueva Ecija, Philippines.

CLSU-FAC (1981b). *Technical Report No. 20*. Freshwater Aquaculture Center, Central Luzon State University, Munoz, Nueva Ecija, Philippines.

FAO (1981). *Proc. Second Expert Consultation on Environmental Criteria for Registration of Pesticides*. Food and Agriculture Organization, Plant Production and Protection Paper No. 28.

Gavarra, M. R., and Raros, R. S. (1973). Studies on the biology of the predatory wold spider, *Lycosa pseudoannulata* Boes. Et. Str. (Araneae, Lycosidae). *Philipp. Ent.*, **2** (6), 427–444.

Gorbach, S., Haarring, R., Knauf, W., and Werner, H. J. (1971a). Residue analysis in the water system of East-Java (River Brantas, Ponds, Sea-water) after continued large-scale application of Thiodan in rice. *Bull. Env. Contam. Toxicol.*, **6** (1), 40–47.

Gorbach, S., Haarring, R., Knauf, W., and Werner, H. J. (1971b). Residue analyses and biotests in rice fields of East Java treated with Thiodan. *Bull. Env. Contam. Toxicol.*, **6** (3), 163–199.

Guenzi, W. D., and Beard, W. E. (1968). Anaerobic conversion of DDT to DDD and anaerobic stability in soil. *Proc. Soil Sci. Soc. Am.*, **32**, 522–534.

Guerrero, L. A., and Guerrero, R. D. III. (1980). Preliminary studies on the culture of edible freshwater snails in Central Luzon, Philippines. *CLSU Sci. J.*, **1** (1), 11–14.

Hashimoto, Y. (1982). Effect of pesticides on aquatic organisms and their environment. *J. Pestic. Sci.*, **7** (3), 281–287.

Hill, D. W., and McCarty, P. K. (1967). Anaerobic degradation of selected chlorinated hydrocarbon pesticides. *J. Wat. Pollut. Control Fed.*, **39**, 1259–1277.

Hirano, C., and Yushima, T. (1969). *Jap. J. Appl. Ent. Ecol.*, **13**, 174–184 (from Masuda, 1979).

IRRI (1981). *Annual Report for 1980*. International Rice Research Institute, Los Banos, Laguna, Philippines.

IRRI (1983a). *Annual Report for 1981*. International Rice Research Institute, Los Banos, Laguna, Philippines.

IRRI (1983b). *Annual Report for 1982*. International Rice Research Institute, Los Banos, Laguna, Philippines.

IRRI (1984). *Annual Report for 1983*. International Rice Research Institute, Los Banos, Laguna, Philippines.

JICA (1981). *Contributions to the Development of Integrated Rice Pest Control in Thailand*. Japan International Cooperation Agency.

Kanauchi, M., Uchida, M., and Tsuchiya, K. (1982). Persistence of isoprothiolane in paddy water and rice plants after submerged applications. *J. Pestic. Sci.*, **7** (3), 377–383.

Kanazawa, J. (1981). Bioconcentration potential of pesticides by aquatic organisms. *Japan Pestic. Info.*, **39**, 12–16.

Knauf, W., and Schulze, E. F. (1973). New findings on the toxicity of endosulfan and its metabolites to aquatic organisms. *Mededelingen Fakulteit Landbouwwetenschappen*, Gent., **38**, 717.

Mabbayad, B. B., Obias, R. O., and Calendacion, R. T. (1983). Rice culture systems in the Philippines. In: *Rice Production Manual*, Philippines. College of Agriculture, University of the Philippines at Los Banos College, Laguna, Philippines.

MacRae, I. C., Raghu, K., and Castro, T. F. (1967). Persistence and biodegradation of four common isomers of benzene hexachloride in submerged soils. *J. Agr. Fd. Chem.*, **15**, 911–914.

MacRae, I. C., Raghu, K., and Bautista, E. M. (1969). Anaerobic degradation of the insecticide lindane by *Clostridium* sp. *Nature*, **221**, 859–860.

Magallona, E. D. (1980). *Pesticide Management*. Fertilizer and Pesticide Authority, Philippines.

Magallona, E. D. (1983). Persistent insecticides in the tropical agroecosystem. *Philipp. Ent.*, **6** (5–6), 567–595.

Magallona, E. D., and Mercado, B. L. (1978). Pesticide use in the Philippines. In: *Pesticide Management Southeast Asia*; *Proc. of the Southeast Asian Workshop in Pesticide Management*. BIOTROP Special Publication No. 7, pp. 71–77.

Magallona, E. D., Bajet, C. M., and Barredo, M. J. V. (1985). Bound residues of isoprocarb in some components of the rice paddy ecosystem. In: *Quantification, Nature and Bio-availability of Bound 1C-Pesticide Residues in Soils, Plants and Food*. International Atomic Energy Agency, Vienna.

Malik, M. A., and Khan, D. U. (1982). The process (testing, clearances, etc.) involved in recommending pesticides in Bangladesh. *Proc. FAO Working Group Meeting on Pesticide Use and Specificity*, Bangkok, Thailand, Nov. 23–26.

Masuda, T. (1979). Behaviour of pesticides applied to rice seedlings and paddy water. In: *Sensible Use of Pesticides*, pp. 35–48. Food and Fertilizer Technology Center Book Series No. 14, Taiwan, ROC.

Matsumura, F., Benezet, H. J., and Patil, K. C. (1976). Factors affecting microbial metabolism of v-BHC. *J. Pestic. Sci.*, **1** (1), 3–8.

Medina-Lucero, C. (1980). *The Dynamics of Transport and Distribution of Two Organochlorine Insecticides (Lindane and Endosulfan) in a Lowland Rice Field Ecosystem*. PhD Dissertation in Chemistry, University of the Philippines at Los Banos, College, Laguna, Philippines.

Medrana, G. T. (1983). Rice Production Statistics. In: *Rice Production Manual Philippines*, pp. 1–26. College of Agriculture, University of the Philippines at Los Banos, College, Laguna, Philippines.

Meier, P. G., Fook, D. C., and Lagler, K. F. (1983). Organochlorine pesticide residues in rice paddies in Malaysia, 1981. *Bull. Environ. Contam. Toxicol.*, **30** (3), 351–357.

Mikami, N., Takahashi, N., Hayashi, K., and Miyamoto, J. (1980). Photodegradation of denvalerate (Sumicidin) in water and on soil surface. *J. Pestic. Sci.*, **5** (2), 225–236.

Muralikrishna, P. V. G., and Venkateswarlu, K. (1984). Effect of insecticides on soil algal population. *Bull. Environ. Contam. Toxicol.*, **33** (2), 241–245.

Nagata, T., and Mochida, O. (1984). Development of insecticide resistance and tactics for prevention. *Proc. FAO/IRRI Workshop on Judicious and Efficient Use of Insecticides on Rice*, pp. 93–106. International Rice Research Institute, Los Banos, Laguna, Philippines.

NCPC (1982). *Annual Report for 1981*. National Crop Protection Center, College, Laguna, Philippines.

Nebeker, A. V., McCrady, J. K., Shar, R. M., and Mcauliffe, C. K. (1983). Relative sensitivity of *Daphnia magna*, rainbow trout and fathead minnows to endosulfan. *Env. Toxicol. Chem.*, **1**, 69–72.

Obcemea, W. N., Mikkelsen, D. S., and de Datta, S. K. (1977). *Factors Affecting Ammonia Volatilization Losses from Flooded Environment Rice*. Paper presented at the 8th Annual Meeting of the Crop Science Society of the Philippines, May 5–7.

Ogawa, K., Tsuda, M., Yamauchi, F., Yamauchi, I., and Misato, T. (1976). Metabolism of 2-sec-butylphenyl N-methylcarbamate (Bassa, BPMC) in rice plants and its degradation in soils. *J. Pestic. Sci.*, **1** (3), 219–229.

Ogawa, K., Tsuda, M., Yamauchi, F., and Misato, I. (1977). Metabolism of 2-isopropylphenyl N-methylcarbamate (Mipcin, MIPC) in rice plants and its degradation in soils. *J. Pestic. Sci.*, **2** (1), 51.

Oyamada, M., Igarashi, K., and Kuwatsuka, S. (1980). Degradation of the herbicide naproanilide, 1-(2-naphthoxy) propioanilide, in flooded soils under oxidative and reductive conditions. *J. Pestic. Sci.*, **5** (4), 495–501.

Partoatmodjo, S., and Alimoeso, S. (1982). *Pesticide Use and Specificity in Indonesia*. Country paper presented during the FAO Working Group meeting on pesticide use and specificity, Bangkok, Thailand, Nov. 23–26.

Perez, G. D. D. (1981). *A Bioassay of the Effects of Pesticides on Rice Field Ostracods (Crustacea) and their Consequent Predation on Blue Green Algae*. B.S. Thesis (unpublished), Dept. of Life Sciences (now Institute of Biol. Sci.), University of the Philippines at Los Banos, College, Laguna.

Peries, I. D. R. (1982). Pesticide use and specificity in Sri Lanka. *Proc. FAO Working Group Meeting on Pesticide Use and Specificity*, Bangkok, Thailand, Nov. 23–26.

Raghu, K., and MacRae, I. C. (1966). Biodegradation of the gamma isomer of benzene hexachloride in submerged soils. *Science*, **154**, 263–264.

Raghu, K., and Drego, J. (1985). *Bound Residues of Lindane: Magnitude, Microbial Release, Plant Uptake and Effects on Microbial Activities*. Paper presented at the 3rd Research Coordination Meeting on Isotopic Tracer-aided Studies on Bound Residues in Soils, Plants and Food, Gainesville, Florida, USA, March 25–29.

Rajak, R. L. (1982). *India*. Country paper presented during the FAO Working Group Meeting on pesticide use and specificity, Bangkok, Thailand, Nov. 23–26.

Ramamoorthy, S. (1985). Competition of fate processes in the bioconcentration of lindane. *Bull. Environment. Contam. Toxicol.*, **34** (3), 349–358.

Reddy, D. B. (1978). Future trends of pesticide use in the world with particular reference to Asia and the Pacific. In: *Pesticide Management Southeast Asia; Proc. Southeast Workshop on Pesticide Management*. BIOTROP Special Publication No. 7, pp. 1–23.

Rice Production Manual (1969). College of Agriculture, University of the Philippines, Philippines.

Rice Production Manual (1983). University of the Philippines at Los Banos, College of Agriculture, College, Laguna, Philippines.

Rumakon, M., Vungsilabutr, P., and Katanyakul, W. (1982). *Thailand*. A country paper presented during the FAO Working Group Meeting on pesticide use and specificity, Bangkok, Thailand, Nov. 23–26.

Sanchez, F. F. (1983). Pest Management: status and potential contribution to Philippine agriculture. *Philipp. Ent.*, **6** (5–6), 595–606.

Santiago, D., and Magallona, E. D. (1982). Toxicity of different insecticides to *Tilapia nilotica*. *1981 Annual Report*. National Crop Protection Center, College, Laguna, Philippines.

Sastrodihardjo, S., Adianto, M., and Yusoh, M. D. (1978). The impact of several insecticides on soil and water communities. *Proc. Southeast Asian Workshop on Pesticide Management*, pp. 117–125. Bangkok, Thailand.

Seiber, J. N., Heinrichs, E. A., Aquino, G. B., Valencia, S. L., Andrade, P., and Argente, A. M. (1978). Residues of carbofuran applied as a systemic insecticide in irrigated wetland rice: implications in insect control. *IRRI Research Paper 17* (May).

Sethunathan, N. (1973). Microbial degradation of insecticides in flooded soil and in anaerobic cultures. *Residue Reviews*, **47**, 143–165.

Sethunathan, N., and MacRae, I. C. (1969). Persistence and biodegradation of diazinon in submerged soils. *J. Agr. Fd. Chem.*, **17** (2), 221–225.

Sethunathan, N., and Yoshida, T. (1969). Fate of diazinon in submerged soil. Accumulation of hydrolysis product. *J. Agr. Fd. Chem.*, **17**, 1192–1195.

Sethunathan, N., and Pathak, M. D. (1972). Increased biological hydrolysis of diazinon after repeated application to rice paddies. *J. Agr. Fd. Chem.*, **20**, 586–589.

Sethunathan, N., Bautista, E. M., and Yoshida, T. (1969). Degradation of benzenehexachloride by a soil bacterium. *Can. J. Microbiol.*, **15**, 1349–1354.
Sethunathan, N., Siddaramappa, R., Gowda, T. K., and Rajarah, K.P. (1975). Pesticide residue problems in flooded rice ecosystem. *Proc. FAO/UNEP Expert Consultation on Impact Monitoring of Residues from the Use of Agricultural Pesticides in Developing Countries*. Rome, Sept. 29–Oct. 3.
Shea, T. B., and Berry, E. S. (1983). Toxicity of carbaryl and 1-naphthhol to goldfish (*Carassius auratus*) and killifish (*Fundulus heteroclitus*). *Bull. Environment. Contam. Toxicol.*, **31** (5), 526–529.
Siddaramappa, R., Tirol, A. C., Seiber, J. N., Heinrichs, E. A., and Watanabe, I. (1977). *Loss of insecticide carbofuran (Furadan) from standing water and in flooded soil.* Paper presented at the 8th Annual Convention of the Pest Control Council of the Philippines, Bacolod City, Philippines.
Siddaramappa, R., Tirol, A., and Watanabe, I. (1979). Persistence in soil and absorption and movement of carbofuran in rice plants. *J. Pestic. Sci.*, **4** (4), 473–479.
Staring, W. D. E. (1984). *Pesticides: Data Collection Systems and Supply, Distribution and Use in Selected Countries of the Asia-Pacific Region*. Economic and Social Commission for Asia and the Pacific, Bangkok, Thailand.
Stephenson, R. R., Choi, S. Y., and Olmos-Jerez, A. (1984). Determining the toxicity and hazard to fish of a rice insecticide. *Crop Prot.*, **3** (2), 151–165.
Takase, I., and Nakamura, H. (1974). *Nippon Nogeikagaku Kaishi*, **48**, 27–32 (from Masuda, T. 1979).
Takase, I., Tsuda, H., and Yoshimoto, Y. (1971). *Jap. J. Appl. Ent. Zool.*, **15**, 63–69 (from Masuda, T. 1979).
Takashi, I., and Masui, A. (1974). *Jap. J. Appl. Ent. Zool.*, **18**, 171–176 (from Masuda, 1979).
Tanaka, F., Usuda, Y., Shibata, Y., Yakumaru, K., Yamamoto, M., Kondo, T., Hayashi, K., and Aida, S., (1976). *Zenno Hoyakushiken Seiseki*, 501–509 (from Masuda, T. 1979).
Tejada, A. W. (1983). *Fate of carbosulfan in rice paddy-ecosystem*. PhD Dissertation, University of the Philippines at Los Banos, College, Laguna, Philippines.
Than, H. (1976). *Ecology of the Yellow Rice Borer*, Tryporyza incertulas *(Walker) and its Damage to the Rice Plant*. M.S. Thesis (unpublished), Dept. of Entomology, College of Agriculture, University of the Philippines, College, Laguna, Philippines.
Tsuge, S., Nishimura, T., Kazano, H., and Tomizawa, C. (1980). Uptake of pesticides from aquarium tank water by aquatic organisms. *J. Pestic. Sci.*, **5** (4), 585–593.
URARTIP (Unified Rice Action Research and Training Intensification Programme). (1985). *16 Steps Masagana-99 Rice Culture*. Bureau of Agricultural Extension. Diliman, Quezon City.
Varca, L. M., and Magallona, E. D. (1982). Residues of BPMC (2-sec-butyl phenyl N-methylcarbamate) on some components of a paddy rice ecosystem. *Proc. 13th Annual Convention of the Pest Control Council of the Philippines*, Baguio City, May 5–8.
Venkateswarlu, K., and Sethunathan, N. (1984). Degradation of carbofuran by *Azospirillum lipoferum* and *Streptomyces* spp. isolated from flooded alluvial soil. *Bull. Environment. Contam. Toxicol.*, **33**, 556–560.
Villegas, L. M., and Feuer, R. (1970). The 'lowland' or flooded soils. In: *Rice Production Manual*, pp. 68–73. UPLB/IRRI.
Yamato, Y., Kiyonaga, M., and Watanebe, T. (1983). Comparative bioaccumulation and elimination of HCH isomers in shortnecked clam (*Venerupis japonica*) and guppy (*Poecilia reticulata*). *Bull. Environ. Contam. Toxicol.*, **31** (3), 352–359.
Zulkifli, M., Tejada, A. W., and Magallona, E. D. (1983). The fate of BPMC and chlorpyrifos in some components of a paddy rice ecosystem. *Philipp. Ent.*, **6** (5–6), 555–565.

Ecotoxicology and Climate
Edited by P. Bourdeau, J. A. Haines, W. Klein and C. R. Krishna Murti

5.4 Fate and Effects of Aldrin/ Dieldrin in Terrestrial Ecosystems in Hot Climates

I. SCHEUNERT

5.4.1 INTRODUCTION

Aldrin and its epoxide dieldrin are largely used in countries with hot climates, both as agricultural insecticides and for the control of tsetse fly, the vector of human and animal trypanosomiasis, and other insects which are sources of various human and animal diseases. Whereas in industrial countries the use of aldrin and dieldrin has been restricted or banned within the last decade, their use continues in developing countries with hot climates.

It is self evident that the fate and residue behaviour of both these insecticides in terrestrial ecosystems of hot climates are different from their fate in temperate climates. The fate in temperate climates has been investigated in numerous studies reported in literature (e.g. Elgar, 1966; Kohli *et al.*, 1973; Klein *et al.*, 1973; Stewart and Gaul, 1977). As a consequence of the differences in persistence, differences are also evident in the effects on ecosystems. However, only limited information is available on these differences. This study attempts to review experiments carried out in terrestrial ecosystems under tropical, subtropical, and Mediterranean conditions and to evaluate them with regard to data in temperate climates.

5.4.2 RESIDUE BEHAVIOUR OF ALDRIN/DIELDRIN IN SOILS IN HOT CLIMATES

In general, two climatic factors affect the residue behaviour and persistence of chemicals in terrestrial ecosystems: temperature and humidity; composition of soil to a considerable extent is also directly related to climate. The influence of these climatic factors on each of the mechanisms and pathways involved in residue decline in soil affects the persistence of chemicals in soil. The major mechanisms and pathways involved in residue decline in soil are volatilization, mobility and leaching, and degradation, including primary degradation (mainly conversion), as well as total degradation to carbon dioxide (mineralization). In the following paragraphs, studies on each of these pathways related to aldrin/dieldrin in hot climates will be discussed.

5.4.2.1 Volatilization

Field studies reporting direct measurements of dieldrin in the air above treated areas in temperate climates have shown that volatilization is an important pathway of residue loss (Caro and Taylor, 1971; Willis *et al.*, 1972; Taylor *et al.*, 1976; Turner *et al.*, 1977). Climatic factors have a considerable influence on this pathway. Since higher temperature results in higher vapour pressure of aldrin and dieldrin, volatilization of pure as well as of adsorbed substances increases with increasing temperature. Kushwaha *et al.* (1976) reported a rapid loss of aldrin from a glass plate at higher temperatures. Similarly, temperature affects volatilization of aldrin and dieldrin from moist soils.

Today, it is assumed that volatilization from moist soils occurs mainly in the liquid phase (Hamaker, 1972). Therefore, a decrease in adsorption due to climatic or soil factors will result in an increase in residue loss from soil. Since adsorption is negatively correlated to water solubility (Kenaga and Goring, 1978; Chiou *et al.*, 1979; Felsot and Dahm, 1979; Briggs, 1981) which, in turn, is positively related to temperature, an increase in temperature will result in an increase in the portion of chemical desorbed in soil solution. The volatilization of chemicals from aqueous solutions, too, depends strongly on temperature, as regards both the temperature dependence of vapour pressure and of liquid-phase transfer velocities (Mackay and Yuen, 1983; Downing and Truesdale, 1955; Wolff and van der Heijde, 1982). For gas-phase controlled substances, such as aldrin and dieldrin, temperature-related increase in vapour pressure results in a considerable increase in volatilization.

Harris and Lichtenstein (1961) found that increased temperature increased the volatilization of aldrin from soil; for a Plainfield sand, an increase in temperature of about 10°C increased the rate of volatilization more than twofold. Farmer *et al.* (1972) demonstrated, for Gila silt loam, that an increase in temperature of 10°C increased the rate of volatilization of dieldrin approximately fourfold.

Higher moisture content of soils also influences volatilization positively (Spencer *et al.*, 1973). This increase in volatilization is not due to 'co-distillation' phenomena but to a displacement of the insecticides by water from the adsorption sites (Igue *et al.*, 1972).

From all these observations reported, it may be assumed that the increase in residue dissipation observed in hot and moist climates, as discussed in paragraph 5.4.2.4, is due more to enhanced volatilization than to enhanced degradation.

5.4.2.2 Mobility and Leaching

As for volatilization from soil, adsorption of chemicals in soil is also a key parameter for their mobility and leaching. Adsorption coefficients of chemicals in soil vary largely depending on soil type, and especially on the organic carbon content of the soil (Lambert *et al.*, 1965; Spencer *et al.*, 1973; Felsot and Dahm, 1979; Rippen *et al.*, 1982). Therefore, adsorption of aldrin/dieldrin in tropical soils will be greater or smaller than in soils of temperate climates, depending on the organic carbon content. In moist tropical and subtropical regions, weathering and mineralization are very intensive due to high temperatures and high moisture contents. For the same reason, bioactivity in soil and hence degradation of dead plant material are more intensive, often resulting—in spite of higher plant growth and plant decay—in a lower humus content than in central European soils. However, there exist also tropical soils with high organic carbon contents (Scheffer and Schachtschabel, 1982).

Baluja *et al.* (1975) demonstrated, for three soils from Spain with an organic matter content between 1.7 and 7.9%, a high adsorption rate of aldrin and a nearly zero desorption rate. Thin-layer chromatography on seven Brazilian soils, with organic matter contents of between 0.6 and 13.1%, demonstrated that aldrin was strongly adsorbed and did not move from the point of application for any soil (Lord *et al.*, 1978).

Temperature may also influence the adsorption which is usually exothermic. Higher temperature probably decreases adsorption and releases insecticides. Solubility of insecticides is also temperature dependent, thus leading to a decrease in adsorption when the temperature rises and more of the adsorbed insecticide becomes dissolved in soil water (Edwards, 1966).

Under environmental conditions, rainfall is another important factor affecting mobility of chemicals in soil. Therefore, mobility of aldrin/dieldrin in soil is greater in moist climates than in dry ones. A comparative outdoor lysimeter study in a moderately moist climate (Germany) and a warm dry climate (Spain) showed that, within one vegetation period, 21% of the aldrin recovered after application to a depth of 10 cm had moved to deeper layers in the moist climate, whereas only 3.5% had done so in the Mediterranean climate. These differences were mostly due to marked differences in rainfall. This conclusion was confirmed by further experiments in other countries (Weisgerber *et al.*, 1974).

However, at least for unchanged aldrin mobility in soil is of minor importance for residue decline in soil. Conversion to dieldrin and other, more polar compounds, followed by leaching of these, is a more important pathway of residue loss from soils. This probably applies to both temperate and hot climates.

5.4.2.3 Degradation (Conversion and Mineralization)

Primary degradation of a chemical is the disappearance of the parent compound by chemical reactions of every kind, including small alterations in the molecule as well as total mineralization to carbon dioxide. For aldrin in the soil-plant

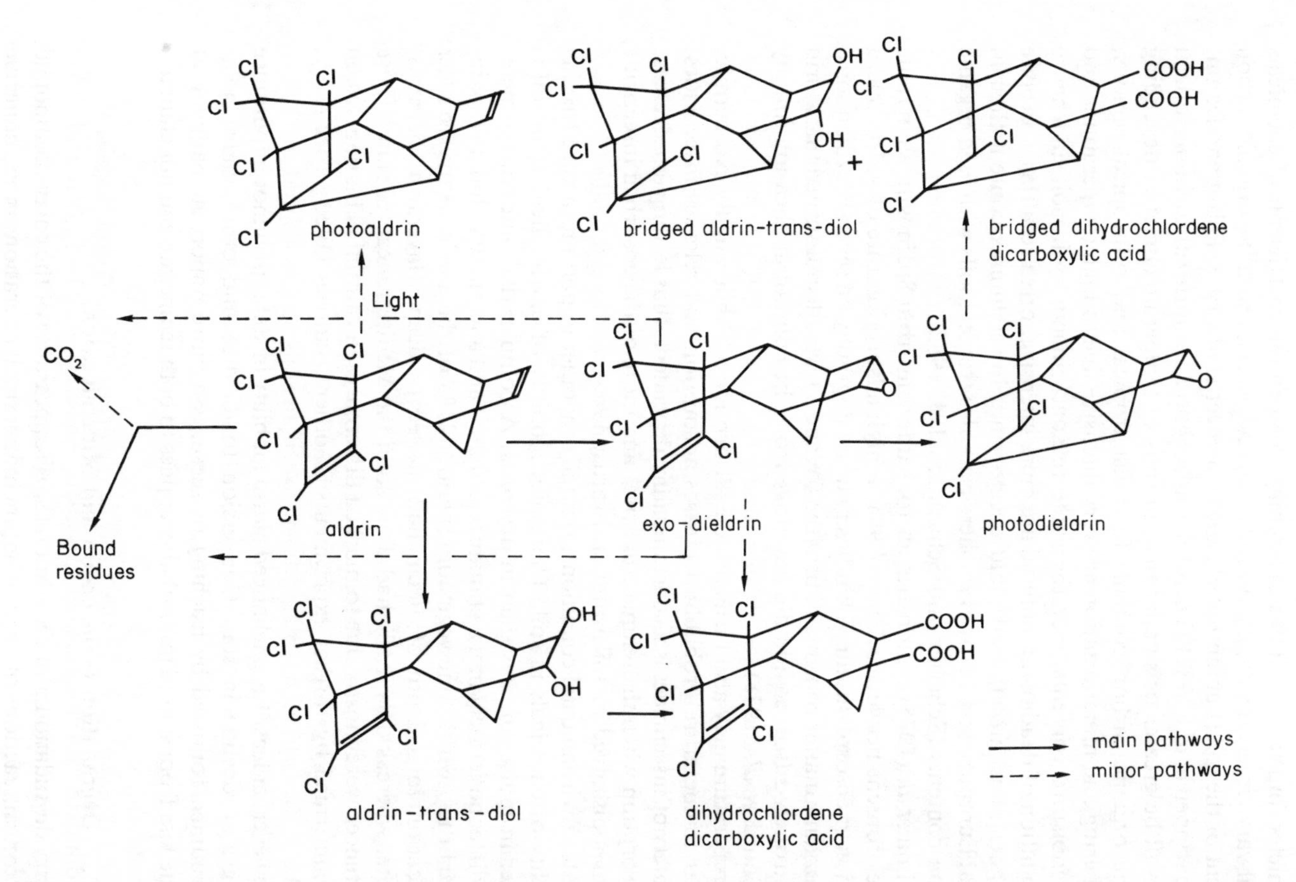

Figure 5.4.1 Conversion pathways of aldrin in plants and soil under outdoor conditions (Scheunert *et al.*, 1977). Reproduced with permission of Academic Press

system, primary degradation comprises all pathways shown in Figure 5.4.1 (Scheunert *et al.*, 1977).

Studies in a temperate climate with ^{14}C-aldrin have shown that the only soluble conversion products of aldrin which were quantitatively significant ($>1\%$ of total ^{14}C-residues) were dieldrin, photodieldrin, and the ring cleavage product, dihydrochlordene dicarboxylic acid. For dieldrin, the only significant soluble conversion product was photodieldrin—a compound which does not represent to any relevant degree a step towards smaller molecules since it was shown to be metabolized only by less than 2% within one year (Weisgerber *et al.*, 1975). Residues bound in soil were formed both from aldrin and dieldrin—from aldrin about 11%, from dieldrin about 1% of total recovered residues within one growing period (Sotiriou *et al.*, 1981). For aldrin, about half of the soil-bound residues could be released by dilute alkali solution and were found to be dihydrochlordene dicarboxylic acid. The chemical nature of the remaining portion of bound residues is not known.

Since microbial processes are normally accelerated by higher temperatures, it is assumed that conversion reactions are higher at higher temperatures. In a comparative outdoor lysimeter study with ^{14}C-aldrin for one vegetation period, unchanged aldrin constituted about 50% of the total ^{14}C residue in soil (0–10 cm depth) in a cool temperature climate, whereas it was 38% in a warm Mediterranean climate. This corresponds to 59 or 58% aldrin, respectively, based on the sum of aldrin and dieldrin, irrespective of other metabolites. A corresponding figure from India for the same time period was 50% (Agnihotri *et al.*, 1977). In another Indian study, aldrin represented more than 90% of the sum of aldrin and dieldrin after 91 days under beet cover (Gupta and Kavadia, 1979). According to these authors, soil moisture content is responsible for differences in aldrin epoxidation.

Other authors (Lichtenstein and Schulz, 1959a; Kushwaha *et al.*, 1978) found a positive temperature-dependence for the conversion of aldrin to dieldrin—i.e. increase with increasing temperature. The conversion of aldrin to polar soluble and soil-bound products—i.e. conversion steps initiating a real degradation—was higher in a warm climate (Weisgerber *et al.*, 1974).

The ability of fungi isolated from Brazilian soils to adsorb and metabolize ^{14}C-aldrin and its metabolites was assayed in a culture growth medium after 76 days of incubation (Musumeci *et al.*, 1982). All the 14 isolates incorporated the radiocarbon as demonstrated by wet combustion of the mycelium. Four of the fungi were able to further metabolize one of the compounds added to the medium.

Total mineralization (utilization) of dieldrin by fungi from Sudanese soils, without detectable intermediate products, was concluded by El Beit *et al.* (1981) from a substantial difference in dieldrin recovery between inoculated samples and controls. However, the utilization of dieldrin as a carbon source could not be demonstrated in carbon-free mineral-salt media since fungi did not grow there.

It may be concluded that an acceleration of conversion and mineralization of aldrin/dieldrin by soils in tropical or subtropical regions could take place, but that an unequivocal demonstration of this is still lacking due to an insufficiency in exact comparative studies.

5.4.2.4 Total Residue Losses from Soils

Total residue losses of aldrin/dieldrin from soil, which represent the sum of losses by volatilization, mobility and leaching, conversion, and degradation, were dependent upon temperature, as measured by several authors.

Kiigemagi *et al.* (1958) found a more rapid disappearance in summer than in winter of aldrin and dieldrin residues. Lichtenstein and Schulz (1959b) found that no aldrin was lost in frozen soils, whereas considerable losses were observed at 7°C, 26°C, and 46°C, and the losses increased with increasing temperature (Figure 5.4.2). In contrast to these findings, Kushwaha *et al.* (1978) reported a shorter half-life of aldrin residues at lower temperatures (25°C) than at higher temperature (35–45°C). Probably in this case, soil microorganisms had optimal conditions at 25°C as compared to higher temperatures. Furthermore, in the bags used in the laboratory, volatilization of aldrin was largely suppressed. Under environmental conditions, however, temperatures of 45°C or more are very rare in agricultural soils, and volatilization is efficient. Therefore, in field studies, an increase in residue loss was generally observed for hot climates as compared to temperate climates.

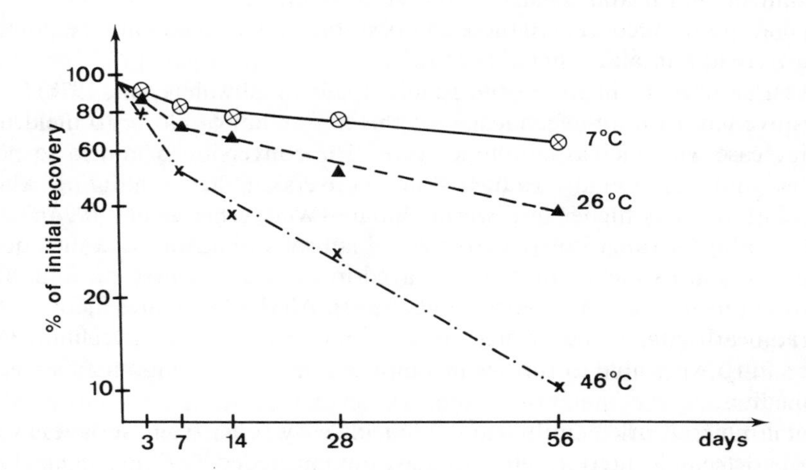

Figure 5.4.2 Loss of aldrin residues from a Plainfield sand as affected by temperature. Application rate: 100 lb/6″ acre (Lichtenstein and Schulz, 1959b). Reproduced with permission of the Entomological Society of America

Atabaev *et al.* (1970) conducted field experiments under arid hot climatic conditions in Uzbekistan (South Russia). They found that aldrin disappeared from the upper soil layer (0–30 cm depth) after two years, and from deeper layers (70–100 cm depth) after five years or more.

Agnihotri *et al.* (1976, 1977) reported a high loss of aldrin and its metabolite dieldrin under field conditions in India: 87.9% after 100 days, 98.7% after 180 days. Under different Indian climatic conditions, Kathpal *et al.* (1981) registered an aldrin dissipation of 89–93% within three months and 92–100% within 8.5 months (aldrin only), corresponding to 69–77% after 8.5 months when the sum of aldrin and its metabolite dieldrin was considered. Similarly, Chawla *et al.* (1981) found an aldrin reduction of 89% in soil during the potato-growing period in India. For soil under the cover of sugar beets, the reduction of aldrin and dieldrin was 65% within 91 days (Gupta and Kavadia, 1979). Within 84 or 120 days, about 76–93% of aldrin were lost from soil under field conditions in Udaipur, India (Kushwaha *et al.*, 1981).

Under subtropical conditions in Taiwan, a long-term experiment was undertaken to investigate the persistence of dieldrin following its repeated seasonal application to soil (Talekar *et al.*, 1977). Dieldrin was sprayed or broadcast uniformly and rototilled immediately to a depth of 15 cm; soil samples were taken also to a depth of 15 cm. The decline in the concentration of dieldrin was 25% in the fall and winter; additional treatment during the following spring did not lead to an accumulation of dieldrin residues in soil, and the concentration at the end of summer was virtually identical with that immediately before the treatment. The persistence of dieldrin in this subtropical area thus appeared to be much shorter than under temperate conditions.

In contrast to this study where dieldrin was incorporated into the soil, another study carried out in the Sudan (El Zorgani, 1976) investigated the persistence of aldrin and dieldrin after surface application. In Figure 5.4.3, the disappearance of soil surface residues of aldrin and dieldrin is presented. The figure shows remarkably fast rates of loss of surface deposits of both insecticides. The residues of aldrin are expressed as the sum of aldrin plus its metabolite dieldrin. The faster disappearance rate of aldrin-derived residues as compared to that of dieldrin-derived residues is probably due, on the one hand, to the higher vapour pressure of aldrin and, on the other hand, to additional degradation pathways of aldrin through routes not involving dieldrin formation (*see* Figure 5.4.1).

Elgar (1975) carried out a comprehensive comparative investigation of the dissipation and accumulation of aldrin-derived residues (aldrin plus dieldrin) in soil at 12 sites in different climatic zones. Aldrin was incorporated immediately after treatment with a rotovator to a maximum depth of 15 cm, and samples were taken also to a depth of 15 cm. Five years of study were reported. The results demonstrated that the difference between the rate of loss at the cool (Northern and Central Europe) and warm (Mediterranean) temperate sites was

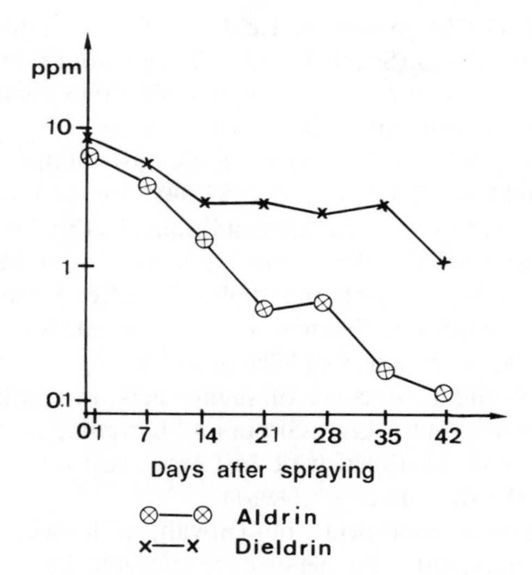

Figure 5.4.3 Disappearance of aldrin and dieldrin residues from soil after surface application in Sudan (El Zorgani, 1976). Reproduced by permission of Springer-Verlag, New York

small, but that the rate of loss at the tropical sites was greater in the first year. There was no correlation between the rate of loss and any of the soil parameters.

The field studies described confirm the conclusions drawn from laboratory experiments reported in paragraphs 5.4.2.1–5.4.2.3, namely that some of the routes of residue disappearance are affected positively by higher temperatures and/or by higher soil moisture contents. It is assumed that the enhanced residue loss under subtropical and tropical conditions is due largely to an increase in volatilization, but degradation is probably also faster in subtropical and tropical soils.

5.4.3 UPTAKE BY PLANTS AND PERSISTENCE ON PLANT SURFACES

Uptake of chemicals by plants is a complex process comprising separate routes, such as root uptake, foliar uptake of vapours in the air or of deposits of sprays or of dust, or uptake through oil cells of lipid-containing plants (Topp *et al.*, 1986; Hulpke and Schuphan, 1970). In view of the assumed negative temperature-dependence of soil adsorption discussed in paragraphs 5.4.2.1 and 5.4.2.2, both root and foliar uptake should be positively influenced by temperature. However, comparative studies for aldrin and dieldrin under both temperate and tropical or subtropical conditions, have not been reported thus far.

In a comparative outdoor lysimeter study, maize root concentration factors, expressed as concentration of aldrin residues in maize roots divided by concentration in soil, were higher in a temperate climate (Central Europe) than in a warm (Mediterranean) climate (Weisgerber *et al.*, 1974). Like the differences in soil mobility discussed in paragraph 5.4.2.2, observed in the same experiment, these differences in plant uptake are probably due to the high differences in rainfall.

Chawla *et al.* (1981) observed a concentration of 0.09–0.135 ppm aldrin and dieldrin in potatoes in India, under field conditions, when 1.875 kg/ha aldrin was applied. The corresponding figures from German and English lysimeter experiments (2.9 kg/ha) were 0.14 or 0.22 ppm (Scheunert *et al.*, unpublished). However, much higher residue levels of these insecticides in potatoes following soil treatments have been reported from other Indian sites.

It may be concluded that the influence of climate on the uptake of aldrin/dieldrin by plants is not yet clarified. It appears that the differences observed are due to differences in soil properties rather than to climate. Further studies with different crops under hot climatic conditions are needed.

The persistence of dieldrin deposits on plant leaves after spray application in hot climates is another important aspect. The situation might be different from that in temperate climates, e.g. due to higher vapour pressures of dieldrin at higher temperatures. Figure 5.4.4 shows the persistence of dieldrin on leaves from the tree canopy (Koeman *et al.*, 1978) after helicopter applications in Nigeria. A rapid decline in the first two weeks after application is followed by a slower decline between day 14 and day 60 and an even slower one thereafter. This three-stage decline is similar to that observed for aldrin in a long-term (13-year) outdoor study in temperate soil (Scheunert, unpublished); however, the rate of decline is much higher on leaves. The formation of photodieldrin from dieldrin on leaves, a mainly photochemical process, should also be more rapid under tropical conditions. Indeed, after only two weeks, up to 20% of the total amount of dieldrin on leaves from tree canopy in Nigeria had changed to photodieldrin. By contrast, the conversion of dieldrin to photodieldrin on cabbage leaves in a temperate climate was only 5% after 4 weeks (Weisgerber *et al.*, 1970).

5.4.4 UPTAKE BY FAUNA AND EFFECTS ON ECOSYSTEMS

A comprehensive field study on the uptake of dieldrin by fauna and of its ecological effects has been carried out in Adamaoua, Cameroon, by Müller *et al.* (1981). Dieldrin was sprayed by a helicopter as an oil formulation with 18% active ingredient at a rate of 5 l/ha on a gallery forest, about 40 m wide and 3 km long. Some data on residue in faunal species, directly after spraying and one year afterwards, are listed in Table 5.4.1. Side-effects observed on the non-target fauna were summarized as follows. With a single treatment, at least

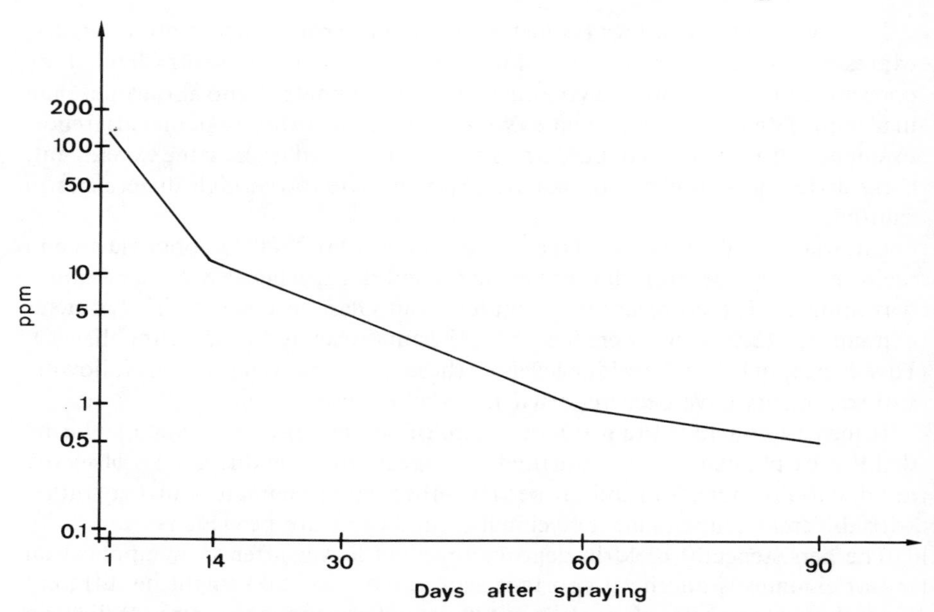

Figure 5.4.4 Disappearance of dieldrin from leaves of tree canopy in Nigeria (Koeman *et al.*, 1978). Reproduced by permission of Elsevier Applied Science Publishers Ltd

Table 5.4.1 Dieldrin content (ppm/fresh weight) in fauna after dieldrin spraying in Adamaoua, Cameroon (Müller *et al.*, 1981). Reproduced with permission of Springer-Verlag, Berlin

	Directly after spraying				One year after spraying			
Species	$\bar{X}/\tilde{X}$[1]	min.	max.	n[2]	$\bar{X}$[1]	min.	max.	n[2]
Praomys tullbergi (liver) (rat)	0.37	0.00	1.20	13	0.29	0.04	0.74	3
Micropteropus pusillus (liver) (fruit bat)	136.00	1.48	174.81	11	0.14	0.02	0.20	3
Halcyon malimbicus (liver) (insectivorous bird)	4.30	2.21	4.30	8	6.52	5.44	7.70	2
Nectarinia verticalis (liver) (nectarivorous, insectivorous bird)	1.80	1.03	2.44	5	0.24			1
Turdus pelios (liver) (polyphagous bird)	0.43	0.20	0.91	8	3.00	0.00	6.00	2
Dorylus spp. (Formicidae) (male)	0			(ca.10)	0.15	0.14	0.15	2 (ca.15)[3]
Lumbricidae (earthworms)					0.02	0.02	0.02	2 (ca.50)[3]

[1] $\bar{X}$ = mean value; for $n \geq 4$ the median value $\tilde{X}$ is given
[2] n = number of samples
[3] In brackets: number of individuals in mixed samples

10% of the whole above-ground biomass of invertebrates was destroyed in Adamaoua in order to extinguish one disease vector, i.e. *Glossina morsitans submorsitans*. One year after treatment, the arthropod fauna of the ground surface in the gallery forest as a whole showed a significant reduction in the abundance of individuals. Although these results might indicate a weakening of the stability of the biocoenosis in the face of exogenous influences, the diversity values simultaneously gave results which indicated a complete restoration of the inner stability of the biocoenosis. For the phytophagous insects of the herbaceous and foliage layers of the gallery forest (studies made of the Macroheterocera as an example) it was illustrated that the diversity of some biocoenoses was reduced by dieldrin treatment. It could further be proved that some non-target species in the treated area were destroyed by the pesticide.

In vertebrates, no acute mortality could be established after spraying. On the basis of the chemical residue analyses in connection with the analyses of the food web, it can be stated that fruit-eating birds might be endangered in the long run. The fruit-eating bats, which in some cases showed an especially high concentration of the noxious agent directly after treatment, could be found to have strongly reduced residue values after one year and moreover did not show any negative changes in their population structure. The residue values of some insect-eating birds indicate the risk of acute as well as of long-term damage, though even one year after spraying no negative population development could be observed. On the other hand, in the selected study area, insectivorous bats and shrews could no longer be recorded.

In spite of the evidence of an extremely strong direct effect of dieldrin on non-target terrestrial organisms, both with regard to the reduction in numbers of individuals and the extirpation of some species within the treated area, the investigations did not reveal any structural or energetic changes of the gallery forest ecosystem as a whole. This statement is valid provided that the acute loss of small insectivorous mammals can be compensated rapidly from uninjured regions.

A similar study was carried out by spraying 900 g/ha active ingredient dieldrin on a Northern Guinea-type savanna zone in Nigeria (Koeman *et al.*, 1978). Before, and at various intervals after spraying, a population census was made of a number of selected bird species. Certain species of fringe forest birds, such as various flycatcher species, appeared to be very vulnerable and either ceased to be recorded or became extremely rare in the treated areas.

The occurrence of residues of a chemical in animals of high trophic levels, such as birds of prey, is often regarded as a measure of its potential for ecological magnification. Frank *et al.* (1977) analysed pectoral muscle samples of 18 species of resident Kenyan raptors of different trophic levels for the presence of dieldrin, and compared the values with reported data of birds of prey in temperate climates. Many raptors from agricultural areas contained dieldrin whereas those from non-agricultural areas did not. While levels were generally low compared to those reported in populations of birds of prey from temperate latitudes, falcons and accipiters from highly agriculturalized areas, such as the region around Nairobi, contained residue levels of more than 2 ppm. The authors suggested that chlorinated hydrocarbon kinetics may not be the same in the tropics as in Northern latitudes inasmuch as terrestrial ecosystems seem to show higher levels than aquatic systems.

5.4.5 OCCURRENCE IN HUMAN FOOD AND HUMAN TISSUES

Müller *et al.* (1981) did not find noxious dieldrin levels in human foodstuffs one year after dieldrin application against tsetse flies in Cameroon. No residues were detected in cow's milk or wild honey, and only 0.03 ppm in beef. However, residues are expected after intentional application of the insecticide on food plants, animal feed, or on animals themselves.

Samples of wheat collected from Bombay markets in India revealed that aldrin residue levels in wheat were in the range of 0.50–0.08 ppm. Of the 18 samples contaminated with aldrin, two had residues above the tolerance limit prescribed by WHO (Krishna Murti, 1984). Aldrin residues in groundnut oil collected from the markets in two Indian districts (Lucknow and Sitapur) averaged 0.290 and 0.892 ppm respectively (Srivastava *et al.*, 1983). Twenty-five egg samples collected from Bombay markets showed a high incidence of contamination from dieldrin and aldrin, the residue levels being 0.61–1.04 ppm and 0.14–0.52 ppm (Banerji, 1979). The content of aldrin in farm eggs from Lucknow in India was 0.93 μg per egg, and in eggs from domestic hens, 0.40 μg per egg (Siddiqui and Saxena, 1983). The concentration of dieldrin and aldrin in muscles of fish from India was 0.03 ppm and 0.03–0.04 ppm respectively (Bhinge and Banerji, 1981). Average concentrations of aldrin in buffalo and goat milk were 0.041 ppm (Saxena and Siddiqui, 1982).

In Brazil, meat from cattle raised in the most developed agricultural regions of the state of Minas Gerais showed the highest residual levels. The mean quantity in the state was 0.02 mg/kg (Maia and Brant, 1980). In 32% of the samples taken from dairy products in South Africa (Luck and van Dyk, 1978), the dieldrin level (mean 0.13 mg/kg) exceeded the international maximum residue limit. The authors suggested that dieldrin is often misused, which probably is the case in other countries also.

The occurrence of pesticide residues in human food is closely related to that in human tissues. Table 5.4.2 (Hunter *et al.*, 1969) presents mean concentrations of dieldrin in adipose tissues of people of various countries, as well as daily intake estimated from these concentrations. The table reveals that concentrations in adipose tissue are lowest in India, with a mostly hot and moist climate, and in Australia, with a mostly subtropical climate, whereas they are higher for the countries in temperate climate zones. However, for Australia, higher values have

Table 5.4.2 Mean concentrations of dieldrin in human adipose tissues and estimated human intake in various countries (Hunter *et al.*, 1969). Reproduced with permision of Heldref Publications

Country	Mean concentration in adipose tissue (mg/kg)	Estimated daily intake (μg/person/day)
India	0.03	1.6*
Australia	0.05	2.7
United States	0.14	7.6*
Canada	0.16	8.6*
United Kingdom	0.23	12.4
New Zealand	0.27	14.6*

*Arithmetical mean

been reported too (0.67 ppm: Wassermann *et al.*, 1968; 0.21 ppm: Brady and Siyali, 1972). Since the residues in human fat depend on many factors, the variability of data is not a good basis for concluding that aldrin/dieldrin is less persistent in the tropics.

Table 5.4.3 shows the concentrations of dieldrin in the adipose tissue of the general population in South Africa (Wassermann *et al.*, 1970). The table indicates that factors such as sex and race exert an influence on the storage levels in a given area. Age is an important factor also. In Nigeria, dieldrin averaged between 0.002 ppm in the adipose tissue of the foetus and 0.18 ppm in the 25–44 year group (Wassermann *et al.*, 1972b). In Brazil, average dieldrin concentrations in adipose tissues were between 0.011 and 0.133 ppm (Wassermann *et al.*, 1972a); in Mexico, between 0.06 and 0.18 (Albert *et al.*, 1980); and in Uganda, between 0.021 and 0.038 (Wassermann *et al.*, 1974a). These data seem to indicate that the storage of dieldrin in these countries is low compared with other countries of Europe, North and South America, and Asia. A positive relationship between *p,p'*-DDT and dieldrin storage was also noted. This finding may be explained by a biochemical interrelationship of the two compounds in the body, the presence of a large amount of DDT interfering with the detoxification of dieldrin, resulting in its accumulation in adipose tissue (Wassermann *et al.*, 1974b).

In India, samples of placenta and accompanying fluid as well as of circulating blood were frequently found to contain aldrin and dieldrin (Saxena *et al.*, 1980a, 1980b, 1981). Breast milk samples from Lucknow (India) contained a mean level of aldrin of 0.03 ppm (Siddiqui *et al.*, 1981).

Due to insufficient information on exposure levels and living habits of the persons in question, these data cannot be interpreted in relation to climatic factors.

5.4.6 CONCLUSIONS

It may be concluded from the review of literature on the fate and effects of aldrin/dieldrin in terrestrial ecosystems of hot climates, that both these

Table 5.4.3 Concentration of dieldrin in the adipose tissue of the general population of South Africa (ppm) (Wassermann *et al.*, 1970). Reproduced with permission from the Medical Association of South Africa

Group	Mean Content
Bantu females	0.034
White females	0.047
Bantu males	0.033
White males	0.048
Females (total)	0.040
Males (total)	0.039
Bantu (total)	0.034
Whites (total)	0.047
General population	0.039

insecticides are less persistent than in temperate climates; however, in aquatic systems the difference between climates is probably even greater. Increased volatilization due to higher temperatures and to higher soil moisture content is probably a major reason for these differences; degradation is also possibly enhanced. Although accumulation in organisms and effects on ecosystems after dieldrin application do occur, birds of prey in general do not have such high residues as those in temperate climates. Residues in human food and storage in human tissues have been observed to be lower in some cases. Since the use of aldrin and dieldrin continues in developing countries, influences of climate on their fate and their effects should be investigated further.

5.4.7 REFERENCES

Agnihotri, N. P., Pandey, S. Y., and Jain, H. K. (1976). Persistence of BHC and aldrin in soil and translocation in mung (*Phaseolus aureus* L.) and lobia (*Vigna sinensis* Siva). *Indian J. Ent.*, **36**, 261–267.

Agnihotri, N. P., Pandey, S. Y., Jain, H. K., and Srivastava, K. P. (1977). Persistence of aldrin, dieldrin, lindane, heptachlor and *p,p'*-DDT in soil. *J. Ent. Res.*, **1**, 89–91.

Albert, L., Mendez, F., Cebrian, M. E., and Portales, A. (1980). Organochlorine pesticide residues in human adipose tissue in Mexico: Results of a preliminary study in three Mexican cities. *Archs. Envir. Hlth.*, **35**, 262–269.

Atabaev, Sh. T., Khasanov, Yu. U., and Nazarova, L. S. (1970). Stability of the pesticide aldrin under hot climate conditions. *Gig. Sanit.*, **35**, 108–109 (in Russian).

Baluja, G., Murada, M. A., and Tejedor, M. C. (1975). Adsorption/desorption of lindane and aldrin by soils as affected by soil main components. In: Coulston, F., and Korte, F. (eds), *Pesticides. Environmental Quality and Safety*, supplement vol. III, pp. 243–249. Georg Thieme Publishers, Stuttgart.

Banerji, S. A. (1979). Organochlorine pesticides residues. *Scavenger*, **9**, 2–7.

Bhinge, R., and Banerji, S. A. (1981). *Estimation of Organochlorine Pesticide-Residue Analysis*. Ludhiana; Punjab Agricultural University and Indian Council of Agricultural Research.

Brady, M. N., and Siyali, D. S. (1972). Hexachlorobenzene in human body fat. *Med. J. Aust.*, **59**, 158–161.

Briggs, G. G. (1981). Theoretical and experimental relationships between soil adsorption, octanol-water partition coefficients, water solubilities, bioconcentration factors, and the parachor. *J. Agric. Fd. Chem.*, **25**, 1050–1059.

Caro, J. H., and Taylor, A. W. (1971). Pathways of loss of dieldrin from soils under field conditions. *J. Agric. Fd. Chem.*, **19**, 379–384.

Chawla, R. P., Kalra, R. L., and Joia, B. S. (1981). Absorption of residues of soil applied aldrin and heptachlor in potatoes. *Indian J. Ent.*, **43**, 266–271.

Chiou, C. T., Peters, L. J., and Freed, V. H. (1979). A physical concept of soil-water equilibria for non-ionic organic compounds. *Science*, **206**, 831–832.

Downing, A. L., and Truesdale, G. A. (1955). Some factors affecting the rate of solution of oxygen in water. *J. Appl. Chem.*, *5*, 570–581.

Edwards, C. A. (1966). Insecticide residues in soils. *Residue Rev.*, **13**, 83–132.

El Beit, I. O. D., Wheelock, J. V., and Cotton, D. E. (1981). Pesticide–microbial interaction in the soil. *Int. J. Envir. Studies*, **16**, 171–180.

Elgar, K. E. (1966). Analysis of crops and soils for residues of the soil insecticides aldrin and 'telodrin'. *J. Sci. Fd. Agric.*, **17**, 541–545.

Elgar, K. E. (1975). The dissipation and accumulation of aldrin and dieldrin residues in soil. In: Coulston, F., and Korte, F. (eds), *Pesticides. Environmental Quality and Safety*, supplement vol. III, pp. 250–257. Georg Thieme Publishers, Stuttgart.

El Zorgani, G. A. (1976). Persistence of organochlorine insecticides in the field in the Gezira soil under cotton. *Bull. Envir. Contam. Toxicol.*, **15**, 378–382.

Farmer, W. J., Igue, K., Spencer, W. F., and Martin, J. P. (1972). Volatility of organochlorine insecticides from soil: I. Effect of concentration, temperature, air flow rate, and vapour pressure. *Soil Sci. Soc. Amer. Proc.*, **36**, 443–447.

Felsot, A., and Dahm, P. A. (1979). Sorption of organophosphorus and carbamate insecticides by soil. *J. Agric. Fd. Chem.*, **27**, 557–563.

Frank, L. G., Jackson, R. M., Cooper, J. E., and French, M. C. (1977). A survey of chlorinated hydrocarbon residues in Kenyan birds of prey. *E. Afr. Wildl. J.*, **15**, 295–304.

Gupta, H. C. L., and Kavadia, V. S. (1979). Dissipation of aldrin residues in clay loam soil under the cover of root crops. *Ind. J. Plant Protection*, **VII**, 43–49.

Hamaker, J. W. (1972). Diffusion and volatilization. In: Goring, C. A. I., and Hamaker, J. W. (eds), *Organic Chemicals in the Soil Environment*, vol. 2, pp. 341–397. Marcel Dekker Inc., New York.

Harris, C. R., and Lichtenstein, E. P. (1961). Factors affecting the volatilization of insecticidal residues from soils. *J. Econ. Ent.*, **54**, 1038–1045.

Hulpke, H., and Schuphan, W. (1970). Contributions to the metabolism of the pesticide aldrin in food plants. 3. Communication: Distribution of residues of ^{14}C-labelled aldrin in storage roots of two carrot cultivars (*Daucus carota* L. ssp. *sativus* Hoffm.) *Qualitas Pl. Mater. veg.*, **XIX**, 347–358 (in German).

Hunter, C. G., Robinson, J., and Roberts, M. (1969). Pharmacodynamics of dieldrin (HEOD). Ingestion by human subjects for 18 to 24 months and postexposure for 8 months. *Archs. Envir. Hlth*, **18**, 12–21.

Igue, K., Farmer, W. J., Spencer, W. F., and Martin, J. P. (1972). Volatility of organochlorine insecticides from soil: II. Effect of relative humidity and soil water content on dieldrin volatility. *Soil Sci. Soc. Amer. Proc.*, **36**, 447–450.

Kathpal, T. S., Yadav, P. R., and Kushwaha, K. S. (1981). Residues of some organochlorine insecticides in soils under different agroclimatic conditions of India. *Indian J. Ent.*, **43**, 420–427.

Kenaga, E. E., and Goring, C. A. I. (1978). Relationship between water solubility, soil-sorption, octanol-water partitioning, and bioconcentration of chemicals in biota. Paper given at: *American Society for Testing and Materials Third Aquatic Toxicology Symposium*, October 17-18, New Orleans, Louisiana.

Kiigemagi, U., Morrison, H. E., Roberts, J. E., and Bollen, W. B. (1958). Biological and chemical studies on the decline of soil insecticides. *J. Econ. Ent.*, **51**, 198-204.

Klein, W., Kohli, J., Weisgerber, I., and Korte, F. (1973). Fate of aldrin- ^{14}C in potatoes and soil under outdoor conditions. *J. Agric. Fd. Chem.*, **21**, 152-156.

Koeman, J. H., Den Boer, W. M. J., Feith, A. F., de Iongh, H. H., Spliethoff, P. C., Na'isa, B. K., and Spielberger, U. (1978). Three years' observation on side effects of helicopter applications of insecticides used to exterminate *Glossina* species in Nigeria. *Environ. Pollut.*, **15**, 31-59.

Kohli, J., Zarif, S., Weisgerber, I., Klein, W., and Korte, F. (1973). Fate of ^{14}C-aldrin in sugar beets and soil under outdoor conditions. *J. Agric. Fd. Chem.*, **21**, 855-857.

Krishna Murti, C. R. (1984). *Pesticide Residues in Food and Biological Tissue—A Report of the Situation in India*. Indian National Science Academy, New Delhi, pp. 10-12 residues in the environment; pp. 15-31 cultivated food crops; pp. 32-35 cultivated cash crops; pp. 36-37 stored grains, pp. 38-41 random sample of foodgrains; pp. 43-66 human tissues.

Kushwaha, K. S., Gupta, H. C. L., and Singh, R. (1976). Effect of temperature on the degradation of aldrin. *Indian J. Ent.*, **38**, 134-137.

Kushwaha, K. S., Gupta, H. C. L., and Kavadia, V. S. (1978). Effect of temperature on the degradation of aldrin residues in sandy loam soil. *Ann. of Arid Zone*, **17**, 200-206.

Kushwaha, K. S., Gupta, H. C. L., and Kavadia, V. S. (1981). Dissipation of aldrin residues from soils under maize/bajra-fallow rotation. *Indian J. Ent.*, **43**, 76-79.

Lambert, S. M., Porter, P. E., and Schieferstein, H. (1965). Movement and sorption of chemicals applied to the soil. *Weeds*, **13**, 185-190.

Lichtenstein, E. P., and Schulz, K. R. (1959a). Breakdown of lindane and aldrin in soils. *J. Econ. Ent.*, **52**, 118-124.

Lichtenstein, E. P., and Schulz, K. R. (1959b). Persistence of some chlorinated hydrocarbon insecticides as influenced by soil types, rate of application and temperature. *J. Econ. Ent.*, **52**, 124-131.

Lord, K. A., Helene, C. G., de Andrea, M. M., and Ruegg, E. F. (1978). Sorption and movement of pesticides on thin-layer plates of Brazilian soils. *Arg. Inst. Biol. Sáo Paulo*, **45**, 47-52.

Luck, H., and van Dyk, L. P. (1978). Chlorinated insecticide residues in dairy products. *S. Afr. J. Dairy Technol.*, **10**, 179-182.

Mackay, D., and Yuen, A. T. K. (1983). Mass transfer coefficient correlations for volatilization of organic solutes from water. *Environ. Sci. Technol.*, **17**, 211-217.

Maia, R., and Brant, P.C. (1980). Comparative study of contamination of beef by organochlorine pesticide residues in various regions of Minas-Gerais state, Brazil. *Revsta. Inst. Adolfo Lutz*, **40**, 15-22 (in Portuguese).

Müller, P., Nagel, P., and Flacke, W. (1981). Ecological side effects of dieldrin application against tsetse flies in Adamaoua, Cameroon. *Oecologia* (Berl.), **50**, 187-194.

Musumeci, M. R., Caienelli, V. C. B., and Ruegg, E. F. (1982). Adsorption of aldrin and its metabolites by soil fungi *in vitro*. *Cienc. Cult.* (Sáo Paulo), **34**, 381-385 (in Portuguese).

Rippen, G., Ilgenstein, M., Klöpffer, W., and Poremski, H. -J. (1982). Screening of the adsorption behaviour of new chemicals: natural soils and model adsorbents. *Ecotox. Environ. Safety*, **6**, 236-245.

Saxena, M. C., and Siddiqui, M. K. J. (1982). Pesticide pollution in India: Organochlorine pesticides in women, buffalo and goat milk. *J. Dairy Sci.*, **65**, 430–434.

Saxena, M. C., Seth, T. D., and Mahajan, P. L. (1980a). Organochlorine pesticides in human placenta and accompanying fluid. *Int. J. Env. Analyt. Chem.*, **7**, 245–251.

Saxena, M. C., Siddiqui, M. K. J., Bhargava, A. K., Seth, T. D., Krishna Murti, C. R., and Kutty, D. (1980b). Role of chlorinated hydrocarbon pesticides in abortions and premature labour. *Toxicology*, **17**, 323–331.

Saxena, M. C., Seth, T. D., Krishna Murti, C. R., Bhargava, A. K., and Kutty, D. (1981). Organochlorine pesticides in specimens from women undergoing spontaneous abortion, premature of full term delivery. *J. Analyt. Toxic.*, **5**, 6–9.

Scheffer, F., and Schachtschabel, P. (1982). *Textbook of Soil Science*, 11th edn. F. Enke (publisher), Stuttgart (in German).

Scheunert, I., Kohli, J., Kaul, R., and Klein, W. (1977). Fate of ^{14}C-aldrin in crop rotation under outdoor conditions. *Ecotox. Environ. Safety*, **1**, 365–385.

Siddiqui, M. K. J., and Saxena, M. C. (1983). Biological monitoring of environmental contaminants: Chlorinated hydrocarbons in eggs of hens. *Sci. Total Environ.*, **32**, 29–34.

Siddiqui, M. K. J., Seth, T. D., Krishna Murti, C. R., and Kutty, D. (1981). Agrochemicals in the maternal blood, milk and cord blood, a source of toxicants for prenates and neonates. *Environ. Res.*, **24**, 24–32.

Sotiriou, N., Klein, W., and Scheunert, I. (1981). Organochlorine compounds in terrestrial ecosystems. In: Kotzias, D., Mansour, M., Politzki, G., Lahaniatis, E. S., Seltzer, H., and Bergheim, W. (eds), *First Internat. Mtg. on Environ. Pollution in the Mediterranean Reg. Proceedings*, pp. 287–317. Mediterranean Scientific Association of Environmental Protection, Munich.

Spencer, W. F., Farmer, W. J., and Cliath, M. M. (1973). Pesticide volatilization. *Residue Rev.*, **49**, 1–47.

Srivastava, S., Siddiqui, M. K. J., and Seth, T. D. (1983). Organochlorine pesticide residues in groundnut oil. *J. Fd. Sci. Technol.* **20**, 25–27.

Stewart, D. K. R., and Gaul, S. D. (1977). Dihydrochlordene dicarboxylic acid residues in soil treated with high rates of aldrin. *Bull. Envir. Contam. Toxicol.*, **17**, 712–713.

Talekar, N. S., Sun, L. -T., Lee, E. -M., and Chen, J. -S. (1977). Persistence of some insecticides in subtropical soil. *J. Agric. Fd. Chem.*, **25**, 348–352.

Taylor, A. W., Glotfelty, D. E., Glass, B. L., Freeman, H. P., and Edwards, W.M. (1976). Volatilization of dieldrin and heptachlor from a maize field. *J. Agric. Fd. Chem.*, **24**, 625–631.

Topp, E., Scheunert, I., and Korte, F. (1986). Factors affecting the uptake of ^{14}C-labelled organic chemicals by plants from soil. *Ecotox. Environ. Safety*, **11**, 219–228.

Turner, B. C., Glotfelty, D. E., and Taylor, A. W. (1977). Photodieldrin formation and volatilization from grass. *J. Agric. Fd. Chem.*, **25**, 548–550.

Wassermann, M., Curnow, D. H., Forte, P. N., and Groner, Y. (1968). Storage of organochlorine pesticides in the body fat of people in Western Australia. *Ind. Med. Surg.*, **37**, 295–300.

Wassermann, M., Wassermann, D., and Lazarovici, S. (1970). Present state of the storage of the organochlorine insecticides in the general population of South Africa. *S. Afr. Med. J.*, **44**, 646–648.

Wassermann, M., Nogueira, D. P., Tomatis, L., Athie, E., Wassermann, D., Djavaherian, M., and Guttel, C. (1972a). Storage of organochlorine insecticides in people of Sáo Paulo, Brazil. *Ind. Med.*, **41**, 22–25.

Wassermann, M., Sofoluwe, G. O., Tomatis, L., Day, N. E., Wassermann, D., and Lazarovici, S. (1972b). Storage of organochlorine insecticides in people of Nigeria. *Environ. Physiol. Biochem.*, **2**, 59–67.

Wassermann, M., Tomatis, L., Wassermann, D., Day, N. E., and Djavaherian, M. (1974a). Storage of organochlorine insecticides in adipose tissue of Ugandans. *Bull. Environ. Contam. Toxicol.*, **12**, 501–508.

Wassermann, M., Tomatis, L., Wassermann, D., Day, N. E., Groner, Y., Lazarovici, S., and Rosenfeld, D. (1974b). Epidemiology of organochlorine insecticides in the adipose tissue of Israelis. *Pestic. Monit. J.*, **8**, 1–7.

Weisgerber, I., Klein, W., and Korte, F. (1970). Contributions to ecological chemistry XXVI. Conversion and residue behaviour of ^{14}C-aldrin and ^{14}C-dieldrin in white cabbage, spinach, and carrots. *Tetrahedron*, **26**, 779–789 (in German).

Weisgerber, I., Kohli, J., Kaul, R., Klein, W., and Korte, F. (1974). Fate of ^{14}C-aldrin in maize, wheat, and soils under outdoor conditions. *J. Agric. Fd. Chem.*, **22**, 609–612.

Weisgerber, I., Bieniek, D., Kohli, J., and Klein, W. (1975). Isolation and identification of three unreported photodieldrin-^{14}C-metabolites in soil. *J. Agric. Fd. Chem.*, **23**, 873–877.

Willis, G. H., Parr, J. F., Smith, S., and Carroll, B. R. (1972). Volatilization of dieldrin from fallow soil as affected by different soil water regimes. *J. Environ. Qual.*, **1**, 193–196.

Wolff, C. J. M., and van der Heijde, H. B. (1982). A model to assess the rate of evaporation of chemical compounds from surface waters. *Chemosphere*, **11**, 103–177.

Ecotoxicology and Climate
Edited by P. Bourdeau, J. A. Haines, W. Klein and C. R. Krishna Murti

5.5 The Use of Insecticides in the Onchocerciasis Control Programme and Aquatic Monitoring in West Africa

C. LÉVÊQUE

5.5.1 INTRODUCTION

Human onchocerciasis is a dermal filariasis widespread throughout tropical Africa. The disease is particularly serious in clinical, social, and economic terms in the Guinean and Sudanian savanna areas, where it causes irreversible blindness among exposed human populations.

The filaria *Onchocerca volvulus*, strictly limited to man, is transmitted in West Africa by the female blackfly of the *Simulium damnosum* complex (Philippon, 1977). The larvae of these flies are aquatic and occur only in fast-flowing water, requiring a minimum flow of about 50 cm/s for survival. Thus, onchocerciasis is prevalent along most watercourses, so that people have tended to leave the river valleys and move to the uplands.

In 1970, the United Nations Development Programme funded the preparation of a strategy for the control of onchocerciasis in West Africa, where about one million people have the disease. The Onchocerciasis Control Programme (OCP) was launched in December 1974 under the aegis of WHO and was planned for 20 years (Davies *et al.*, 1978). It covers a vast area of 764 000 km^2 and includes Upper Volta and parts of Ivory Coast, Ghana, Togo, Benin, Niger, and Mali (Figure 5.5.1). Up to 18 000 km of rivers with potential breeding sites for *Simulium damnosum sensu lato* were investigated and partly treated.

In the absence of any effective treatment (prophylaxis, chemotherapy) suitable for mass application, vector control is the only way to prevent the spread of the disease. Adult control being difficult, it was decided to use chemicals for controlling larval stages whose distribution is restricted to rapids.

The first insecticide treatments (Phase I) started in February 1975 in the central parts of the OCP area (Figure 5.5.2) and progressively extended eastwards (Phase II, March 1976; Phase III, July 1977), westwards (Phase III, March 1977), and southwards (Phase IV, March 1977 and April 1978). A southeastern extension was planned for 1986 in Togo and Benin as well as a western extension.

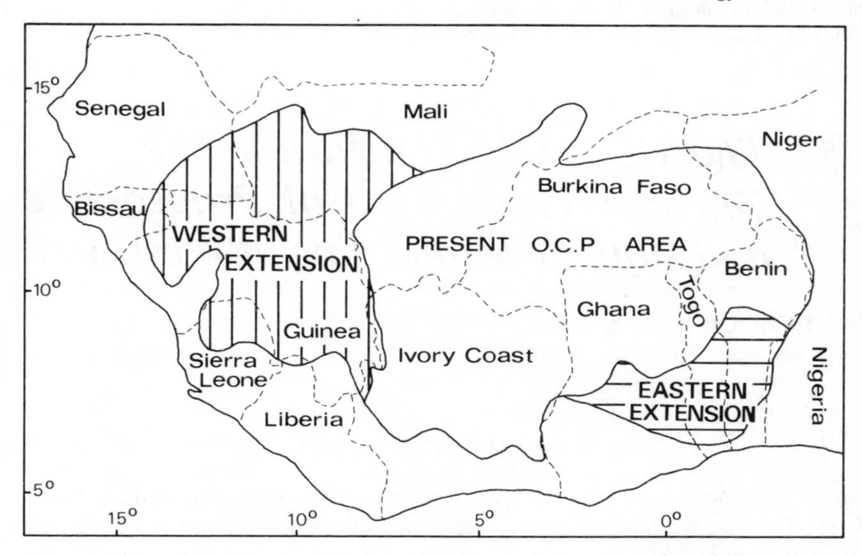

Figure 5.5.1 Present Onchocerciasis Control Programme area and extension zones planned

The OCP includes two main operational units:

(i) the Vector Control Unit (VCU) undertakes all the vector control, evaluation, and research activities and comprises three divisions: Aerial Operations; Entomological Evaluation of Vector Control; Applied Research and Staff Training. The latter coordinates and supervises studies on blackfly vectors, trials of new larvicides, and activities of the different teams in charge of the hydrobiological monitoring (see below);

(ii) The Epidemiological Evaluation Unit (EPI) responsible for the medical and parasitological assessment of the results of the VCU activities.

Several committees (Joint Programme Committee, Committee of Sponsoring Agencies, National Onchocerciasis Committees, Expert Advisory Committee) are in charge of evaluating results, providing funds for operational activities, and giving advice to WHO and the Programme Director. The Expert Advisory Committee (EAC) has attached to it a permanent Ecological Group, composed of five members, which is in charge of assessing the impact on the environment of the use of insecticides in the Programme. The Ecological Group proposes to EAC such measures as may be needed to supplement the ecological studies undertaken under the Programme and makes recommendations to ensure effective protection of the environment. The Ecological Group meets independently at least once a year.

5.5.2 TYPE OF ECOSYSTEM

The OCP covers major river systems in West Africa, such as the Volta system, part of the Niger basin and its tributaries, the northern part of Sassandra, Bandama, Comoé, Mono, and Ouémé.

Most of these rivers are savanna type with a water regime characterized by a flood period from July to December with a peak in September, and a low water period from January to March.

Many of the rivers in the central part of OCP are intermittent and dry up during the dry season. For the permanent ones, discharge is very low during that time and the upper course is sometimes reduced to a series of pools. Thus there are severe seasonal variations in flow which result in major ecological changes for species inhabiting rivers.

When the monitoring programme started, little was known about the biology of African rivers and even less about their biology when polluted. This knowledge was gradually improved and specific research was conducted for a better understanding of the results obtained in monitoring stations.

In order to help the different teams for identification of species, a catalogue of aquatic insects was produced (Dejoux *et al.*, 1981a) as well as a catalogue of fishes (Lévêque and Paugy, 1985).

Details will be found in different papers published on physico-chemistry of rivers (Iltis and Lévêque, 1982), hydrology (Moniod *et al.*, 1977), aquatic insects ecology and biology (Dejoux *et al.*, 1981a, 1981b; Elouard, 1983; Elouard and Lévêque, 1977; Gibon *et al.*, 1983; Gibon and Statzner, 1985; Statzner, 1982, 1984), fish ecology and biology (Albaret, 1982; Lévêque and Herbinet, 1980, 1982; de Mérona, 1980, 1981; de Mérona & Albaret, 1978; de Mérona *et al.*, 1977, 1979; Paugy, 1978, 1980a, 1980b), phytoplankton (Iltis, 1982a, 1982b, 1982c, 1983), river biology (Lévêque *et al.*, 1983).

5.5.3 THE AQUATIC MONITORING PROGRAMME

Since prolonged and intensive use of insecticides could present risks, it was necessary to evaluate the possible short-term and long-term effects of applications on the present organisms of the treated watercourses.

Consequently an aquatic environmental monitoring programme was devised before the beginning of OCP, so as to be sure that the insecticide released did not excessively disturb the functioning of the treated ecosystems and to provide warning to those carrying out treatments, should toxic effects be noted.

When setting up the monitoring programme, several important considerations had to be kept in mind (Lévêque *et al.*, 1979):

(i) the monitoring work was to deal with a long-term regular sampling aimed at investigating the ecological effects of treatment over the duration of the programme, combined with shorter duration research programmes looking at specific short-term problems.

(ii) the periodicity of sampling, the sites selected for monitoring, and the field methods used had to combine reliability of sampling technique with reliability of access in both wet and dry seasons, over many kilometres of roads or tracks which are not yet hard surfaced.

(iii) the monitoring techniques had to work equally well in shallow, slow-flowing rivers in the dry season and in the same rivers flowing fast and deep in the wet season.

(iv) the best possible use had to be made of the available manpower and of local facilities. The monitoring was, therefore, based on national teams of scientists from the countries concerned with OCP, with the help of foreign specialists. Many of these scientists were trained in the ORSTOM Hydrobiological Laboratory in Bouaké (Ivory Coast).

(v) in order to ensure reasonable comparability of results all teams had to use the same methods.

The monitoring programme was primarily concerned with two major categories of organisms:

(i) the benthic invertebrates that abound in the watercourses and that are directly threatened by the insecticide in the same way as *Simulium damnosum* larvae.

(ii) fishes, by virtue of their economic interest for the people living along the rivers, but also for psychological reasons to show the villages occupied in fishing that care was taken about the risks of pollution.

Shorter duration research was also conducted on water quality, phytoplankton, zooplankton, etc.

The selection of sampling stations was based on a preliminary field investigation in order to cover a wide range of river types. Some of the stations were on untreated rivers in order to act as permanent controls. Unfortunately it was not possible to collect enough ecological data to serve as reference for fauna before spraying began. The progressive extension of the OCP (Phases II–III–IV, Figure 5.5.2), however, provided an opportunity to remedy this omission, insofar as some monitoring stations were selected on rivers which had remained untreated for some years until subjected to repeated applications of temephos.

Details of monitoring and sampling methods are given in Lévêque *et al.* (1979), Dejoux *et al.* (1979), and Dejoux (1980). Only a brief summary will be given here.

For invertebrates, three main sampling methods were used:

(a) Drift net sampling using 2 m long nets, 20 × 20 cm aperture, 300 μm mesh size. Three samples were taken approximately 1½ hours before sunset (day drift) and six samples 1½ hours after sunset (night drift). River flow was measured at the time of the sampling in order to evaluate the actual numbers of animals per cubic metre filtered. The basic techniques of drift sampling used in the programme are described in Elouard and Lévêque (1977).

(b) Surber samples using a 15 × 15 Surber sampler. This simple method, which allows rocky substrates to be sampled, cannot be used in deep waters and

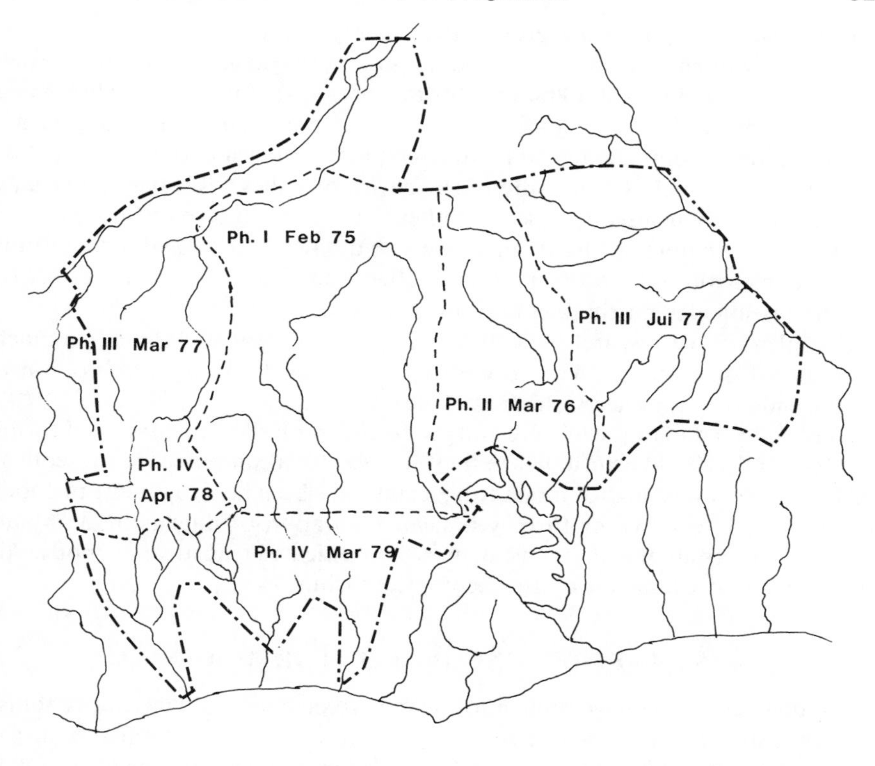

Figure 5.5.2 Chronology and extension of the different operational phases

was, therefore, limited to the low-water period. For comparative work, the results are expressed as number of animals per square metre.

(c) Artificial substrates. A special apparatus was designed for the monitoring programme. It consisted of small concrete blocks which were left immersed on the bottom for one month (Dejoux and Venard, 1976; Dejoux *et al.*, 1983). Later, other types of artificial substrates were used, such as an artificial floating substrate made of a bunch of plastic fibres (Elouard, 1983).

(d) The use of gutters *in situ* was introduced late (Dejoux, 1975; Troubat, 1981). It seems to be the most precise method for short-term study of the toxicity of an insecticide. It gives a relatively accurate picture of the mortality of organisms because the number of individuals tested is known. Moreover, the use of multiple gutters makes it possible to compare toxicity of different insecticides, or different concentrations of the same substance, with that of a control, under similar conditions.

For fishes the monitoring programme mainly concerned:

(a) the study of changes in the catch composition of experimental fishing carried out at regular intervals (3 months for the three years following the beginning of treatment, then at 6-monthly intervals) with a standardized set of gill nets.

(b) the study of some biological parameters, more especially the coefficient of condition, which is a measure of the health of fishes. This parameter may consequently be used to assess whether fishes are still able to find the food that they require in the treated rivers and whether ecological conditions remain favourable to them. It also reflects any possible adverse effect of the insecticides on the metabolism of fishes.

Complementary research was also conducted on the analysis of stomach contents of selected species, fecundity, and impact of organophosphorus compounds on brain acetylcholinesterase activity.

Data collected in the field were and are recorded on specially designed forms and sent to WHO at Geneva, where they are fed into a computer for subsequent analysis. Yearly meetings of monitoring teams have been held to discuss methods and results. Every two or three years, an evaluation of data is made by an independent group of experts, using more sophisticated statistical methods. All results are also examined by the Ecological Group.

5.5.4 SCREENING OF NEW LARVICIDES

The sequence for the development of acutely toxic chemical larvicides against *S. damnosum s.l.* follows a screening fitting into WHO's general system for the development of insecticides (WHOPES) adapted to conditions of the OCP area. These are different from those of temperate countries, especially as regards the physico-chemical characteristics of the river waters.

After different laboratory tests carried out in troughs with different concentrations, larvicides giving 100% mortality for *S. damnosum* larvae at 0.5 mg/l over 10 min (or at least 95% at 0.2 mg/l over 10 min) are tested in rivers to determine the possible operational dose. Afterwards, the impact on non-target fauna is tested in troughs at the operational dose and at twice the dose, taking temephos and chlorphoxim as controls. If the larvicide appears promising, river tests are conducted at assumed operational dose, with ecological monitoring of the river using Surber sampler, drift nets, and gutters.

The decision for operational trials depends on toxicity as regards non-target fauna, mammalian toxicity, and various technical aspects, such as ease of handling and corrosive effect of formulation. The Ecological Group proposed criteria applicable to the selection of alternative compounds for *Simulium* control:

(i) the acute effects of a candidate pesticide, in the formulation and dose rate as appropriate for its use against *Simulium*, should not reduce the numbers of invertebrate species to a level at which their survival at a given locality would be endangered.

(ii) the pesticide should not give rise to the regional loss of any invertebrate species; the temporary seasonal local disappearance of some invertebrate species at the breeding sites of *Simulium* may have to be accepted.
(iii) the pesticide should not cause a long-term (i.e. extending beyond the next season) imbalance under normal conditions of application, e.g. marked shifts in the relative abundance of species should not occur.
(iv) The use of the pesticide should have neither any direct impact on fish nor any effect on the life cycle of fish.
(v) compounds likely to accumulate in the food web should be avoided.
(vi) in the process of selecting pesticides for *Simulium* control in an area, full account should be taken of human activities which either by themselves or in combination with the vector control operations might cause adverse effects on the environment.

A hundred insecticide formulations have been tested by OCP during recent years. Many of them were not completely effective against *Simulium* larvae. Few were tested with non-target entomofauna (Dejoux 1983b; Dejoux and Troubat, 1982; Paugy *et al.*, 1984; Troubat and Lardeux, 1982; Yameogo *et al.*, 1984) but were not selected due to their toxic effects. Many others are still under screening.

It should be noted that different methoxychlor formulations were not completely successful in *Simulium* larvae control, whereas good results were obtained in Canada and the USA. The reason should be a decrease in activity where temperature increases. The inverse phenomenon is observed for organophosphorus compounds, temephos giving feeble results in temperate zones.

5.5.5 POLLUTANT INPUT

The insecticide selected for a large-scale campaign of this type, due to last for about 20 years, must have properties that allow it to meet often contradictory requirements, such as effective action against the larvae of *S. damnosum*, ease of application, lowest possible cost, little residue but far-reaching effect, harmlessness for man and mammals, and lowest toxicity possible for the rest of the aquatic environment.

An organophosphate, temephos (or Abate®), was selected according to the above criteria, after numerous laboratory and field tests, on account of its efficacy against the larvae of the vector and its low toxicity for the non-target fauna (Dejoux and Troubat, 1973, 1974; Lauzanne, 1973; Lauzanne and Dejoux, 1973).

A 20% emulsionable concentrate is used for operational activities. The dosage of 0.05 ppm/l over 10 min is effective for about 40 km in the wet season. In the dry season, treatments tend to be made to each riffle and the dosage is normally 0.1 ppm/l over 10 min.

Temephos was the only insecticide used between 1974 and 1980. In December 1979 temephos resistance developed in the species *S. soubrense* and *S. sanctipauli* from the *S. damnosum* complex, in breeding sites of the lower Bandama (Ivory Coast). This resistance spread rapidly to the southern forest zone and part of the humid savanna zone in the rainy season, but has not appeared so far in the savanna species *S. damnosum* and *S. sirbanum*, which are encountered in the main part of OCP. Consequently, insecticides other than temephos have been used for controlling the resistant species so as to maintain efficacy. They are:

(i) *Bacillus thuringiensis* serotype H-14, a spore-forming bacterium. At sporulation, each bacterium produces a crystal of toxic protein, lethal to larvae upon ingestion. B.t. H-14 is highly host-specific, unlike broad-spectrum insecticides. Unfortunately it can be used until now only in the dry season because of the low concentration of active ingredient in available formulations, which makes it unusable in the rainy season with the logistic resources available to OCP. The commercial formulation used in the OCP area is Teknar, but the search for better B.t. H-14 formulations would appear to be the best approach to find an insecticide as a real replacement for temephos.

(ii) Chlorphoxim is another organophosphate. A 20% emulsionable concentrate is used at the dosage of 0.025 ppm/l over 10 min. This pesticide is more toxic than temephos but a resistance developed in the forest species already resistant to temephos, around July 1981. Fortunately, the resistance to chlorphoxim is less stable than resistance to temephos, and regresses when treatment is stopped.

In 1984, temephos was still used in three-quarters of the OCP area. In the south-west where strains resistant to this larvicide appeared, Teknar is used at river discharge of up to 200 m^2/s. Above this level, chlorphoxim is substituted.

Since larval development is short (around 10 days) weekly spraying has proved necessary for effective control of vector populations in breaking development cycles of blackflies. To cover the area, vector control has been carried out by aerial applications of larvicides, using six to nine helicopters and one or two specially equipped fixed-wing aircraft, depending on the season.

An estimate of the amount of insecticides used in OCP from 1975 to 1983 is given in Table 5.5.1. It is clear that B.t. H-14, the less toxic larvicide, is increasingly used in the OCP area.

5.5.6 EFFECTS NOTED

5.5.6.1 Invertebrates

Temephos

A routine spraying operation (0.05–0.1 mg/l over 10 min) produces a massive detachment of insect fauna, reflected by a rise in the drift, after a 15–45 min

Table 5.5.1 Amount of insecticides (10^3 l used in the OCP area from 1975 to 1983. Sources: OCP, 1983)

	1975	1976	1977	1978	1979	1980	1981	1982	1983
Abate 200 C E*	75	130	156	216	263	184	132	163	120
Chlorphoxim*						6	81	7	22
B.t. H-14 (Teknar)[†]						0.5	8	233	290

*For Abate and chlorphoxim an emulsionable concentrate (20% active ingredients) was used throughout the period
[†]For Teknar, concentration was 3000 units *Aedes aegypti* per mg

period of latency (Dejoux, 1983a). Regular evaluation of the mortality rate of drifting organisms has shown that within 5 hours of treatment nearly 100% were dead. The mortality rate was reduced to 75% during the following hours and to nearly zero 24 hours after application of insecticide (Dejoux, 1982). Generally speaking, the first applications of temephos in rivers have a fairly strong effect, and 30 to 50% of the invertebrate population (experimental data) release their hold on their substrate at low-water period. Subsequent applications have a less quantitative impact because there is some selection of the least susceptible species and the most resistant ones (Dejoux, 1983a).

Although all taxonomic groups are affected, some of them, such as the Tricorythidae and some Batidae (Ephemeroptera), some Philopotamidae and Leptoceridae (Trichoptera), are particularly susceptible to temephos. The taxa with moderate susceptibility include the Hydropsychidae (Trichoptera), Caenidae (Ephemeroptera), and *Simulium* species other than *S. damnosum s.l.* The chironomids display little susceptibility to this insecticide (Dejoux, 1983a; Dejoux *et al.*, 1980; Elouard, 1983, 1984a; Elouard and Jestin, 1982, 1983; Samman and Pugh Thomas, 1978).

It has also been observed that there are variations in susceptibility during the various larval stages. Early stages are much more seriously affected by temephos than older organisms (Elouard, 1983).

Evaluation of the quantitative variations produced by temephos in the long term gave an overall value of 40% reduction in the quantity of fauna. But this value should be regarded as relative because it includes both insects which have proliferated and those which have greatly diminished due to the insecticide. In fact, in a monitoring programme applied on such a vast scale, with such wide seasonal fluctuations and variations in distribution, it seems difficult to quantify the long-term effects of temephos in terms of variations of population densities. On the other hand, long-term structural variations in population are more easily identified. This is due to the establishment of stable biocoenotic structures which are typical of temephos treatment periods and different from those observed in untreated rivers or during periods without treatment (Elouard and Jestin, 1982, 1983).

The most obvious indicators of long-term modifications are the disappearance of *Simulium adersi*, the rarefaction of Trichorythidae and of some species of Batidae (*Pseudocleon* sp., *Centropilum* sp.), and the proliferation of *S. schoutedeni* and the Chironomidae.

It should be noted that the nycthemeral pattern of drift does appear to be affected by temephos, being less marked and becoming even patternless after several months of treatment.

B.t. H-14

All the tests and field experiments carried out in the OCP area shown that B.t. greatly affects all *Simulium* larvae but is safe for most of the non-target invertebrate fauna with the exception of a few taxa, such as *Orthotrichia* (Trichoptera) (Dejoux *et al.*, 1985; Gibon *et al.*, 1980; Dejoux, 1979; Elouard and Gibon, 1984). According to Rishikesh *et al.* (1983), this larvicide is an exceptionally safe agent for non-target organisms, including man and other vertebrates.

Field experiments showed that drift of invertebrates after application of B.t. behaved very differently from the drift observed when organophosphorus insecticides were applied. The maximum drift following application was very low, below the night peak for the control drift. That is another proof of the low toxicity of this larvicide.

Chlorphoxim

All the observations show that this insecticide is much more toxic in the short term than temephos for the non-target invertebrate fauna. All insect taxa are affected except for the Orthocladiinae (Chironomidae) and the Caenidae (Dejoux *et al.*, 1981c, 1982; Elouard and Gibon, 1984; Gibon and Troubat, 1980).

Alternation of insecticides

An interesting question was to know if alternation of insecticides would permit the recolonization of treated stretches by groups partly eliminated by the use of a single insecticide or if, conversely, the alternation would have an even more catastrophic effect on the non-target fauna.

The results obtained during the fairly intensive study conducted in the lower Maraoue river (Ivory Coast) are quite reassuring. The river has been monitored since 1975 and treated with temephos from March 1979 to August 1980. Then since November 1980, as a consequence of the appearance of temephos resistance in the complex *Simulium soubrense—S. sanctipauli*, three insecticides have been used alternately: temephos, chlorphoxim, and B.t. H-14.

After four years of larvicide treatment, it cannot be said that repeated weekly insecticide applications have an effect on the population densities for the taxonomic groups as a whole (Elouard and Gibon, 1984). For the major groups, the seasonal variations in numbers observed during the pre-treatment years are in most cases of the same order of magnitude as those observed after larvicide treatment. But that does not mean that there has been no impact. As regards the Chironomidae, their densities increased substantially whatever insecticide was used. For the Batidae and possibly for the Tricorythidae, it would seem that the use of insecticides, particularly chlorphoxim, was producing a gradual reduction in their numbers. But the wide fluctuations in density of these taxa which were observed during the pre-treatment period prevent any definitive conclusion on this point.

With all the results being expressed at family level, it must also be pointed out that an increase in the density of one or more species may mask a decrease or even the disappearance of other species. Thus, the very high densities of Hydropsychidae (Trichoptera) observed in 1981–1982 are due to the genus *Cheumatopsyche*, while all the available data (Statzner and Gibon, 1984) indicate that since larvicide treatment began, there has been a substantial regression of the Macroematinae, particularly the genera *Macronema* and *Protomacronema*.

The overall conclusion of this study is that alternation of the insecticides as practised by OCP in the lower Maraoue does not appear to disrupt the population of aquatic insects to a greater or lesser extent than each of the insecticides alone.

5.5.6.2 Fish

From the results obtained in the course of monitoring rivers treated with temephos in the OCP area, the overall conclusion was that temephos had no detectable effect on the fish populations (Abban *et al.*, 1982; Lévêque *et al.*, 1982). There were no major changes in the size of the experimental fishing catches in monitoring stations as illustrated in Figure 5.5.3. Some changes observed in the composition of catches were not ascribable to the insecticide but rather to year-to-year changes in river discharge. This is the case, for instance, for *Schilbe mystus*, which disappeared almost completely after the flood in 1976, both from treated and untreated rivers, and reappeared in abundance by the end of 1979 (Lévêque and Herbinet, 1980).

The analysis of stomach contents carried out in 1975 on different species in treated and untreated rivers did not provide evidence of an influence exerted by the insecticide since the diet was appreciably the same in composition whatever the provenance of fishes (Vidy, 1976). This result was confirmed in 1976 and 1977 (unpublished data).

For the coefficient of condition, results obtained in various basins show that values are relatively random, fluctuating around a mean, for each species concerned; they did not seem appreciably altered after five years of monitoring.

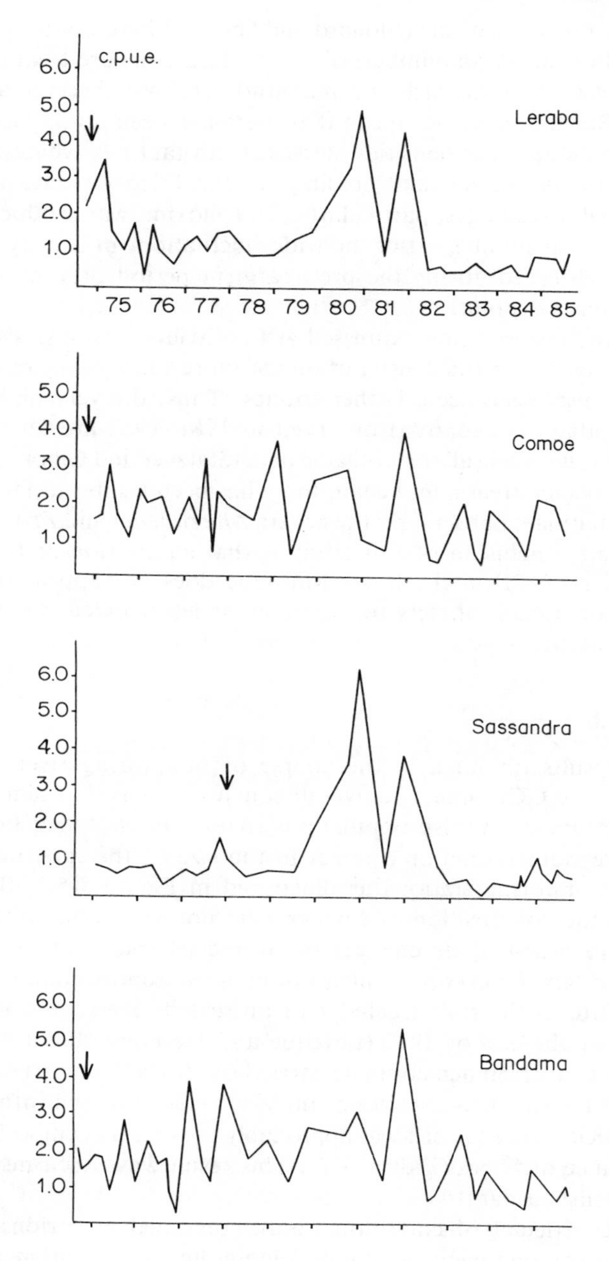

Figure 5.5.3 Changes in total catch per 100 m of gill net per night in experimental fishing for different rivers treated. The arrow indicates the beginning of insecticide application

Studies on fecundity of the principal species showed no differences between fishes from treated or untreated basins.

But even if temephos does not seem to affect fish, it can accumulate in tissues, as shown by some results obtained during the dry season in the OCP area (Quelennec *et al.*, 1977): the concentration of temephos (in ppm) one day and six days after spraying was, respectively, 14.3 and 7.1 for *Tilapia zillii*, 0.77 and 0.25 for *Alestes nurse*, and 1.3 and 0.96 for *Labeo parvus*. Much lower values were obtained for *Tilapia zillii* during flood.

Since toxicity of organophosphorus compounds is due mainly to their inhibitory effect on cholinesterase activity, a study on the effect of temephos on brain acetylcholinesterase (AChE) activity of fish was conducted in the OCP area (Antwi, 1983, 1984). No inhibitory effect was found in the brain AChE activity of *Alestes nurse*, *Schilbe mystus*, and *Tilapia* spp. in rivers treated for many years with temephos. But some *Tilapia galilaeus* and *Alestes nurse*, caught 24 hours after chlorphoxim application in the Maraoue river, showed a 20% reduction in the brain AChE activity.

5.5.6.3 Others

Studies on phytoplankton in Ivory Coast (Iltis 1982a, 1982b, 1982c; 1983) did not show any noticeable changes in species composition or biomass in rivers treated with temephos.

5.5.7 RECOVERY AND REVERSIBILITY— THE RECOLONIZATION POTENTIAL

The results obtained in the monitoring programme for treated rivers in the OCP area clearly demonstrated that on a long-term basis, the impact of insecticides was far from drastic. A good illustration is given by results obtained during the detailed study of the Lower Maraoue (see above). But it appears also that recolonization potential is high among invertebrates in rivers treated with temephos. That is the case, for instance, with the Red Volta, an intermittent river treated from 1976 to 1981. The structure of invertebrate populations during that period was typical of the treated rivers, namely with predominance of Oligochaeta and chironomids. The river was not treated in 1982 and the population structure was therefore similar to that found in untreated sites (Guenda, 1985).

Some data are also available for rivers treated with chlorphoxim. The Bandama basin was subjected to six to eight cycles of chlorphoxim treatment, depending on breeding sites, from 18 November 1980 to 6 January 1981. In the course of the campaign, this insecticide reduced the density of the fauna colonizing rocky substrates by 75 to 98% (Dejoux *et al.*, 1981c). Nevertheless, a week after the larvicide was discontinued there was a spectacular rise in densities, but the community structure was different from that customarily observed on these rivers, whether untreated or treated with temephos. The situation seems to return to normal within one or two months after suspension of treatment.

The recovery capacity of the invertebrate fauna in the rivers under study appears to depend on the existence of refuge zones and their capacity to fuel the recolonization of stretches of water that have been depopulated by insecticides. The problem of recolonization is probably highly complex, with different mechanisms being involved (Elouard, 1984b):

(i) a number of tributaries and some of the upper reaches of large watercourses are free of breeding sites for *S. damnosum* and are not treated with insecticide. They may, therefore, serve as refuge areas or rather as nurseries for recolonization by non-target fauna. A study carried out on the tributaries of the N'zi in Ivory Coast (Gibon *et al.*, 1983) demonstrated that the fauna found in the small tributaries was taxonomically very similar—as far as the running-water fauna is concerned—to the fauna found on the treated stretches of water. These small untreated watercourses could, therefore, act as reservoirs and ensure the survival of the non-target fauna.

(ii) in large rivers, there are some stretches of water flowing too slowly to permit the development of *S. damnosum s. l.* On the other hand, neither do they permit development of most of the non-target fauna. These stretches are generally not treated by OCP and constitute excellent potential reservoirs.

(iii) as far as insects are concerned, treated stretches as well as temporary watercourses could be recolonized from the air by imagoes from untreated rivers, either inside or outside the OCP area.

Moreover, in the central parts of OCP, where so far control has been very successful, there has been considerable reduction in treatment over the last few years. Systematically, weekly treatment has been replaced by 'opportune' treatment carried out only when *S. damnosum s.l.* are present. Walsh (1981) evaluated the length of rivers in the central parts of OCP (Phases I, II, III) as approximately 8000 km in the dry season and 23 000 km in the wet season. During the first years, respectively 5500 and 14 000 km of rivers were treated, but in 1980 only 4500 and 11 400.

In 1983, owing to a particularly favourable dry season and better utilization of the existing knowledge, Teknar and temephos treatments were progressively curtailed so that only 600 km of the Bandama river system, in the Control Programme area, were treated using Teknar only at the time when the water level was lowest.

The lightening of insecticide pressure on the rivers is most favourable to the maintenance of the non-target fauna.

5.5.8 CONCLUSIONS

Throughout the history of OCP, considerable attention has been given to the possible effects of larvicides on the non-target fauna and river biology. It is probably the only major programme to date with such an environmental monitoring element within its own structure.

The results obtained after many years' treatment lead us to assume that the larvicides employed had little effect on the non-target fauna. Although the first applications of temephos and chlorphoxim had a fairly strong impact on invertebrate communities in the short term, it would seem that these situations disappear fairly quickly after a year or less of successive applications. In operational conditions, the treated rivers seem to have fairly strong resilience, and at any rate a great capacity of recovery.

The situation is still improving with the reduction of treated rivers, resulting from the success of vector control, and increasing utilization of B.t. H-14, an exceptionally safe pesticide for non-target fauna.

But environmentalists are also concerned with the considerable quantities of pesticide locally available for agricultural purposes, and with the danger of fish poisoning in the Programme area caused by abuse of them. The attention given by OCP to protect the aquatic environment could be completely jeopardized by uncontrolled pesticide practices, which seem to be relatively common within the area.

5.5.9 REFERENCES

Abban, E. K., Fairhurst, C. P., and Curtis, M. S. (1982). *Observations on Fish Populations in Abate Treated Rivers in Northern Ghana.* Unpublished report.

Albaret, J. J. (1982). Reproduction et fécondité des poissons d'eau douce de Côte d'Ivoire. *Rev. Hydrobiol. Trop.*, **15** (4), 347–371.

Antwi, L. A. K. (1983). *The Effect of Abate and Chlorphoxim on the Brain Acetylcholinesterase Activity of Fish from Treated Rivers in the Volta Basin Area.* Report, Institute of Aquatic Biology, Ghana.

Antwi, L. A. K. (1984). *The Effect of Abate on the Brain Acetylcholinesterase Activity of Fish from two Treated Rivers in the Upper Volta: Rivers White Volta and Black Volta.* Report OCP/VCU/HYBIO/84.13.

Davies, J. B., Le Berre, R., Walsh, J. F., and Cliff, B. (1978). Onchocerciasis and *Simulium* control in the Volta River Basin. *Mosquito News*, **38**, 466–472.

Dejoux, C. (1975). Nouvelle technique pour tester *in situ* l'impact de pesticides sur la faune aquatique non cible. *Cah. ORSTOM, sér. Ent. méd. et Parasitol.*, **13** (2), 75–80.

Dejoux, C. (1979). *Recherches préliminaires concernant l'action de Bacillus thuringiensis israelensis de Barjac sur la faune invertébrée d'un cours d'eau tropical.* Report WHO/VBC/79–721.

Dejoux, C. (1980). *Effets marginaux de la lutte chimique contre* Simulium damnosum. *Techniques d'études.* Rapport du Centre ORSTOM de Bouaké, No. 35.

Dejoux, C. (1982). Recherches sur le devenir des invertébrés dérivant dans un cours d'eau tropical à la suite de traitements antisimulidiens au temephos. *Rev. fr. Sci. eau*, **1**, 267–283.

Dejoux, C. (1983a). Utilisation du temephos en campagne de lutte contre *Simulium damnosum* en Afrique de l'Ouest. Impact des premiers cycles de traitement sur le milieu aquatique. *Rev. Hydrobiol. trop.*, **16** (2), 165–179.

Dejoux, C. (1983b). Toxicité pour la faune aquatique de quelques nouveaux insecticides—III—La deltaméthrine. *Rev. Hydrobiol. trop.*, **16** (3), 263–275.

Dejoux, C., and Troubat, J. J. (1973). *Etude en laboratoire de la toxicité sur la faune non-cible de nouveaux insecticides employés en lutte anti-simulies—1ère partie: action sur les insectes, les batraciens et sur* Bulinus forskalii *(mollusque)*. Rapport du Centre ORSTOM de N'Djamena.

Dejoux, C., and Troubat, J. J. (1974). *Action in situ de l'Abate sur la faune aquatique non-cible. Toxicité à moyen terme en milieu tropical.* Rapport du Centre ORSTOM de N'Djamena.

Dejoux, C., and Venard, P. (1976). *Efficacité comparée de deux types de substrats artificiels.* Rapport du Centre ORSTOM de Bouaké, No. 1.

Dejoux, C., and Troubat, J. J. (1982). Toxicité pour la faune aquatique non-cible de quelques larvicides antisimulidiens. II—L'actellic R M 20. *Rev. Hydrobiol. Trop.*, **15** (2), 151–156.

Dejoux, C., Elouard, J. M., Lévêque, C., and Troubat, J. J. (1979). La lutte contre *Similium damnosum* en Afrique de l'ouest et la protection du milieu aquatique. *C. R. du Congrès de Marseille sur la lutte contre les insectes en milieu tropical*, **II**, 873–883.

Dejoux, C., Elouard, J. M., Jestin, J. M., Gibon, F. M., and Troubat, J. J. (1980). *Action du temephos (Abate) sur les invertébrés aquatiques—VIII—Mise en évidence d'un impact à long terme après six années de surveillance.* Rapport du Centre ORSTOM de Bouaké, No. 36.

Dejoux, C., Elouard, J. M., Forge, P., and Maslin, J. L. (1981a). *Catalogue iconographique des insectes aquatiques de Côte d'Ivoire.* Rapport OCP/VCU/HYBIO 83-2.

Dejoux, C., Elouard, J. M., Forge, P., and Jestin, J. M. (1981b). Mise en évidence de la microdistribution des invertébrés dans les cours d'eau tropicaux. Incidence méthodologique pour la recherche d'une pollution à long terme par insecticides. *Rev. Hydrobiol. Trop.*, **14** (3), 253–262.

Dejoux, C., Gibon, F. M., and Troubat, J. J. (1981c). *Impact de six semaines de traitement au chlorphoxim sur les invertébrés du bassin du Bandama.* Rapport du Centre ORSTOM de Bouaké, No. 41.

Dejoux, C., Gibon, F., Lardeux, F., and Ouattara, A. (1982). *Estimation de l'impact du traitement au chlorphoxim de quelques rivières de Côte d'Ivoire durant la saison des pluies de 1981.* Rapport du Centre ORSTOM de Bouaké, No. 47.

Dejoux, C., Jestin, J. M., and Troubat, J. J. (1983). Validité de l'utilisation d'un substrat artificel dans le cadre d'une surveillance écologique des rivières tropicales traitées aux insecticides. *Rev. Hydrobiol. Trop.*, **16** (2), 181–193.

Dejoux, C., Gibon, F. M., and Yameogo, L. (1985). Toxicité pour la faune non cible de quelques insecticides nouveaux utilisés en milieu aquatique tropical—IV—*Bacillus thuringiensis* var. *israelensis* H-14. *Rev. Hydrobiol. Trop.*, **18** (1), 31–49.

Elouard, J. P. (1983). *Impact d'un insecticide organophosphoré (le temephos) sur les entomocénoses associées aux stades préimaginaux du complexe* Simulium damnosum *Theobald (Diptera: Simuliidae)*. 10 microfiches (TDM 13) (0735–6), ORSTOM, Paris.

Elouard, J.M. (1984a). *General conclusions with regard to the effects of an organophosphorous insecticide (Temephos) on entomocoenoses associated with preimaginal stages of the* Simulium damnosum *Theobald complex (Diptera: Simuliidae)*. Report OCP/VCU/HYBIO/84.2.

Elouard, J. M. (1984b). *Some views on the recolonization potential of stretches of water treated with insecticides.* Report OCP/VCU/HYBIO/84.3

Elouard, J. M., and Lévêque, C. (1977). Rythme nycthéméral de dérive des insectes et des poissons dans les rivières de Côte d'Ivoire. *Cah. ORSTOM, sér. Hydrobiol.*, **11** (2), 179–183.

Elouard, J. M., and Jestin, J. M. (1982). Impact of temephos (Abate) on the non-target invertebrate fauna. A—Utilization of correspondence analysis for studying surveillance data collected in the Onchocerciasis Control Programme. *Rev. Hydrobiol. Trop.*, **15** (1), 23–31.

Elouard, J. M., and Jestin, J. M. (1983). Impact du temephos (Abate) sur les invertébrés non cibles. B—Un indice biocénotique pour mesurer l'action du temophos sur la faune lotique non-cible des rivières traitées dans le cadre du Programme de Lutte contre l'Onchocercose. *Rev. Hydrobiol. Trop.*, **16** (4), 341–351.

Elouard, J. M., and Gibon, F. M. (1984). *Incidence on non-target insect fauna of the alternate use of three insecticides (temephos, chlorphoxim and B.t. H-14) for the control of the larvae of* Simulium damnosum *S.I.* Report OCP/VCU/HYBIO/84.4.

Gibon, F. M., and Troubat, J. J. (1980). *Effets d'un traitement au chlorphoxim sur la dérive des invertébrés benthiques.* Rapport du Centre ORSTOM de Bouaké, No. 37.

Gibon, F. M., and Statzner, B. (1985). Longitudinal zonation of lotic insects in the Bandama river system (Ivory coast). *Hydrobiologia*, **122**, pp. 61–64.

Gibon, F. M., Elouard, J. M., and Troubat, J. J. (1980). *Action du* Bacillus thuringiensis *var.* israelensis *sur les invertébrés aquatiques. I—Effet d'un traitement expérimental sur la Maraoué.* Rapport du Centre ORSTOM de Bouaké, No. 38.

Gibon, F. M., Troubat, J. J., and Bihoum, M. (1983). *Recherches sur la faune invertébrée benthique des cours d'eau non traités aux larvicides antisimulidiens.* Rapport du Centre OSTOM de Bouake, No. 50.

Guenda, W. (1985). *Hydrobiologie d'un cours d'eau temporaire en zone soudanienne: la Volta Range (Burkina Faso-Ghana)—Relation avec les traitements chimiques antisimulidiens.* Thèse Doct. 3ième cycle, Univ. Aix-Marseille.

Iltis, A. (1982a). Peuplements algaux des rivières de Côte d'Ivoire. I—Stations de prélèvement, méthodologie, remarques sur la composition qualitative et biovolumes. *Rev. Hydrobiol. Trop.*, **15** (3), 231–240.

Iltis, A. (1982b). Peuplements algaux des rivières de Côte d'Ivoire. II—Variations saisonnières des biovolumes, de la composition, de la diversité spécifique. *Rev. Hydrobiol. trop.*, **15** (3), 241–251.

Iltis, A. (1982c). Peuplements algaux des rivières de Côte d'Ivoire. III—Etude du périphyton. *Rev. Hydrobiol. Trop.*, **15** (4), 303–312.

Iltis, A. (1983). Peuplements algaux des rivières de Côte d'Ivoire. IV—Remarques générales. *Rev. Hydrobiol. Trop.*, **16** (3), 235–240.

Iltis, A., and Lévêque, C. (1982). Caractéristiques physico-chimiques des rivières de Côte d'Ivoire. *Rev. Hydrobiol. Trop.*, **15** (2), 115–130.

Lauzanne, L. (1973). *Etude au laboratoire de la toxicité sur la faune non-cible de nouveaux insecticides employés en lutte antisimulies. 2ème partie: action sur les poissons.* Rapport du Centre ORSTOM de N'Djamena.

Lauzanne, L., and Dejoux, C. (1973). *Etude de terrain de la toxicité sur la faune aquatique non-cible de nouveaux insecticides employés en lutte antisimulies.* Rapport du Centre ORSTOM de N'Djamena.

Lévêque, C., and Herbinet, P. (1980). Caractères méristiques et biologie des *Schilbe mystus* (Pisces, Schilbeidae) en Côte d'Ivoire. *Cah. ORSTOM, sér. Hydrobiol.*, **13**, 161–170.

Lévêque, C., and Herbinet, P. (1982). Caractères méristiques et biologie des *Eutropius mentalis* (Pisces, Schilbeidae) en Côte d'Ivoire. *Rev. Zool. Afr.*, **96**, 366–392.

Lévêque, C., and Paugy, D. (1985). *Guide des poissons d'eau douce de la zone du Programme de lutte contre l'onchocercose en Afrique de l'Ouest.* Rapport ORSTOM, Paris.

Lévêque, C., Odei, M., and Pugh Thomas, M. (1979). The Onchocerciasis Control Programme and the monitoring of its effects on the riverine biology of the Volta

River Basin. In: Perring, F. H., and Mellanby, K. (eds), *Ecological Effects of Pesticides*, pp. 133–143. Linnaean Society Symposium series, 5.

Lévêque, C., Paugy, D., and Jestin, J. M. (1982). *Fish Communities of Ivory Coast Rivers Treated by Temephos* (Unpublished report).

Lévêque, C., Dejoux, C., and Iltis, A. (1983). Limnologie du fleuve Bandama. Côte d'Ivoire. *Hydrobiologia*, **100**, 113–141.

Mérona, B. de (1980). Ecologie et biologie de *Petrocephalus bovei* (Poissons; Mormyridae) dans les rivières de Côte d'Ivoire. *Cah. ORSTOM, sér. Hydrobiol.*, **13**, 117–127.

Mérona, B. de (1981). Zonation ichtyologique du bassin du Bandama. *Rev. Hydrobiol. Trop.*, **14**, 63–75.

Mérona, B. de, and Albaret, J. J. (1978). *Répartititon spatiale des poissons dans les radiers des rivières de Côte d'Ivoire*. Rapport du Centre ORSTOM de Bouaké, No. 17.

Mérona, B. de, Lévêque, C., and Herbinet, P. (1977). *Observations préliminaires sur les peuplements ichtyologiques des radiers. Résultats des pêches électriques effectuées dans les stations du programme de surveillance de l'environnement aquatique*. Rapport du Centre ORSTOM de Bouaké, No. 9.

Mérona, B. de, Abban, E. K., Herbinet, P., and Sape, E. R. (1979). *Peuplements ichtyologiques de biotopes d'eaux peu profondes des rivières du nord du Ghana*. Rapport du Centre ORSTOM de Bouaké, No. 31.

Moniod, F., Pouyard, B., and Sechet, P. (1977). Le bassin du fleuve Volta. *Monographies hydrologiques*, ORSTOM, No. 5.

Paugy, D. (1978). Ecologie et biologie des *Alestes baremoze* (Pisces. Characidae) des rivières de Côte d'Ivoire. *Cah. ORSTOM, sér. Hydrobiol.*, **12**, 245–275.

Paugy, D. (1980a). Ecologie et biologie des *Alestes imberi*. (Pisces. Characidae) des rivières de Côte d'Ivoire. Comparaison méristique avec *A. nigricauda*. *Cah. ORSTOM, sér. Hydrobiol.*, **13**, 129–141.

Paugy, D. (1980b). Ecologie et biologie des *Alestes nurse* (Pisces, Characidae) des rivières de Côte d'Ivoire. *Cah. ORSTOM, sér. Hydrobiol.*, **13**, 143–159.

Paugy, D., and Coulibaly, B. (1983). *Effects of treatment with Propoxur on Benthic Invertebrate Drift in the Wawa*. Report OCP/VCU/HYDBIO/83.3.

Paugy, D., Yameogo, L., Bihoum, M., and Coulibaly, B. (1984). *Short-term Impact of Permethrin on the Non-target Aquatic Fauna*. Report OCP/VCU/HYBIO/84.14.

Philippon, B. (1977). Etude de la transmission d' *Onchocerca volvulus* (Leuckart, 1983) Nematoda, Onchocercidae) par *Simulium damnosum* (Theobald, 1903) (Diptera, Simuliidae) en Afrique tropicale. *Travaux et Documents ORSTOM*, No. 63.

Quelennec, G., Miles, J. W., Dejoux, C., and Mérona B. de (1977). *Chemical Monitoring for Temephos in Mud, Oysters and Fish from a River within the Oncocerchiasis Control Programme in the Volta Basin Area*. Report WHO/VBC/683.

Rishikesh, N., Burges, H. D., and Vandekar, M. (1983). *Operational use of* Bacillus thuringiensis *serotype H-14 and environmental safety*. Report WHO/VBC/83.871.

Samman, J., and Pugh Thomas, M. (1978). Effect of an organophosphorous insecticide, Abate, used in the control of *Simulium damnosum* on non-target benthic fauna. *Int. J. Environ. Stud.*, **12**, 141–144.

Statzner, B. (1982). Population dynamics of Hydropsychidae (Insecta, Trichoptera) in the N'zi River (Ivory Coast), a temporary stream partly treated with the insecticide Chlorphoxim. *Rev. Hydrobiol. Trop.*, **15** (2), 157–176.

Statzner, B. (1984). Keys to adult and immature Hydropsychidae in the Ivory Coast (West Africa) with notes on their taxonomy and distribution (Insecta, Trichoptera). *Spixiana*, **7**, 23–50.

Statzner, B., and Gibon, F. M. (1984). Keys to adult and immature Macronematinae (Insecta: Trichoptera) from the Ivory Coast (West Africa) with notes on their taxonomy and distribution. *Rev. Hydrobiol. Trop.*, **17** (2), 129–151.

Troubat, J. J. (1981). Dispositif à gouttières multiples destiné à tester *in situ* la toxicité des insecticides vis-à-vis des invertébrés benthiques. *Rev. Hydrobiol. trop.*, **14** (2), 149–152.

Troubat, J. J., and Lardeux, F. (1982). Toxicité pour la faune aquatique de quelques larvicides antisimulidiens. I—Le G H 76®. *Rev. Hydrobiol. Trop.*, **15** (1), 15–21.

Vidy, G. (1976) *Etude du régime alimentaire de quelques poissons insectivores, dans les rivières de Côte d'Ivoire. Recherche de l'influence des traitements insecticides effectués dans le cadre de la lutte contre l'onchocercose.* Rapport du Centre ORSTOM de Bouaké, No. 2.

Yameogo, L., Bihoum, M., and Coulibaly, B. (1984). *Short-Term Effects of Azametiphos on Non-Target Aquatic Invertebrates.* Report OCP/VCU/HYBIO/84-15.

Walsh, J. F. (1981). The untreated component of river systems in the OCP area and its environmental implications. Unpublished report to OCP.

Ecotoxicology and Climate
Edited by P. Bourdeau, J. A. Haines, W. Klein and C. R. Krishna Murti

5.6 Herbicides in Warfare: The Case of Indochina

A. H. WESTING

5.6.1 INTRODUCTION

The Second Indochina War (or Vietnam Conflict) of 1961–1975 is noted for the widespread and severe environmental damage inflicted upon its theatre of operations, especially in the former South Vietnam (Westing, 1976, 1980, 1982a, 1984b). The US strategy in South Vietnam, *inter alia*, involved massive rural area bombing, extensive chemical and mechanical forest destruction, large-scale chemical and mechanical crop destruction, wide-ranging chemical anti-personnel harassment and area denial, and enormous forced population displacements. In short, this US strategy represented the intentional disruption of both the natural and human ecologies of the region. Moreover, this war was the first in military history in which massive quantities of anti-plant chemical warfare agents (herbicides) were employed (Buckingham, 1982; Cecil, 1986; Lang *et al.*, 1974; Westing, 1976, 1984b).

The Second Indochina War was innovative in that a great power attempted to subdue a peasant army through the profligate use of technologically advanced weapons and methods. One can readily understand that the outcome of more than a decade of such war in South Vietnam and elsewhere in the region resulted not only in heavy direct casualties, but also in long-term medical sequelae. By any measure, however, its main effects were a widespread, long-lasting, and severe disruption of forestlands, of perennial croplands, and of farmlands—that is to say, of millions of hectares of the natural resource base essential to an agrarian society.

This section first reviews the history of the use of herbicides in warfare. It goes on to summarize the employment of these agents during the Second Indochina War against forest trees and crop plants, and then describes their immediate effect on flora and fauna. Following a brief treatment of the persistence of the agents used, it concludes with a summary of the long-term ecological effects, primarily with reference to South Vietnam.

5.6.2 HISTORICAL BACKGROUND

Humans, in common with all animals, are dependent upon the food and shelter they derive from the plant kingdom. The intentional military destruction during war of vegetation in territory under actual or potential enemy control is a recognition of this fundamental relationship. Indeed, crop destruction has been a continuing part of warfare for millennia (Westing, 1981a), and the military importance of forests has also long been recognized (Clausewitz, 1832–1834, pp. 426, 530).

Vegetational destruction can be accomplished via high explosives, fire, tractors, and other means. The account below describes the employment of chemical agents for this purpose. Indeed, the sporadic use of plant-killing chemicals during both peace and war is thousands of years old. Abimelech, the son of Jerubbaal and an ancient prophet and king of Israel, is recorded in the Bible as having sowed the conquered city of Shechem (at or near the modern Nablus, Jordan; approx. 50 km north of Jerusalem) with salt as the final, perhaps symbolic, act in its destruction (*Judges* 9:45). The ancient Romans seem also to have employed salt in this way (Scullard, 1961).

Various inorganic herbicides (including arsenicals) have been in routine agricultural and horticultural use since the late 19th century and a number of organic ones since the mid-1930s. However, the most important advance to date in the development of herbicides was the discovery of the remarkable utility of the phenoxy and other plant-hormone-mimicking chemicals. It is thus interesting to note that their development as herbicides was tied to the then secret chemical warfare research carried out by the British and US governments during World War II (Peterson, 1967).

More than a thousand chemicals were tested in the USA during World War II in the hopes of perfecting militarily usable crop-destroying chemicals (Merck, 1946, 1947; Norman, 1946). Clearly the single most important herbicidal compound developed during this period was 2,4-D, still the most widely used herbicide in the world. Its less used and more controversial cousin, 2,4,5-T, was developed in the same way during this period. Although the possibility was considered, herbicides were not used for military purposes during World War II.

It fell to the United Kingdom in its attempt to suppress an insurgency in Malaya to be the first to employ modern herbicides for military purposes, primarily during the mid-1950s (Clutterbuck, 1966; Henderson, 1955; Henniker, 1955). The chemical anti-plant agents were used for two different purposes in this desultory decade-long war. Some of the herbicidal attacks were for defoliation along lines of communication in order to reduce the possibility of ambushes, whereas others were for the destruction of crops which were presumably being grown by or for the insurgents. These applications (both by air and from the ground) were relatively modest and rather short-lived.

The major agent employed appears to have been a mixture of 2,4,5-T, 2,4-D, and trichloroacetic acid (Connor and Thomas, 1984).

The only really large-scale military use of herbicides was by the USA in pursuing the Second Indochina War. This programme, the details of which are presented below, began on a very small scale in 1961, increased to a crescendo in 1969, and finally ended during 1971. Although the major US effort was directed against forests, a continuing aspect of the programming from beginning to end was crop destruction and food denial. Through the years, the US herbicide spraying was confined largely to South Vietnam, but a modest fraction of eastern Kampuchea was also involved once in 1969 (Westing, 1972). Moreover, the USA also carried out a series of herbicide missions against Laos (Westing, 1981b) and possibly a few against North Vietnam as well (Agence France Presse, 1971).

Other than the above-noted instances, herbicides have been associated with other theatres of war and with the armed forces of other nations to only a very limited extent. For example, in 1972 the Israeli Army used 2,4-D on one occasion for crop destruction in Aqaba, Jordan (approx. 40 km north-northeast of Jerusalem) (Holden, 1972) and thus amazingly close to the Shechem mentioned earlier, where one of the first military uses of herbicides may have occurred some three thousand years ago.

5.6.3 THE SECOND INDOCHINA WAR OF 1961–1975

5.6.3.1 Technical Background

During the Second Indochina War, the USA carried out a massive herbicidal programme that stretched over a period of a decade. Although the USA was neither the first nor the only nation to employ chemical anti-plant agents as weapons of war, the magnitude of this programme was without precedent. It was aimed for the most part at the forests of South Vietnam and, to a lesser extent, at its crops. Herbicidal attacks upon the other nations of Indochina were modest in comparison. Using a variety of agents, the USA eventually expended a volume of more than 72 million litres (91 million kilograms), containing almost 55 million kilograms of active herbicidal ingredients.

The major anti-plant agents that were employed by the USA in Indochina were colour-coded 'Orange', 'White', and 'Blue' (Table 5.6.1). Agents Orange and White consist of mixtures of plant-hormone-mimicking compounds which kill by interfering with the normal metabolism of treated plants; Agent Blue, on the other hand, consists of a desiccating compound, which kills by preventing a plant from retaining its moisture. Agents Orange and White are particularly suitable for use against dicotyledonous plants, whereas Agent Blue is relatively more suitable for use against monocotyledonous plants. At the high levels that were used for military application—28 l/ha, averaging 21 kg/ha in terms of active ingredients (i.e. 20 to 40 times higher than normal civil usage)—these

Table 5.6.1 Major chemical anti-plant agents employed by the USA in the Second Indochina War. Source: Westing (1976) adjusted in accord with Westing (1982b)

Type	Composition	Physical properties	Application
Agent Orange*	A 1.124:1 mixture (by weight) of the *n*-butyl esters of 2,4,5-trichlorophenoxyacetic acid (2,4,5-T) (545.4 kg/m^3 acid equivalent) and 2,4-dichlorophenoxyacetic acid (2,4-D) (485.1 kg/m^3 acid equivalent); also containing 2,3,7,8-tetrachloro-p-dioxin (dioxin) (an estimated 3.83 g/m^3)	Liquid; oil-soluble; water-insoluble; weight 1285 kg/m^3	Applied undiluted at 28.06 l/ha, thereby supplying 15.31 kg/ha of 2,4,5-T and 13.61 kg/ha of 2,4-D, in terms of acid equivalent, and also an estimated 107 mg/ha of dioxin
Agent White	A 3.882:1 mixture (by weight) of the tri-iso-propanolamine salts of 2.4-dichlorophenoxyacetic acid (2,4-D) (239.7 kg/m^3 acid equivalent) and 4-amino-3,5,6-trichloropicolinic acid (picloram, 'Tordon') (64.7 kg/m^3 acid equivalent)	Aqueous solution; oil-insoluble; weight 1150 kg/m^3	Applied undiluted at 28.06 l/ha, thereby supplying 6.73 kg/ha of 2,4-D and 1.82 kg/ha of picloram in terms of acid equivalent
Agent Blue	A 2.663:1 mixture (by weight) of Na dimethyl arsenate (Na cacodylate) and dimethyl arsinic (cacodylic) acid (together 371.46 kg/m^3 acid equivalent)	Aqueous solution; oil-insoluble; weight 1310 kg/m^3	Applied undiluted at 28.06 l/ha, thereby supplying 10.42 kg/ha, in terms of acid equivalent (of which 5.66 kg/ha is elemental arsenic).

*Numerous herbicidal formulations have been tested by the USA as chemical anti-plant agents, several of which were assigned a colour code during their more or less ephemeral existence: 'Orange II' was similar to the 'Orange' above, except that its 2,4,5-T moiety was replaced by the iso-octyl ester of 2,4,5-T; 'Pink' was a mixture of the *n*-butyl and iso-butyl esters of 2,4,5-T; 'Green' consisted entirely of the *n*-butyl ester of 2,4,5-T; and 'Purple' was a mixture of the *n*-butyl ester of 2,4-D and the *n*-butyl and iso-butyl esters of 2,4,5-T herbicides are, however, not as selective as one might expect on the basis of civil experience.

Of the several agents used, Agent Orange represented 61% of the total volume expended over the years (Table 5.6.2). The three peak years of herbicide spraying—1967 to 1969—were about equal in magnitude and together accounted for over three-quarters of the volume of total US wartime expenditures. These were also very active war years in other respects, as evidenced, for example, by the heavy US munition expenditures during this period and the high numbers of US fatalities.

Forest destruction was generally accomplished through the use of Agents Orange or White. Conversely, Agent Blue was usually the agent of choice for the destruction of rice and other crops, although Agent Orange was much used for this purpose as well (Table 5.6.3). All told, about 86% of the missions were

Table 5.6.2 US herbicide expenditures in the Second Indochina War: a breakdown by agent and year* (in $m^3 = 10^3$ l). Source: Westing (1976). Reproduced with permission

Year	Agent Orange†	Agent White‡	Agent Blue§	Total¶
1961	?	0	?	?
1962	56	0	8	65
1963	281	0	3	283
1964	948	0	118	1066
1965	1767	0	749	2516
1966	6362	2056	1181	9599
1967	11 891	4989	2513	19 394
1968	8850	8483	1931	19 264
1969	12 376	3572	1309	17 257
1970	1806	697	370	2873
1971	0	38	?	38
Total‖	44 338	19 835	8182	72 354

*To convert any of the herbicide volume data given to area covered in hectares (not considering overlap), multiply by 35.6

†To convert any of the Agent Orange volume data given to total weight in kilograms, multiply by 1285; similarly for 2,4,5-T content in kilograms, multiply by 545; similarly for 2,4-D, multiply by 485; similarly for dioxin estimate, multiply by 0.003 83

‡To convert any of the Agent White volume data given to total weight in kilograms, multiply by 1150; similarly for 2,4-D content in kilograms, multiply by 240; similarly for picloram, multiply by 64.7

§To convert any of the Agent Blue volume data given to total weight in kilograms, multiply by 1310; similarly for dimethyl arsinic (cacodylic) acid in kilograms, multiply by 371; similarly for elemental arsenic, multiply by 202

¶To convert any of the total herbicide volume data given to average total weight in kilograms, multiply by 1251; similarly for average kilograms of active ingredients, multiply by 757

‖The following amounts were sprayed in terms of active ingredients: 2,4-D, 26 million kilograms; 2,4,5-T, 24 million kilograms (containing about 170 kg dioxin); picloram, 1.3 million kilograms; dimethyl arsinic (cacodylic) acid, 3.0 million kilograms (of which elemental arsenic represents 1.7 million kilograms): total active ingredients, 55 million kilograms directed primarily against forest and other woody vegetation and the remaining 14% primarily against crop plants.

Total geographical coverage of the spray missions was less than one might expect on the basis of the total expenditure of herbicides, since about 34% of the target areas were chemically attacked more than once during the course of the war (Table 5.6.4). Thus the total area subjected to spraying one or more times came to an estimated 1.7 million hectares, this area being treated 1.5 times on average, thereby receiving an average dose of approx. 42 l/ha, or approx. 32 kg/ha in terms of active ingredients.

Most of the anti-plant chemicals—in the neighbourhood of 95%—were dispersed from C-123 (UC-123) transport aircraft equipped to deliver somewhat over 3000 l onto 130 ha or so. The high-pressure nozzles which were used delivered droplets having an approximate median diameter of 350 μm, so that there was reasonably little drift as long as wind speeds exceeding 5 m/s were

Table 5.6.3 US herbicide expenditures in the Second Indochina War: a breakdown by type of mission and agent* (in m^3). Source: Westing (1976). Reproduced with permission

Type of mission	Agent Orange	Agent White	Agent Blue	Total
Forest	39 816	19 094	1684	60 594
Misc. woody vegetation	709	529	312	1550
Crop	3813	212	6185	10 210
Total	44 338	19 835	8182	72 354

*The same conversions provided in Table 5.6.2 footnotes are also applicable to this table

Table 5.6.4 US herbicide expenditures in the Second Indochina War: a breakdown by number of repeat sprayings within the area covered. Source: Westing (1976). Reproduced with permission

Number of sprayings of one area	Ultimate herbicide expenditure* (m^3)	Area involved† (10^3 ha)
One	31 572	1125
Two	21 431	382
Three	11 412	136
Four	5335	48
Five or more	2603	19
Total	72 354	1709

*To convert any of the herbicide volume data given to average total weight in kilograms, multiply by 1251; similarly for average kilograms of active ingredients, multiply by 757

†Based on the standard application rate of 28.1 l/ha. Had no area been sprayed more than once, then the total coverage would have been 2578×10^3 ha. As it was, the areas which were sprayed received an overall average of 42.3 l/ha, that is, they were sprayed an average of 1.51 times

avoided, as they usually were. Normal spray time for an aircraft was just over 2 minutes, although the entire payload could, if the need arose, be ejected (dumped) in about 30 seconds, and thus onto approximately 30 hectares. Of the order of 50 such dumpings occurred during the war, in which the dose level became approx. 120 l/ha, or about 90 kg/ha in terms of active ingredients. One aircraft (one sortie) sprayed a strip roughly 150 m wide and 8.7 km long. A mission generally consisted of 3–5 aircraft flying in lateral (side-by-side) formation. Most of the 5% of the anti-plant chemicals not dispensed from fixed-wing aircraft was dispensed from helicopters, although small amounts were also dispensed from trucks and boats.

It is impossible to provide an accurate regional breakdown of herbicide expenditures for the whole of Indochina inasmuch as the necessary information has never been made public by the USA. It is known that about 10% of South Vietnam, the hardest hit nation, was sprayed (Table 5.6.5). Within South Vietnam, it was a rather large region surrounding Saigon (Ho Chi Minh City)—so-called Military Region III (Figure 5.6.1)—that was singled out for the most intensive coverage, both on a per-unit-area or per-capita basis. Indeed, it appears that essentially 30% of the land area of Military Region III was sprayed one or more times.

Table 5.6.5 US herbicide expenditures in the Second Indochina War: a breakdown by region. Source: Westing (1976) adjusted in accord with Westing (1972) and Westing (1981b)

Region	Herbicide expenditure* (m^3)	Area sprayed once or more (10^3 ha)	Fraction of area sprayed (%)	Spraying in relation to the population (l/capita)
South Vietnam‡	70 720	1670	10	4.0
Military Region I	12 022	284	10	3.9
Military Region II	14 851	351	5	4.8
Military Region III	37 482	885	29	7.7(15.9)†
Military Region IV	6365	150	4	1.0
North Vietnam	?	?	?	?
Kampuchea	34	1	—	—
Laos	1600	38	0.2	0.6
Total	72 354	1709	2	1.6

*To convert any of the herbicide volume data given to average total weight in kilograms, multiply by 1251; similarly for average kilograms of active ingredients, multiply by 757

†The parenthetical value is based on the regional population less that of Saigon

‡The former Military Regions are depicted, and their areas and mid-war populations provided, in Figure 5.6.1

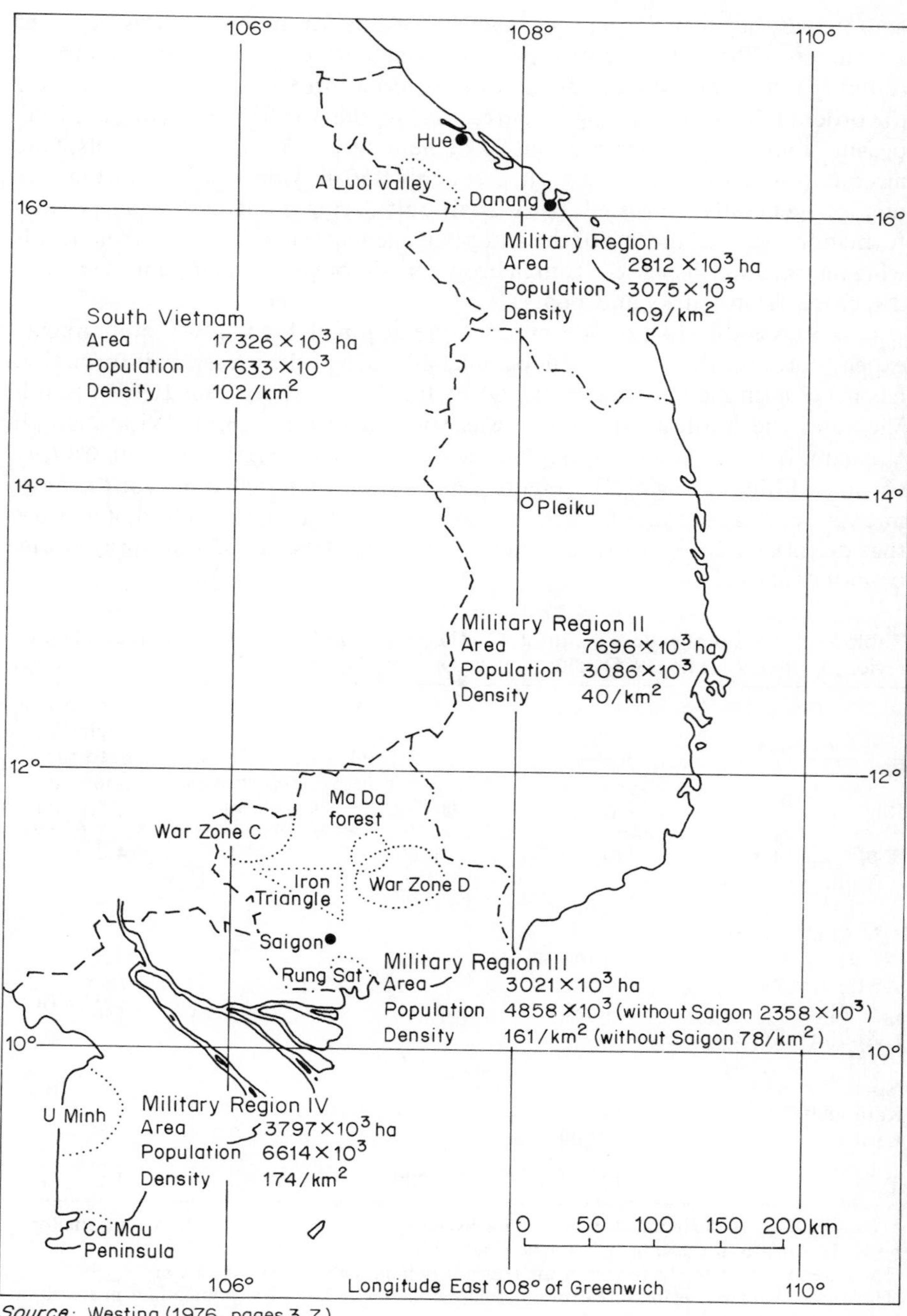

Source: Westing (1976, pages 3,7).

5.6.3.2 Immediate Effects on Flora and Fauna

Dense Inland (Upland) Forest

Woody vegetation covers about 10.4 million hectares, or 60%, of South Vietnam, the largest single category of which is dense inland (upland) forest. The dense inland forest, extending over about 5.8 million hectares, is composed of a complex and variable floristic conglomeration. It includes a bewildering diversity of dicotyledonous trees, lianas, epiphytes, and herbs, as well as some monocotyledons, ferns, etc. The tree species vary in height, usually forming two and occasionally three rather indistinct strata (storeys). The upper canopy usually attains a height of 20–40 m. The dominant plant family is the Dipterocarpaceae, which is represented by at least 30 major species in the genera *Dipterocarpus*, *Anisoptera*, *Hopea*, and *Shorea*. Another important genus is *Lagerstroemia* in the family Lythraceae. There are also a number of important genera of Leguminosae (e.g. *Erythrophleum*), Guttiferae, and Meliaceae. Moreover, this dense inland forest supports some 200 commercial tree species, a dozen of which are of exceedingly high quality and suitable for the world market. Chief among these so-called luxury woods are three rosewood species (*Dalbergia bariensis*, *D. cochinchinensis*, and *Pterocarpus pedatus*; all Leguminosae), an ebony (*Diospyros mun*; Ebenaceae), and a false mahogany (*Melanorrhea laccifera*; Anacardiaceae).

The dense inland forest was the militarily most important of South Vietnam's land categories. To begin with, it can be estimated that about 1.4 million hectares, or 14% of the total extent of South Vietnam's woody vegetation was sprayed one or more times (Table 5.6.6). Of this, perhaps 1.1 million hectares occurred in the dense inland forest type, which represents about 19% of that vegetational category. The dense forest lands within so-called War Zones C and D (Figure 5.6.1) were particularly hard hit.

Following herbicidal attack of an inland forest, fairly complete leaf abscission (as well as flower and fruit abscission) occurred within two or three weeks. The surviving trees usually remained bare until the onset of the next rainy season, often, therefore, for a period of several or more months. To achieve total defoliation of the lower storeys of a multiple-canopy forest necessitated one or more follow-up sprayings.

Virtually all of the many dicotyledonous tree species subjected to spraying were defoliated at the intensity of treatment employed. Then, at the time of refoliation, a spectrum of sensitivity became evident among the many hundreds of tree species which comprise this vegetational type. Among the most sensitive of the dense inland forest species of South Vietnam are *Pterocarpus pedatus* (Leguminosae) and *Lagerstroemia* spp. (Lythraceae). Among the most resistant

---

Figure 5.6.1 *(opposite)* South Vietnam during the Second Indochina War (populations shown are estimates for 1969). Source: Westing (1976). Reproduced by permission

Table 5.6.6 US herbicide expenditures, in 10^3 ha, in South Vietnam in the Second Indochina War: a breakdown by vegetational type. Source: Westing (1976) adjusted in accord with Table 5.6.5

Vegetational type	Area	Area sprayed once or more
Dense forest	5800	1077
Primary	4500	836
Primary plus secondary	600	111
Secondary	700	130
Open (clear) forest	2000	100
Bamboo brake	800	40
Mangrove forest (swamp)	500	151
True	300	124
Rear (back)	200	27
Rubber plantation	100	30
Pine forest	100	0
Miscellaneous woody vegetation	1100	36
Woody subtotal	10 400	1434
Paddy (wet) rice	2500	59
Field crops (upland rice, etc.)	500	177
Agricultural subtotal	3000	236
Miscellaneous	3926	0
Total	17 326	1670

are *Cassia siamea* (Leguminosae) and *Sandoricum indicum* (Meliaceae). And among those intermediate between these two extremes are *Hopea odorata*, *Dipterocarpus alatus*, and *Shorea cochinchinensis* (all Dipterocarpaceae).

Only about 10% (some observers have suggested more) of the trees were killed outright by a single military spraying, a situation true for perhaps 66% of the total sprayed area (Table 5.6.4). The survivors displayed various levels of injury, as evidenced by differing severities of crown (branch) dieback, temporary sterility, and other symptoms. Such injury in time resulted in some further delayed mortality among the survivors. Moreover, once the understorey was deprived of the protection of the overstorey, some fraction of the understorey trees in time die of environmental conditions too harsh for their existence. When the military situation led to more than one herbicidal attack, as occurred on about 34% of all sprayed lands, the level of tree mortality increased more or less exponentially with each subsequent spraying, more steeply so with briefer intervals between sprayings. It might also be noted that in economic terms the spraying of South Vietnam's dense inland forest resulted in a loss of commercial timber estimated to be of the order of 20 million m^3 (Westing, 1980, 1982a).

When the trees were sprayed, causing the leaves to fall and decompose, the soil was for the most part unable to hold the released nutrients so that these were lost to the local ecosystem, a phenomenon referred to as nutrient dumping, which is especially acute in the tropics. Further nutrient losses and other site debilitation occurred, since the death of the vegetation led to accelerated soil erosion. Both erosion and nutrient dumping continued until the area in question was again stabilized by the establishment of a replacement (pioneer) vegetation, usually during the subsequent growing season.

The principal impact on the wildlife of sprayed sites was via a diminution in the food and cover (shelter) afforded by the vegetation. Here it must be noted that a significant majority of the animal life in a tropical forest is found in—and depends upon—the upper vegetational storeys, precisely the portion of the ecosystem most seriously impaired by massive herbicidal attack. There was also a more or less modest level of damage to wildlife via the direct toxic action of the herbicides sprayed. Some birds appear to have succumbed in this fashion, and probably several kinds of invertebrates as well, including some aquatic invertebrates and some terrestrial insects.

Thus it can be seen that the immediate ecological impact of military spraying of an inland forest site with anti-plant chemicals can be severe, especially if repeated. The primary producers (green plants) of an ecosystem are knocked back drastically, to be replaced by a new community of significantly lesser biomass, smaller nutrient-holding capacity, and reduced primary productivity. A poorer soil results from the attack, with a lesser fraction of humus (organic matter), and often exhibiting a chronic shortage of nitrogen. Fire subsequent to herbicidal attack would aggravate the situation. Particularly in those inland areas that were sprayed several times—some 200 thousand hectares (Table 5.6.4)—the overstorey destruction was sufficient to permit the release of, or invasion by, certain tenacious pioneer grasses. These included both herbaceous types, such as *Imperata cylindrica* (Gramineae), and woody types, such as frutescent (shrubby) bamboos.

Coastal Mangrove Habitat

The mangrove habitat, scattered primarily along the southerly coastline of South Vietnam, occupies approximately 500 thousand hectares of inhospitable, seemingly impenetrable, and outwardly unimportant swamps. The dominant mangrove vegetation consists of several species of small trees, mostly 3 m to 15 m high, primarily in the genera *Rhizophora* (Rhizophoraceae), *Avicennia* (Verbenaceae), and *Bruguiera* (Rhizophoraceae)—all so-called mangroves. As soil deposition extends the coastline slowly out into the sea, the most common pioneer of this virgin land is *Sonneratia* (Lythraceae), which in turn prepares the site for a subsequent invasion by *Avicennia*. *Rhizophora* is likely to be next in this succession, to be followed by *Bruguiera*, which extends back to the limit of daily tidal wash (and the limit of the true mangrove type). Through time, the soil level builds up beyond the reach of flood tide and a new community (the rear (or back) mangrove type) develops, which is dominated by a tree species of *Melaleuca* (Myrtaceae). The mangrove habitat provides substantial amounts of firewood, charcoal, tannin, thatch (from *Nipa fruticans*; Palmae), and other secondary forest products.

The mangrove habitat is singled out here owing to the widespread and peculiarly drastic herbicidal damage it suffered during the Second Indochina War. An estimated 124 thousand hectares of true mangrove—41% of that entire sub-type—plus another 27 thousand hectares of rear mangrove (13% of that sub-type) were subjected to military herbicide spraying during the war. Unlike the relatively modest degree of kill resulting from such action in inland forest types, even a single spraying in this coastal lowland type most often destroyed essentially the entire plant community (the herbicidal sensitivity probably being related to its osmotic versatility, i.e. to its wide tolerance to changes in salinity). Virtually nothing remained alive after a single herbicidal attack and the resulting scene was weird and desolate. It subsequently became even worse when exacerbated by the usual salvage harvesting of the killed trees and by the inevitable erosion.

The taxonomically diverse plant species that make up the mangrove community all displayed great sensitivity to the hormone-mimicking herbicides, with *Rhizophora*—the economically most important genus of trees—being especially sensitive, and *Avicennia* somewhat less so (Truman, 1961–1962; Walsh *et al.*, 1973). Little if any immediate recolonization occurred on the herbicide-obliterated sites. With the primary producers essentially wiped out, their energy-capturing function was lost, along with all else that built on this. Both food and cover were eliminated by the US attacks, affecting not only the enemy forces but the indigenous forest creatures as well. For example, an enormous reduction was reported in numbers of birds (Orians and Pfeiffer, 1970).

Less obviously, the reticulation of channels throughout a mangrove swamp—roughly one-quarter of the surface area of the mangrove type in South Vietnam—supports a rich variety of aquatic fauna during all or part of their life cycle. These organisms depend directly or indirectly on a steady and enormous supply of nutrients dropped, flushed, and leached out of the terrestrial part of the system. Numerous species of fish and crustaceans that spend their adult lives offshore, and some that migrate and live up the rivers, utilize the mangrove estuaries as breeding and/or nursery grounds. There were early indications of post-war declines in South Vietnam's offshore fishery (involving both true fish and shellfish (crustaceans)) attributed to the wartime spraying.

It is important to note that the mangrove type is a transitional zone between land and sea and thus appears to serve the added important function of stabilizing the shoreline. As the coastline accretes, mangroves invade the virgin lands and their roots hold the soil against the action of wind, wave, current, and tide. The unprotected channel banks and barren mud flats created by the spraying were seen to erode at rapid rates.

With so much of the mangrove habitat destroyed, it was clearly the ecosystem most seriously affected by the Second Indochina War. With the promise of long-term conversion of a significant fraction of this habitat to other vegetational types (whether by natural or anthropogenic means) the question arises to what extent such long-term habitat loss will lead to species extinctions. It is known that the number of species within any particular taxonomic group that an isolated habitat can support is related to its area. If a habitat is reduced in size, as has been the case with South Vietnam's mangroves, the resulting excess of species will in due course die out. It has been estimated by the author that a 10% reduction in this mangrove habitat (a likely situation) will in time lead to a 3 or 4% loss in the indigenous plant and animal species (Westing, 1982a).

Agriculture

During the Second Indochina War the USA carried out a routine military policy of systematic large-scale crop destruction in South Vietnam. Chemical crop destruction from the air made up the greatest fraction of this major resource denial programme. Chemical crop destruction is estimated to have affected some 236 thousand hectares of agricultural lands in South Vietnam one or more times (approx. 8% of the total) (Table 5.6.6). In all, herbicides were sprayed over some 356 thousand hectares of agricultural lands in South Vietnam, although this larger value does not take into account sprayings of the same fields during different years. (At least 8 thousand hectares of crop lands were additionally sprayed elsewhere in Indochina, especially in Laos.) The crop spraying is estimated to have resulted in the immediate destruction of more than 300 million kilograms of food.

Additionally, perhaps 30% of South Vietnam's 135 thousand hectares of rubber plantations were destroyed by herbicides during the war (Westing, 1980).

5.6.4 SOUTH VIETNAM TODAY

5.6.4.1 Herbicidal Persistence

Important in any consideration of the long-term impact of herbicides is their persistence and mobility, that is, how long they will remain present and active in the soil and biota, and whether they will move up in food chains, perhaps even concentrating in the process (so-called ecological amplification). For 2,4-D, representing 48% of the chemicals sprayed (Table 5.6.2), a level of environmental insignificance (as determined by lack of obvious effect on all but the most highly sensitive of subsequently planted test species) is reached within a month or so. For 2,4,5-T, representing 44% of the chemicals sprayed, this occurs within five months or so; for picloram, representing 2% of the chemicals sprayed, within perhaps 18 months; and for dimethyl arsinic acid, representing 6% of the chemicals sprayed, after about a week. Chemical analysis would, of course, reveal traces of all these substances for some time beyond the durations just given, as might the sowing of particularly sensitive indicator species. Some relevant examples from the literature follow.

In Hawaiian field trials, it was found that 2,4-D applied at the rate of 11 kg/ha continues to affect subsequently planted indicator plants such as bean (*Phaseolus*; Leguminosae) or tomato (*Lycopersicon*; Solanaceae) for 2–14 weeks (Akamine, 1950–1951). In Puerto Rican field trials, it was found that a 1:1 mixture of 2,4-D and 2,4,5-T applied at 27 kg/ha exerted a significant residual effect on various herbaceous monocotyledons and dicotyledons for 2 months; and picloram at 7 kg/ha for 3 months (Bovey *et al.*, 1968). In further trials there, the effects of picloram at 3 kg/ha could be detected on cucumber (*Cucumis*; Cucurbitaceae) for 12 months (Bovey *et al.*, 1969).

In South Vietnamese field trials, it was found that when Agent Orange was sprayed onto cleared inland forest soils at military dose levels, such treatment ceased to cause a reduction in survival and growth of, *inter alia*, subsequently planted corn (*Zea*; Gramineae) after 4 weeks, of upland rice (*Oryza*; Gramineae) or peanut (*Arachis*; Leguminosae) after 10 weeks, and of bean (*Phaseolus*; Leguminosae) after 18 weeks. In an equivalent trial with Agent White the intervals to insignificant damage were 10 weeks for both corn and upland rice, 31 weeks for the bean used, and 24 weeks for peanut (although some presumably trivial effects were still discernible on the peanuts for at least 34 weeks, the time at which the observations were terminated) (Blackman *et al.*, 1974).

When either Agent Orange or Agent White was similarly applied to cleared mangrove forest soils, the herbicidal rate of disappearance was reported to be similar to that found in the inland forest soils. Moreover, when seedlings of two common mangroves (*Rhizophora* and *Ceriops*; both Rhizophoraceae) were transplanted into such treated soils 40 days after spraying, their rate of survival appeared to be equivalent to that in control soils.

In temperate-zone field trials with dimethyl arsinic acid, it has been found that for such horticultural operations as lawn renovation, grass can be sown without harm to it immediately following an application of 22 kg/ha since sufficient inactivation occurs during the few days prior to seed germination (Ehman, 1965). It was further reported that subsequently harvested alfalfa (*Medicago*; Leguminosae) and ryegrass (*Lolium*; Gramineae) that had been sown 3 days after pasture treatment at 6 kg/ha contained arsenic levels no higher than in control plants.

The dioxin contaminant of the 2,4,5-T in Agent Orange turned out to be considerably more persistent than its carrier agent. A conservatively estimated total of 170 kg was applied to South Vietnam, primarily during 1966 to 1969 and largely in the former Military Region III (Table 5.6.7). The dioxin, once incorporated into the local ecosystem, can be assumed to disappear from the environment following first-order kinetics and can be calculated to have an environmental half-life of the order of 3.5 years (Westing, 1982a; *see also* Olie, 1984). If one makes the simplifying assumptions that the estimated 170 kg of applied dioxin had all been introduced into the South Vietnamese environment in 1968 and that half of it had become incorporated into the soil and biota (the other half presumably having been rather rapidly photodecomposed), then perhaps 8 kg remained present in 1980, 3 kg in 1985, and 1 kg will presumably be present in 1990. The action of wind and water is likely to have enlarged (and continues to be enlarging) the original area of application of 1.0 million hectares, a matter that is disturbing in the sense that the area of contamination is expanding, but reassuring in the sense that the severity of contamination in any one locality is declining, not only via decomposition but also through scattering.

Table 5.6.7 Dioxin applications in South Vietnam in the Second Indochina War: a rough approximation*. Source: calculated from the data in Tables 5.6.2 and 5.6.5

Year	Military Region I†	Military Region II†	Military Region III†	Military Region IV†	Total
A. Amount (kg)					
1961	?	?	?	?	?
1962	—	—	0.1	—	0.2
1963	0.2	0.2	0.6	0.1	1.1
1964	0.6	0.8	1.9	0.3	3.6
1965	1.2	1.4	3.6	0.6	6.8
1966	4.1	5.1	12.9	2.2	24.4
1967	7.7	9.6	24.1	4.1	45.5
1968	5.8	7.1	18.0	3.1	33.9
1969	8.1	10.0	25.1	4.3	47.4
1970	1.2	1.5	3.7	0.6	6.9
Total	28.9	35.7	90.1	15.3	170.0
B. Amount per unit area, assuming uniform distribution over the entire region (mg/ha)					
1961	?	?	?	?	?
1962	—	—	—	—	—
1963	0.1	—	0.2	—	0.1
1964	0.2	0.1	0.6	0.1	0.2
1965	0.4	0.2	1.2	0.2	0.4
1966	1.5	0.7	4.3	0.6	1.4
1967	2.8	1.2	8.0	1.1	2.6
1968	2.0	0.9	5.9	0.8	2.0
1969	2.9	1.3	8.3	1.1	2.7
1970	0.4	0.2	1.2	0.2	0.4
Total	10.3	4.6	29.8	4.0	9.8

*The estimated 170 kg of dioxin was directly applied to about 1.0 million hectares, that is, onto about 6% of the surface of South Vietnam (Table 5.6.2). Thus the average dose on this directly sprayed land was about 163 mg/ha. About 155 kg, or 91% of the applied dioxin was sprayed onto forest lands and the remaining 15 kg, or 9%, onto agricultural lands (Table 5.6.3)

†The former Military Regions are depicted, and their areas and mid-war populations provided in Figure 5.6.1

5.6.4.2 Long-term Effects on Flora and Fauna

Recent examination of the inland forests of South Vietnam has established that the wartime herbicidal damage of more than a decade ago is still much in evidence. It was reaffirmed that the severity of original damage and progress towards recovery depend on a variety of complex (and often little understood) factors, including: pre-spray condition of the stand; frequency and season of original spraying; species composition; steepness and other features of the terrain: local climate; areal extent of damage; availability of a seed source; and subsequent fire history (see below). It appears that with one or two original sprayings of a dense inland forest, a sufficient number of understorey trees survived that will grow and provide at least a poor harvest in three to four decades following attack (Ashton, 1984). However, it was estimated to take eight to ten decades following such spraying for a stand comparable to the pre-spray one to become established.

Those inland forests sprayed three or more times were generally damaged sufficiently to result in subsequent site damage from soil erosion and nutrient dumping (loss of nutrients in solution) and for the establishment of a grassy cover, usually herbaceous though sometimes woody (Ashton, 1984; Hiêp, 1984a). It turns out that these now herbaceous-grass-dominated sites have been burned over during many of the annual dry seasons since the war, such fires often being of human origin (Galston and Richards, 1984). These repeated fires have not only essentially prevented the re-establishment of trees, but have even been encroaching on the surrounding forest and have thus been slowly expanding in size. The modest natural forest regeneration in these badly damaged areas has been with trees of poor quality. The very important post-war role of fire in impeding forest recovery, and even in exacerbating the original degree of damage, has been the major revelation of the recent inland forest studies.

Herbicidal decimation of a forest leads to site debilitation for a number of reasons. The nutrients released by the fallen foliage cannot be held to any great extent by the soil and are thus lost to that ecosystem. Such nutrient dumping is especially severe in the tropics and often prominently involves potassium, nitrogen, and phosphorus (Zinke, 1984). As the trees die, the newly unprotected soil is subject to erosional loss—the more so the steeper the terrain—until the re-establishment of a new vegetational cover (a grass cover, which, however, protects the soil less well than the former trees). Indeed, recent soil studies have revealed that soils on steep slopes that had been subjected to the wartime spraying are, more than a decade later, still seriously depleted in nitrogen as well as in total organic matter content (Huây and Cu, 1984).

It has become quite clear that, for vegetational recovery to occur in the seriously damaged inland forests, fire must be excluded and, moreover, that the worst damaged areas will require artificial planting. Indeed, site debilitation has in many instances been sufficiently severe to require pre-planting (or inter-planting) with hardy soil-holding and soil-enriching species, for example, with nitrogen-fixing leguminous trees (Ashton, 1984; Galston and Richards, 1984).

The close association between an animal's geographical distribution (i.e. the animal's presence or absence) and its particular habitat requirements is a fundamental tenet of ecology. Indeed, this relationship is an especially tight one in tropical forests (Leighton, 1984). Recent comparisons in South Vietnam of unsprayed inland forest sites with comparable sites that had been multiply sprayed during the war, have been subjected to subsequent fires, and are now dominated by grasses, abundantly confirm this relationship. For example, in two unsprayed forests 145 and 170 bird species were recorded whereas in the destroyed forest (now grassland) there were only 24 (Westing, 1984b). Similar values for mammal species were 30 and 55 in the two unsprayed sites, but only 5 in the comparable though previously sprayed site. Moreover, an examination of the mammalian species that comprise these numbers reveals that whereas most taxa of wildlife declined, the numbers of undesirable rodent species increased.

To ameliorate the disastrous long-term impact of destroyed habitat on wildlife populations will require an accelerated programme of reforestation, the prohibition of game hunting, and restrictions on fuel-wood gathering (Huỳnh *et al.*, 1984). More sophisticated actions are called for as well (Leighton, 1984).

As noted earlier, the one habitat of South Vietnam which had been most seriously disrupted by the wartime herbicidal attacks was mangrove. Roughly 124 thousand hectares of this highly productive ecosystem (i.e. approx. 40% of it) had been utterly devastated. A rough field survey carried out by the author in 1980 indicated the following current situation regarding these 124 thousand hectares (Westing, 1982a): (a) barren patches of 5–50 ha, approx. 5–10%; (b) natural regeneration of *Rhizophora* adjacent to residual stands (a highly desirable outcome) approx. 1%; (c) artificial planting of *Rhizophora* approx. 10%; (d) conversion to rice and other crops, about 5–6%; and (e) natural regeneration of low-growing locally undesirable species of palms, ferns, poor-quality mangrove species, etc., about 75%. Much site damage by sheet erosion and wave action has occurred since the war (Snedaker, 1984). The lack of natural regeneration by the ecologically and economically desirable *Rhizophora* has to a considerable extent resulted from a lack of seed source so that an accelerated programme of artificial planting is indicated (Hiêp, 1984b; Snedaker, 1984). Where mangrove species have become established (whether by natural or artificial means and whether of inferior or superior species) a closed canopy can be expected within a decade or two of the time of establishment, and a harvestable crop of wood (small timbers and firewood) in perhaps four or five decades.

The offshore marine fishery of South Vietnam is known to have declined since the war, but whether this phenomenon finds its roots in the wartime herbicidal attacks, as is suggested from time to time, has not been demonstrated (Snedaker, 1984). However, one recent study indicates that freshwater fish in inland forest areas that had been attacked with herbicides during the war became and have remained substantially reduced both in species numbers and biomass (Yên and Quýnh, 1984). The reduction was attributed to a long-lasting decline within the affected waters of the algae and invertebrates that provide the food for these fish.

5.6.5 CONCLUSION

Faced during the Second Indochina War with a dispersed and elusive enemy in South Vietnam, the USA sought to deny this foe sanctuary, freedom of movement, and a local civilian economy from which to help derive its sustenance. This strategy was pursued, *inter alia*, through an unprecedentedly massive and sustained expenditure of herbicidal chemical warfare agents against the fields and forests of South Vietnam. The use of these agents resulted in large-scale devastation of crops, in widespread immediate damage to the inland and coastal forest ecosystems, and—it might be added—in a variety of health problems among exposed humans.

The damage to nature involved the death of millions of trees and often their ultimate replacement by grasses, in turn maintained to this day by subsequent periodic fires; deep and lasting inroads into the mangrove habitat; widespread site debilitation via soil erosion and loss of nutrients in solution (nutrient dumping); decimation of terrestrial wildlife, primarily via destruction of their habitat; losses in freshwater fish, largely because of reduced availability of food species; and a possible contribution to declines in the offshore fishery.

A vigorous and sustained research effort is warranted in Vietnam in order to pursue and ameliorate the long-term ecological (and medical) effects of the wartime use of the herbicides. The proposed ecological studies should pursue techniques of fire prevention, soil restoration, tree planting (including pre-planting and inter-planting), and wildlife restoration. Study areas and field stations should be established in both inland and coastal habitats. The ecological studies might best be carried out with the active cooperation of such international agencies as the United Nations Educational, Scientific and Cultural Organization (e.g. with its Regional Coastal and Marine Programme).

Finally, the question arises regarding future employment of herbicides as anti-plant chemical warfare agents and of the potentially ecocidal outcome of their use (Westing, 1980). Military evaluations have been favourable as regards a diversity of potential operational theatres (Engineers, 1972). On the other hand, a widely held interpretation of the Geneva Protocol of 1925 makes illegal their use in war (Westing, 1985). Moreover, their impact—especially as demonstrated by the Second Indochina War—makes it illegal to use them in the light of the Environmental Modification Convention of 1977 (Westing, 1984a).

5.6.6 REFERENCES

Agence France Presse (1971). Hanoi issues list of raids: defoliation flights alleged. *New York Times*, 21 Jan., p. 13.

Akamine, E. K. (1950–1951). Persistence of 2,4-D toxicity in Hawaiian soils. *Bot. Gaz.*, Chicago, **112**, 312–319.

Ashton, P. S. (1984). Long-term changes in dense and open inland forests following herbicidal attack. In: Westing, A. H. (ed.), *Herbicides in War: the Long-term Ecological and Human Consequences*, pp. 33–37. Taylor & Francis, London.

Blackman, G. E., Fryer, J. D., Lang, A., and Newton, M. (1974). *Effects of Herbicides in South Vietnam: Persistence and Disappearance of Herbicides in Tropical Soils*. National Academy of Sciences, Washington.

Bovey, R. W., Miller, F. R., and Diaz-Colon, J. (1968). Growth of crops in soils after herbicidal treatments for brush control in the tropics. *Agron. J.*, Madison, Wis., **60**, 678–679.

Bovey, R. W., Dowler, C. C., and Merkle, M. G. (1969). Persistence and movement of picloram in Texas and Puerto Rican soils. *Pestic. Monit. J.*, Washington, **3**, 177–181.

Buckingham, W. A. Jr. (1982). *Operation Ranch Hand: the Air Force and Herbicides in Southeast Asia 1961–1971*. US Air Force Office of Air Force History, Washington.

Cecil, P. F. (1986). *Herbicidal Warfare: the Ranch Hand Project in Vietnam*. Praeger, New York.

Clausewitz, C. von. (1832–1834). *On War*. Translated from the German by O. J. M. Jolles (1950), Infantry Journal Press, Washington.

Clutterbuck, R. L. (1966). *Long, Long War: Counterinsurgency in Malaya and Vietnam*. Praeger, New York.

Connor, S., and Thomas, A. (1984). How Britain sprayed Malaya with dioxin. *New Scientist*, London, **101** (1393), 6–7.

Ehman, P. J. (1965). Effect of arsenical build-up in the soil on subsequent growth and residue content of crops. *Proc. of the Southern Weed Conference*, USA, **18**, 685–687.

Engineers, US Army Corps of (1972). *Herbicides and Military Operations*. Engineer Strategic Group Rept. No. TOPOCOM 9022300. 3 vols. US Army Corps of Engineers, Washington.

Galston, A. W., and Richards, P. W. (1984). Terrestrial plant ecology and forestry: an overview. In: Westing, A. H. (ed.), *Herbicides in War: the Long-term Ecological and Human Consequences*, pp. 39–42. Taylor & Francis, London.

Henderson, G. R. G. (1955). Whirling wings over the jungle. *Air Clues*, London, **9** (8), 239–243.

Henniker, M. C. A. (1955). *Red Shadow over Malaya*. Wm. Blackwood, Edinburgh.

Hiêp, Dinh (1984a). Long-term changes in dense inland forest following herbicidal attack. In: Westing, A. H. (ed.), *Herbicides in War: the Long-term Ecological and Human Consequences*, pp. 31–32. Taylor & Francis, London.

Hiêp, Dinh (1984b). Long-term changes in the mangrove habitat following herbicidal attack. In: Westing, A. H. (ed.), *Herbicides in War: the Long-term Ecological and Human Consequences*, pp. 89–90. Taylor & Francis, London.

Holden, D. (1972). Israelis admit army killed Arab crops with chemical sprays. *Sunday Times*, London, 16 Jul., p. 2.

Huây, Hoàng Van, and Cu, Nguyên Xuan (1984). Long-term changes in soil chemistry following herbicidal attack. In: Westing, A. H. (ed.), *Herbicides in War: the Long-term Ecological and Human Consequences*, pp. 69–73. Taylor & Francis, London.

Huyǹh, Dang Huy; Cân, Dan Ngoc; Anh, Quôc; and Thàng, Nguyên Van (1984). Long-term changes in the mammalian fauna following herbicidal attack. In: Westing, A. H. (ed.), *Herbicides in War: the Long-term Ecological and Human Consequences*, pp. 49–52. Taylor & Francis, London.

Lang, A., Thói, L. V., Åberg, E., Bethel, J. S., Blackman, G. E., Chandler, R. F. Jr., Drew, W. B., Fraser, F. C., Fryer, J. D., Golley, F. B. Jr., Hô, P. H., Kunstadter, P., Leighton, A. H., Odum, H. T., Richards, P. W., Tschirley, F. H., and Zinke, P. J. (1974). *Effects of Herbicides in South Vietnam: Summary and Conclusions.* National Academy of Sciences, Washington.

Leighton, M. (1984). Terrestrial animal ecology: an overview. In: Westing, A. H. (ed.), *Herbicides in War: the Long-term Ecological and Human Consequences*, pp. 53–62. Taylor & Francis, London.

Merck, G. W. (1946). Biological warfare. *Milit. Surg.* (now *Milit. Med.*), Washington, **98**, 237–242.

Merck, G. W. (1947). Peacetime benefits from biological warfare research studies. *J. Am. vet. med. Ass.*, Schaumburg, Ill., **110**, 213–216.

Norman, A. G. (ed.) (1946). Studies on plant growth-regulating substances. *Bot. Gaz.*, Chicago, **107**, 475–632.

Olie, K. (1984). Analysis for dioxin in soils of southern Viet Nam. In: Westing, A. H. (ed.), *Herbicides in War: the Long-term Ecological and Human Consequences*, pp. 173–175. Taylor & Francis, London.

Orians, G. H., and Pfeiffer, E. W. (1970). Ecological effects of the war in Vietnam. *Science*, Washington, **168**, 544–554.

Peterson, G. E. (1967). Discovery and development of 2,4-D. *Agric. Hist.*, Davis, Cal., **41**, 243–253.

Scullard, H. H. (1961). *History of the Roman World from 753 to 146 B.C.*, 3rd edn Methuen, London.

Snedaker, S. C. (1984). Coastal, marine and aquatic ecology: an overview. In: Westing, A. H. (ed.), *Herbicides in War: the Long-term Ecological and Human Consequences*, pp. 95–107. Taylor & Francis, London.

Truman, R. (1961–1962). Eradication of mangroves. *Aust. J. Sci.*, Sydney, **24**, 198–199.

Walsh, G. E., Barrett, R., Cook, G. H., and Hollister, T. A. (1973). Effects of herbicides on seedlings of the red mangrove, *Rhizophora mangle* L. *BioScience*, Washington, **23**, 361–364.

Westing, A. H. (1972). Herbicidal damage to Cambodia. In: Neilands, J. B. *et al.*, *Harvest of Death: Chemical Warfare in Vietnam and Cambodia*, pp. 177–205. Free Press, New York.

Westing, A. H. (1976). *Ecological Consequences of the Second Indochina War*. Almqvist & Wiksell, Stockholm.

Westing, A. H. (1980). *Warfare in a Fragile World: Military Impact on the Human Environment*. Taylor & Francis, London.

Westing, A. H. (1981a). Crop destruction as a means of war. *Bull. atom. Scient.*, Chicago, **37** (2), 38–42.

Westing, A. H. (1981b). Laotian postscript. *Nature*, London, **294**, 606.

Westing, A. H. (1982a). Environmental aftermath of warfare in Viet Nam. *SIPRI Yearbook*, London, pp. 363–389.

Westing, A. H. (1982b). Vietnam now. *Nature*, London, **298**, 114.

Westing, A. H. (ed.) (1984a). Environmental Warfare: a Technical, Legal and Policy Appraisal. Taylor & Francis, London.

Westing, A. H. (1984b). Herbicides in war: past and present. In: Westing, A. H. (ed.), *Herbicides in War: the Long-term Ecological and Human Consequences*, pp. 3–24. Taylor & Francis, London.

Westing, A. H. (1985). Towards eliminating the scourge of chemical war: reflections on the occasion of the sixtieth anniversary of the Geneva Protocol. *Bull. of Peace Proposals*, Oslo, **16**, 117–120.

Yên, Mai Dinh, and Quyńh, Nguyên Xuân. (1984). Long-term changes in the freshwater fish fauna following herbicidal attack. In: Westing, A. H. (ed.), *Herbicides in War: the Long-term Ecological and Human Consequences*, pp. 91–93. Taylor & Francis, London.

Zinke, P. J. (1984). Soil ecology: an overview. In: Westing, A. H. (ed.), *Herbicides in War: the Long-term Ecological and Human Consequences*, pp. 75–81. Taylor & Francis, London.

Ecotoxicology and Climate
Edited by P. Bourdeau, J. A. Haines, W. Klein and C. R. Krishna Murti

5.7 Fate and Undesirable Effects of Pesticides in Egypt

A. H. EL-SEBAE

5.7.1 INTRODUCTION

An overview of the status of the pesticides used in Egypt is presented in a case study. The fate, distribution, and adverse effects of the widely used pesticides are discussed. Chlorinated hydrocarbon insecticides and structurally related derivatives are found to be highly persistent and biomagnified in the environment. However, the OP insecticide Leptophos was shown to be quite persistent and liable to be stored in lipoid tissues. In addition, the hazards resulting from acute, semi-chronic toxicity of the insecticides used are reported.

5.7.2 ECOLOGICAL FACTORS AFFECTING AGRICULTURAL ACTIVITIES IN EGYPT

Egypt is a semi-arid country where the six million acres (2.4 million hectares) of arable land lie in the Nile River delta and narrow valley. This irrigated land is only 5–7% of the total area of the country, while the rest is a mere desert. Additional agricultural activities in the vast deserts are limited to some oases depending on underground water sources, and the only rain helpful for agriculture is confined to the northwestern coast along the Mediterranean, where only about 100 000 acres (40 000 hectares) are cultivated.

The annual River Nile input is 60 billion cubic metres. It is estimated that above one-third of that total flows to the Mediterranean Sea. Another third is used for irrigation, and the rest is lost in vaporization, runoff, and leaching down to the water table.

The Nile water originates from the African plateau and crosses the following eight countries before reaching Egyptian territory: Sudan, Ethiopia, Uganda, Tanzania, Kenya, Zaire, Rwanda and Burundi. While flowing through these countries, the Nile River is loaded with various types of pesticides and many other contaminants. Thus it arrives in Egypt after already being polluted with different pollutants, including the persistent chlorinated pesticides.

The majority of the 50 million inhabitants of Egypt live in crowded cities and villages along the narrow green strip of land beside the River Nile and its north delta around the capital, Cairo. The Nile watercourse is thus used for irrigation and transportation, as well as for industrial and recreational activities. The river is also partially used for disposal of some agricultural waste water, and some industrial wastes. In this densely populated and limited area, more than 30 000 metric tonnes of formulated pesticides are imported and used annually. More than 70% of these pesticides are insecticides used to control cotton insect pests, especially the leaf and bollworms. This programme is important to protect cotton, which is the main Egyptian cash crop.

To control these insect pests, aerial spraying is used to apply more than 75% of the pesticides, a method which is particularly hazardous to the inhabitants and non-target organisms. Such congestion makes it difficult to implement an evacuation or re-entry programme. Moreover, herbicides, fungicides, fertilizers, molluscicides, food additives, and synthetic dyes and other chemical pollutants are present in the Egyptian environment.

The recent agricultural development plans adopt the horizontal and vertical intensified condensed agriculture, which might require the use of more pesticides and other agrochemicals and which thus might magnify the spectrum and magnitude of environmental pollution and hazards of such chemical agents.

5.7.3 STATUS OF PESTICIDES USED IN EGYPT

In Table 5.7.1, the area of field crops treated with pesticides during the period 1951–1981 is indicated. The area treated to control cotton bollworms was higher than expected because it shows 3–4 sprays per season in the area of cotton

Table 5.7.1 Area of field crops treated with pesticides during the period 1951–1981 in Egypt. Source: Egyptian Ministry of Agriculture Records. Reproduced with permission

	Area ($\times 10^2$ acres) (1 acre = 0.4 hectare)			
Crop Pests	1951	1961	1971	1981
Cotton leafworm	200	1100	1400	300
Cotton bollworm	–	1400	3980	4500
Cotton thrips	2	104	420	200
Cotton spidermite	1	111	171	16
Corn borers	–	300	437	36
Rice pests	–	–	200	500
Vegetable pests	–	–	50	250
Fruit-tree pests	–	100	100	200
Household insects (in tons)	1	10	20	50

Table 5.7.2 Total active ingredient (a.i.) insecticides used in Egyptian agriculture during the 30 year period 1955–1985. Source: Ministry of Agriculture records. Reproduced with permission

Compound	Total a.i. (metric tonnes)	Years of consumption
Toxaphene	54,000	1955–1961
Endrin	10,500	1961–1981
DDT	13,500	1952–1971
Lindane	11,300	1952–1978
Carbaryl	21,000	1961–1978
Trichlorfos	6,500	1961–1970
Monocrotophos	8,300	1967–1978
Leptophos	5,500	1968–1978
Chlorpyriphos	13,500	1969–1985
Phosfolan	5,500	1968–1983
Mephosfolan	7,000	1968–1983
Methamidophos/Azinphos-Me	4,500	1970–1979
Triazophos	5,500	1977–1985
Profenofos	6,000	1977–1985
Methomyl	6,500	1975–1985
Fenvalerate	6,500	1976–1985
Cypermethrin	4,300	1976–1985
Deltamethrin	3,400	1976–1985

cultivation, which is on average 1.2–1.5 million acres (0.4–0.6 million hectares) per year.

In Table 5.7.2, the types and amounts of insecticides used on cotton during the last 30 years are shown. Toxaphene, which had been used extensively since 1955, was stopped in 1961 after its failure due to the build-up of resistance in the cotton leafworm, *Spodoptera littoralis*, leading to an outbreak of this insect which resulted in 50% loss of the cotton yield in that season. Carbaryl and trichlorfos were introduced to replace toxaphene; however they also lost their effectiveness after 4–5 years due to the problem of resistance. DDT/endrin, DDT/lindane and methyl parathion combinations were attempted but were soon found to be ineffective.

From 1967, monocrotophos was widely used in cotton in four sprays per season, but this has suffered since 1973, due to resistance. This opened the way for the introduction of compounds which were not yet registered in the producing countries. Leptophos, the OP phosphothionate ester, was shown to cause the adverse effect of delayed neuropathy in humans and livestock. The 1971 water buffalo episode is quite famous. Then chlorpyrifos, triazophos, profenofos, methomyl, and synthetic pyrethroids were recently introduced.

In Table 5.7.3, the generally used types of pesticides in Egyptian agriculture are listed.

Table 5.7.3 Types of pesticides imported into Egypt in 1985. Source: Ministry of Agriculture Records. Reproduced with permission

Item	Amount in tonnes of formulated materials
Organophosphorus insecticides	1000
Carbamate insecticides	550
Synthetic pyrethroids insecticides	800
Chlordane	50
Agricultural spray oils	5000
Photoxin	100
Acaricide (Kelthane)	500
Rodenticides	9000
Sulphur	20 000
Fungicides	4500
Herbicides	3500

Table 5.7.4 Concentration (ppb) of organochlorine insecticides in soil samples at El-Minia, El-Behera, and Dakahlieh Provinces, August, 1979

Samples/ location	Lindane	Endrin	DDT	DDD	DDE	Chlordane
El-Minia 1	0.15	0.16	1.40	1.20	n.d.	0.25
El-Minia 2	0.24	0.19	1.30	1.10	n.d.	0.30
El-Behera 1	0.70	0.30	0.38	0.56	n.d.	n.d.
El-Behera 2	0.50	0.48	0.40	0.84	n.d.	n.d.
Dakahlieh 1	1.20	1.00	1.54	1.34	n.d.	0.20
Dakahlieh 2	1.25	1.50	1.70	1.50	n.d.	0.24

5.7.4 DATA DEALING WITH FATE AND DEGRADATION OF PESTICIDES IN THE EGYPTIAN ENVIRONMENT

The chlorinated hydrocarbon insecticides were shown to be highly persistent in the soil, sediments, and food-chain organisms. They are also stored in human adipose tissues. The concentration of organochlorine insecticides was determined in soil samples from the three Egyptian Governorates: El-Minia (middle of Egypt); El-Behera (northwest of the delta); and Dakahlieh (northeast), during 1979 by Aly and Badawy (1981). The data are shown in Tables 5.7.4 and 5.7.5. Generally, the levels of chlorinated insecticides were higher in the Dakahlieh followed by El-Behera and then El-Minia. This order is parallel to the actual frequencies of utilization of the insecticides. The relatively high levels of chlorinated insecticides in soil years after cessation of application reflect the high persistence and the long half-life of such compounds in the soil.

The chlorinated insecticides were monitored by El-Sebae and Abo-Elamayem (1978) in municipal water in Alexandria City (Table 5.7.6). Data indicated that classical water treatment might reduce the organochlorine insecticide levels, but

Table 5.7.5 Concentration (ppb) of organochlorine insecticides in soil samples at El-Minia, El-Behera, and Dakahlieh Provinces, October, 1979

Samples location	Lindane	Endrin	DDT	DDD	DDE	Chlordane
El-Minia 1	0.15	0.15	1.30	1.32	n.d.	0.30
El-Minia 2	0.20	0.22	1.35	1.25	n.d.	0.42
El-Behera 1	0.52	0.25	0.40	0.60	n.d.	n.d.
El-Behera 2	0.55	0.40	0.45	0.90	n.d.	n.d.
Dakahlieh 1	1.15	1.10	1.60	1.50	n.d.	0.22
Dakahlieh 2	1.20	1.30	1.40	1.38	n.d.	0.34

Table 5.7.6 Chlorinated pesticides in different water sources in Alexandria City

	Concentration (ppb)			
Pesticide detected	Raw water Mahmoudia	Treated water El-Soyef plant	Tap water	Waste water of slaughter-house
BHC*	0.39	N.D.[†]	0.10	0.19
Lindane	0.34	0.19	0.29	0.63
Heptachlor	0.70	0.10	0.12	0.19
p,p′-DDT	0.65	0.47	0.47	0.95
O,p′-DDT	0.95	N.D.	0.95	0.25

*Calculated as lindane
[†]Not detected

Table 5.7.7 Chlorinated pesticides in water and sediment samples at Lake Mariut, Alexandria

	Mean concentrations (ppb)			
	Lindane		p,p′-DDT	
Lake stations	Water	Sediment	Water	Sediment
I	2.06	142.8	3.85	982
II	2.10	74.7	2.54	512
III	1.93	61.6	2.79	715
IV	1.65	120.3	2.80	920
V	1.75	92.2	5.35	796
VI	1.75	52.8	4.31	910
VII	1.79	54.3	6.39	972
VIII	2.76	114.5	4.86	318

that there is still an appreciable level of these pollutants in the tap drinking water. The chlorinated insecticides were also detected in sediments of the northern brackish lakes, such as Lake Mariut near Alexandria, as shown in Table 5.7.7 (Abo-Elamayem *et al.*, 1979). Similar data were reported by Askar (1980) in

Brullus Lake in the northern part of the Delta. Storage and bioaccumulation of these chlorinated insecticides were shown by levels more than 100-fold higher in sediments and fish than in water. Due to the known long half-lives of these insecticides, they are expected to last for several years to come and to continue to be one of the environmental stresses.

Recently, Ernst *et al.* (1983) reported that organochlorine compounds were monitored in some marine aquatic organisms from the Alexandria area. DDT and its degradation products, gamma-BHC, alpha-BHC, dieldrin, and PCBs, were the major detected compounds. The results indicated that the western coast of Alexandria is polluted with organochlorine compounds. Moving towards the east at Rosetta, the levels of PCBs decrease because of the absence of industrial discharges. Generally, the recorded levels are in the range of the tolerated maximum residue limits. However, these low levels still represent a potential hazard as sources for continuous bioaccumulation and biomagnification in the food chain.

Similar data in fish samples were reported by Macklad *et al.* (1984a), concerning Lake Maryout and Alexandria Hydrodrome. Macklad *et al.* (1984b) detected the chlorinated compounds in different fish samples from two sampling sites at Edku Lake and Abu Quir Bay. BHC, DDE, DDD, endrin, DDT, and polychlorinated biphenyls were the major ones recorded. DDE was the major detected DDT metabolite in fish samples. The ratio of alpha to gamma-BHC isomers in different fish species from Edku Lake was higher than Abu Quir fish samples, suggesting older residues in Edku Lake. The level of chlorinated pesticides in *Tilapia* fish was positively correlated with fat tissue content. PCBs, such as Arochlor 1260, were higher in Abu Quir samples, where most industrial wastes are discharged.

Organophosphorus, carbamate, and synthetic pyrethroid insecticides replaced the organochlorine insecticides. Othman *et al.* (1984) estimated the half-life of residues of a number of these recently introduced insecticides on tomato. These

Table 5.7.8 Residue half-lives of certain insecticides on cabbage and tomato

Insecticide	Plant part	Half-life in days
Flucythrinate	Cabbage leaves	4.0
	Tomato leaves	5.9
	Tomato fruits	3.3
Cypermethrin	Cabbage leaves	13.1
	Tomato leaves	7.3
	Tomato fruits	11.6
Dimethoate	Cabbage leaves	2.95
	Tomato leaves	3.40
	Tomato fruits	2.40
Methomyl	Cabbage leaves	1.40
	Tomato leaves	1.30
	Tomato fruits	0.52

values are shown in Table 5.7.8. Methomyl, the oxime carbamate insecticide, had the shortest half-life, followed by dimethoate, flucythrinate, and cypermethrin.

The type of soil affects the adsorption, leaching, or translocation rate of pesticides. El-Sebae *et al.* (1969) demonstrated the variation in the characteristics of three Egyptian soil types as shown in Table 5.7.9. They also showed that the initial adsorption of the herbicide Dalapon on the three soil types differs widely, being higher on the sandy than the silty type, while the muck clay type retained the lowest level of the herbicide (Table 5.7.10). However, the dissipation rate was higher in the sandy soil due to the high loss through evaporation and leaching. The higher organic matter content and compactness in heavy soil types account for the greater ability of these two types to hold the residues of Dalapon for longer intervals.

Riskallah *et al.* (1979) studied the stability of Leptophos in water samples from different sources in the Egyptian environment. Leptophos proved to be a rather stable compound. After four months a considerable amount of this compound still remained unchanged in different water samples. The same trend was reproduced in samples from the River Nile irrigation and drainage water samples. This high stability was also shown on sprayed plant foliage even under direct sunlight. Thus Leptophos can be considered one of the most persistent OP insecticides in the environment.

El-Zorgani (1980) recorded the DDT content of samples of seven fish species taken at Wadi Halfa on Lake Nubia at the border between Sudan and Egypt. The results are shown in Table 5.7.11. p,p′-DDE found in all ten samples analysed was at relatively high levels, confirming contamination of fish samples with levels of DDT significantly higher than the maximum permissible levels. Such high pollution with DDT was attributed to the continued use of DDT on cotton fields in the Gezera project in central Sudan. DDT was banned in Sudan in 1981.

Table 5.7.9 Characteristics of three soil types

Soil Type	pH	Water saturation (%)	Number of bacteria per gm	Organic matter (%)
Sandy	7.8	18.9	50 000	0.09
Silty	7.1	38.8	112 000	0.70
Clay	8.0	60.7	211 000	0.75

Table 5.7.10 Persistence of Dalapon in three soil types

	ppm after shortage for			
Soil type	0 days	7 days	14 days	21 days
Sandy	29.95	17.5	3.5	0.5
Silty	27.30	20.5	15.5	8.0
Clay	24.85	23.0	17.5	12.0

5.7.5 ECOTOXICOLOGICAL FACTORS AFFECTING PERSISTENCE AND DISTRIBUTION OF PESTICIDES

Plant foliage differs widely in its wettability according to variations in plant species, age, plant part, and upper or lower surface of plant leaves. Data in Table 5.7.12 demonstrate such differentiation between four plant species: maize, broad bean, cotton, and squash (El-Sebae *et al.*, 1982). Such variation was found to affect the deposit toxicity of the two insecticides fenvalerate and tetrachlorvinphos in E.C. formulations against the cotton leafworm (*Spodoptera littoralis*) larvae (Table 5.7.13). Squash, which was shown in Table 5.7.12 to be readily wetted and to have the least thick cuticle, was the most susceptible at the same rate of application due to the higher initial deposit.

Temperature is one of the main extrinsic factors which has a continuous impact on chemical and biological processes in the environment. Increased temperature in tropical and semi-tropical areas is expected to increase the evaporation of the pesticide residues under humid conditions.

Table 5.7.14 presents the physico-chemical properties of some widely used insecticides, including the partition coefficient, the hydrolytic half-life, water solubility and vapour pressure. Such vapours can be transported by wind movement and then reprecipitated with rain to areas which might never have

Table 5.7.11 Residues of organochlorine insecticides in some fishes from Lake Nubia

Type	Residue content (ppb) μg/kg		
	p,p′-DDE	p,p′-DDT	Total as p,p′-DDT*
Barbus bynni (Forsk.)			
1–Muscle	1.0	5.0	6.0
2–Muscle	107.0		119.0
3–Muscle	39.0		43.0
Hydrocynus forskalii (Cuv.)			
1–Muscle	3.0	5.0	8.0
2–Muscle	153.0	14.0	184.0
Labeo coubie (Rupp.)			
1–Muscle	21.0		23.0
Labeo niloticus (Forsk.)			
1–Liver	4.0		4.0
2–Liver	2.0		2.0
3–Liver	12.0		13.0
Lates niloticus (Linn.)			
1–Muscle	6.0		7.0

* DDE content was multiplied by 1.11

Table 5.7.12 Plant species variation in wettability at different leaf development stages

Plant species	% leaf area wetted			
	1st-stage leaves		4th-stage leaves	
	Upper	Lower	Upper	Lower
Maize	8	9	28	47
Broad bean	12	27	38	69
Cotton	71	78	36	39
Squash	72	76	83	84
	Thickness of cuticular layer (μm)			
Maize	12.6	10.4	9.3	6.2
Broad bean	8.9	5.6	5.3	3.8
Cotton	4.7	2.8	2.5	1.8
Squash	2.1	1.2	0.7	0.3

Table 5.7.13 Plant species variation in deposit toxicity of fenvalerate and Tetrachlorvinphos to cotton leafworm (by dipping technique)

Plant species	LC50 (ppm)	
	Phenvalerate	Tetrachlorvinphos
Maize	21	920
Broad bean	19	850
Cotton	13.5	730
Squash	11.0	600

Table 5.7.14 Physico-chemical properties of some widely used insecticides

Pesticide	Partition coefficient (log P) octanol/water	Half-life* (hours)	Water solubility (ppm at 25°C)	Vapour pressure (mm Hg at 15°C)
p,p′-DDT	6.19	–	0.0012	1.9×10^{-7}
Leptophos	6.31	22.8	0.0047	–
Ronnel	4.95	6.7	1.08	8×10^{-4}
Chlorpyriphos	5.08	21.3	2.00	1.87×10^{-5}
Chlorpyriphos-methyl	4.31	–	4.00	4.22×10^{-5}
Parathion	3.82	37.3	11.9	3.78×10^{-5}
Methyl parathion	2.97	6.9	55.00	9.7×10^{-6}
Fenitrothion	3.38	–	30.00	5.4×10^{-5}
Bromophos	5.81	36.6	2.00	4.6×10^{-5}
Me bromophos	5.16	7.1	40.00	1.3×10^{-4}
Dimethoate	–1.71	10.4	25,000.0	8.5×10^{-6}
EPN	4.68	40.9	insoluble	3.0×10^{-4}

*Half-life in hours is the hydrolysis rate at 72°C in ethanol at pH 6.0 buffer solution.

used such an insecticide. This might explain the cyclodiene organochlorine insecticides detected in closed lakes in Sweden and some other European countries where they have never been used.

Temperature variation might affect the level of toxicity of the same compound to the same insect. This is called the temperature coefficient for each compound. Most of the insecticides have positive temperature coefficients, while DDT and some synthetic pyrethroids are generally of negative temperature coefficient.

Increased relative humidity favours insecticidal toxicity (Kamel and El-Sebae, 1968). Lipoid solubility in terms of partition coefficient (Table 5.7.14) reflects the potential of bioaccumulation and biomagnification of persistent insecticides. The data showed that Leptophos is more lipoid soluble than p,p′-DDT.

Photolysis is one of the degradation processes and is intensified under subtropical arid and semi-arid areas. Recently El-Sebae *et al.* (1983) reported that photoperiodism affects fish susceptibility to insecticides. The toxicity of cypermethrin and fenvalerate to mosquito fish (*Gambusia affinis*) was significantly increased under the 12 hr darkness and 12 hr light rotation when compared with the condition of continuous darkness. It was also found that at 0900 fish were more susceptible to poisoning than at 2100, due to circadian rhythmic response effects.

5.7.6 ADVERSE AND HAZARDOUS EFFECTS OF PESTICIDES IN EGYPT

Hegazi *et al.* (1979) studied the effect of the application of a group of pesticides (aldicarb, pendimethalin, dinoseb, chlorpyrifos and simazine) on nitrogenase activity in Giza clay-loam soil under maize cultivation near Cairo. All these pesticides showed different inhibitory effects and these effects were increased with higher doses and longer incubation periods.

Carbamate and organophosphorus insecticides were found to cause harmful inhibition of soil dehydrogenase activity (Khalifa *et al.*, 1980). Similar unfavourable side-effects of other pesticides on different soil enzymes were reported.

Cole *et al.* (1976) demonstrated that sub-surface application of aldrin resulted in reduction of the corn plant's height due to its phytotoxicity. Other compounds and solvents are known to be phytotoxic. Edwards and Thompson (1973) indicated that pesticides in the soil affect its content of non-target and beneficial organisms, including earthworms, collembolans, and insect larvae. This leads to deleterious effects on the texture and fertility of the soil. Organochlorine insecticides tend to accumulate to 9-fold in earthworms and 20-fold in soil snails (Gish, 1970). Such soil fauna can be used as indicators for monitoring levels and effects of pesticides and their degradation products in the soil.

El-Sebae *et al.* (1978) were able to minimize the acute toxicity of methomyl and zinc phosphide through the microcapsulation formulations using sustained gelatine capsule walls.

Delayed neuropathy is one of the adverse effects caused by some organophosphorus insecticides, particularly of the phosphonate type. El-Sebae *et al.* (1977, 1979, 1981) proved that Leptophos, EPN, trichloronate, salithion, and cyanophenphos are delayed neurotoxicants. Evidence was given clinically and biochemically. This effect is irreversible and there is no antidote for their application. Recently, methamidophos, trichlorfon, and DDVP were found to cause delayed neuropathy in man.

Cytotoxic effects, including mutagenesis, carcinogenesis, and teratogenesis, were reported for a number of pesticides. The list in Table 5.7.15 indicates these health hazards (El-Sebae, 1985).

El-Mofty *et al.* (1981) proved that the bilharzia snail molluscicide, niclosamide, is carcinogenic to amphibian Egyptian toads. Raymond and Alexander (1976) indicated that nitrosoamines can be formed in the soil from the reaction between some carbamates and nitrites. Nitrosamines can be translocated to edible plants.

El-Sebae (1985) found also that the response of exposed workers to pesticide poisoning differs widely, depending upon their blood group type. Such wide genetical variation should be taken into consideration when applying a safety factor in setting acceptable daily intakes of such pesticides and related toxic derivatives.

Table 5.7.15 Cytotoxic hazards of pesticides used in Egypt

Compound	Type of cytotoxic hazard expected
Amitraz	Oncogenicity
Azinphos-methyl*	Oncogenicity
Benomyl	Teratogenicity
BHC	Oncogenicity
Captan*	Teratogenicity
Carbaryl	Fetotoxicity
Chlorbenzilate	Oncogenicity
Chlordane	Oncogenicity
Chlordimeform	Oncogenicity
Dimethoate	Oncogenicity
Dichlorovos	Fetotoxicity
Endrin	Oncogenicity
Monuron	Oncogenicity
Niclosamide	Oncogenicity
Permethrin	Oncogenicity
Propanil	Oncogenicity
Trichlorofon	Oncogenicity
Trifluralin*	Oncogenicity
Toxaphene	Oncogenicity
Thiodicarb	Oncogenicity

*Waters *et al.* (1980). Reproduced with permission

5.7.7 REFERENCES

Abo-Elamayem, M., Saad, M. A., and El-Sebae, A. H. (1979). Water pollution with organochlorine pesticides in Egyptian lakes. *Proc. of the Internat. Egyptian-German Seminar on Environment Protection from Hazards of Pesticides*, Alexandria, Egypt. March 24–29, pp. 94–108.

Aly, Osama A., and Badawy, M. I. (1981). Organochlorine insecticides in selected agricultural areas in Egypt. *Proc. of the Internat. Symposium on Mgmt. of Indust. Wastewater in the Developing Nations*, Alexandria, Egypt, March 28–31, pp. 273–281.

Askar, A. (1980). *Monitoring of pesticides in Lake Brullus*. M.Sc. Thesis, Alexandria University.

Cole, L. K., Sanborn, J. R., and Metcalf, R. L. (1976). Inhibition of corn growth by aldrin and the insecticides fate in the soil, air and wildlife of a terrestrial model ecosystem. *Envir. Entomol*, **5**, 538–589.

Edwards, C. A., and Thompson A. R. (1973). *Pesticides and the Soil Fauna, Residue Reviews*, Gunther, F. A. (ed.), **45** : 1.

El-Mofty, M., Reuber, M., El-Sebae, A. H., and Sabry, I. (1981). Induction of neoplastic lesions in toads, *Bufo regularis*, with niclosamide. *Proc. of Internat. Symposium on Prevention of Occupational Cancer*, Helskinki, April 21–24.

El-Sebae, A. H. (1985). Management of Pesticide Residues in Egyptian Environment, Appropriate Waste Management for Developing Countries, Kriton Curi, (ed.), Plenum Publishing Corp., pp. 563–577.

El-Sebae, A. H., and Abo-Elamayem, M. (1978). A survey of expected pollutants drained to the Mediterranean in the Egyptian Region. *Proc. of the XXXVI Congress and Plenary Assembly of the Internat. Comm. of Sci. Explor. of the Mediterranean Sea*, Antalya, Turkey, pp. 149–153.

El-Sebae, A. H., Kassem, E. S., and Gad, A. (1969). Persistence and breakdown of dalapon and 2,4-D in three soil types. *Assiut. J. Agric. Sciences*, **3**, 359–366.

El-Sebae, A. H., Soliman, S. A., Abo-Elamayem, M., and Ahmed, N. S. (1977). Neurotoxicity of organophosphorus insecticides: Leptophos and EPN. *J. Environ. Sci. and Hlth*, **B12** (4), 269–288.

El-Sebae A. H., Ibraheim, S. M., El-Feel, S. A., and Srivastava, S. N. (1978). Microencapsulation of methomyl, zinc phosphide and copper sulphate—methodology and activity. *Fourth Internat. Congress of Pesticide Chem. Proc. (IUPAC)*, Zurich, July, pp. 562–566.

El-Sebae, A. H., Soliman, S. A., and Ahmed, N. S. (1979). Delayed neurotoxicity in sheep by the phosphonothioate insecticide Cyanophenphos. *J. Environ. Sci. and Hlth*, **B14** (3), 247–263.

El-Sebae, A. H., Soliman, S. A., Ahmed, N. S., and Curley, A. (1981). Biochemical interaction of six OP delayed neurotoxicants with several neurotargets. *J. Environ. Sci. and Hlth*, **B16**, 463–474.

El-Sebae, A. H., Morsy, F. A., Moustafa, F. I., and Abo-Elamayem, M. (1982). Impact on plant leaf wettability of insecticidal efficiency. *Proc. of the Second Egyptian–Hungarian Conf. on Plant Protection*, Alexandria University, pp. 295–403.

El-Sebae, A. H., El-Bakary, A. S., Le Paturel, J., Kadous, E., and Macklad, M. F. (1983). Effect of photoperiodism on fish susceptibility to insecticides. *Proc. of the Internat. Conf. on Photochemistry and Photobiology*, vol. 2, pp. 961–966.

El-Zorgani, G. A. (1980). Residues of organochlorine pesticides in fishes in Sudan. *J. Environ. Sci. and Hlth*, **B15** (6), 1090–1098.

Ernst, W., Macklad, F., El-Sebae, A. H., and Halim, Y. (1983). Monitoring of organochlorine compounds. 1. Some marine organisms from Alexandria Region. *Proc. Int. Conf. Env. Haz. Agrochem*, vol. 1, pp. 95–108.

Gish, C. D. (1970). Organochlorine insecticide residues in soils and soil invertebrates from agricultural lands. *Pesticides Monitor J*, **3** (4), 241–252.

Hegazi, N., Menib, M., Belal, M., Amer, H., and Farag, R. S. (1979). The effect of some pesticides on asymbiotic N2-fixation in Egyptian soil. *Archs. Environ. Contam. Toxicol*, **8**, 629–635.

Kamel, A. M., and El-Sebae, A. H. (1968). Effect of temperature and humidity on the effectiveness of insecticide deposits, Alex. *J. Agric. Res.*, **16**, 59–66.

Khalifa, M. A. S., Tag El-Din, A., Komeil, A. A., Desheesh, M. A., and Helwa, L. A. (1980). Pesticides and soil enzymes relationships. Ill effects of some organophosphates and carbamates on soil dehydrogenase activity. *Proc. of Fourth Conf. on Microbiology*, Cairo, Egypt, December 24–28.

Macklad, F., El-Sebae, A. H., Halim, Y., and Barakat, M. (1984a). Monitoring of chlorinated pesticides in fish samples from Lake Maryout and Alexandria Hydrodrome. *Bull. of the High Inst. of Public Hlth*, Alexandria Univ., vol. XIV (2), pp. 161–173.

Macklad, F., El-Sebae, A. H., Halim, Y., and El-Belbesi, M. (1984b). Monitoring of chlorinated hydrocarbons in some fish species from Lake Edku and Abu. Quir Bay. *Bull. High Inst. of Public Hlth.*, Alexandria Univ., vol. XIV (4), pp. 145–157.

Othman, Mohamed A. S., Antonious, G. P., Khamis, A., and Tantawy, G. (1984). Determination of residues of flucythrinate, cypermethrin, dimethoate, and methomyl on tomato and cabbage plants. Abstract in: *Symposium on Integrated Pest Mgmt. and Rationalization of Pesticide Use in the Arab Countries*, Algeria, Sept. 16–20.

Raymond, D. E., and Alexander, M. (1976). Plant uptake and leaching of dimethylnitrosoamine. *Nature*, **262**, 394–395.

Riskallah, M. R., El-Sayed, M. M., and Hindi, S. A. (1979). Study on the stability of Leptophos in water under laboratory conditions. *Bull. Environ. Contam. Toxicol*, **23**, 607–614.

Waters, M. D., Simmon, V. F., Mitchell, A. D., Jorgenson, T. A., and Valencia, R. (1980). Overview of short term tests for the mutagenic and carcinogenic potential of pesticides. *J. Environ. Sci. and Hth*, **B15**, 867–906.

Ecotoxicology and Climate
Edited by P. Bourdeau, J. A. Haines, W. Klein and C. R. Krishna Murti

5.8 Mercury in Canadian Rivers

D. R. MILLER

5.8.1 INTRODUCTION

The general history of the problem of mercury in the environment, and mercury as a toxic chemical generally, has been told many times, and we will not repeat here any details of the tragic poisoning in Japan, Iraq, and elsewhere. It is the purpose of this brief report to describe the environmental mercury problem specifically in Northwestern Ontario, in the English–Wabigoon river system.

In view of what was known about mercury intoxication at the time, and particularly in the light of the observation that its effects are extremely unpleasant and that therapy, at least at the stage of clinical expression, is almost totally unavailable, it is not surprising that the reaction to the discovery that mercury concentrations in certain Canadian foodstuffs were at dangerous levels produced something akin to a panic response. In retrospect, it is generally agreed that the issue was not handled in a particularly effective way; however, some important lessons were learned about the institutional frameworks and procedures that ought to have been in place to deal with such a problem. In addition, the story serves as a good example of how Northern systems may be especially delicate. In a sense that goes beyond what we think of as the ecosystem, it was found that reliance on a single resource, narrowness of the food chain, specialization in food and livelihood at the highest level (in this case, man), and difficulty in dealing with the problem in a holistic fashion led to profoundly serious consequences.

The lessons learned were quite expensive, in the sense that much irrelevant work was done, and in the sense that the victims were not quickly relieved of their most serious problems. The consequences were at the social and economic level; we can at least be satisfied to observe that no actual human deaths caused by clinically demonstrable mercury poisoning have been documented.

5.8.2 BACKGROUND

Historically, mercury has entered the Canadian environment in five ways (Hocking, 1979). They are:

(i) The use of mercury-containing chemicals as fungicides on certain seed grains;
(ii) The use of mercury compounds, most often phenylmercuric acetate, as a general-purpose antibiotic (a so-called 'slimicide') in the pulp and paper industry;
(iii) The use of mercury as an electrode in the chemical preparation of chloralkali products;
(iv) As a by-product of smelting operations directed at the production of other metals; and
(v) The general, diffuse input caused by the presence of mercury as an impurity in fossil fuels.

The first three of these have been discontinued in Canada. The other two are unavoidable in the short term. The fourth is particularly difficult to deal with, since the mercury in stack gases is normally in the form of elemental mercury vapour, and standard scrubbing devices are of limited use. However, some methods are available (Stuart and Down, 1974). The fifth will in fact increase as the fossil fuel consumption continues to rise in North America and, of course, throughout the world. It is not yet clear how soon this will result in a demonstrable danger on a wide scale (see below).

The first indication that mercury was a problem in Canada was the observation that mercury was present in the tissues of birds that had eaten fungicide-treated grains after they had been distributed as seeds. This was a relatively easy situation to correct, particularly since the source was the same as that responsible for the large number of deaths of humans in Iraq, where seed grains had been diverted into the making of bread for human consumption (Bakir *et al.*, 1973). This direct and widespread danger to human life made it very easy to justify the resources to identify and use alternative fungicides, as is now done. This, together with the imposition of more stringent controls on the disposition of such seed grains, should make it possible to avoid the problem in the future.

The second and third activities, although now discontinued, resulted in large quantities of mercury being released into river systems, much of which is still present and is still a source of considerable concern and, indeed, outright danger (Hocking, 1979). It is this collection of reservoirs of mercury, generally contaminated sediments in rivers, with which we are primarily concerned in what follows.

5.8.3 PHYSICAL BEHAVIOUR OF MERCURY

Most of the mercury released into Canadian rivers was in a relatively innocuous form in terms of toxicity, typically elemental mercury from chloralkah plants or phenylmercuric acetate from pulp and paper mills (Miller, 1977). The contaminant generally adhered to fine-grained sediments, especially to richly organic sediments. Initially it was thought that this mercury would be permanently trapped in the sediment, and completely unavailable on purely chemical grounds, since the sediments are rich in sulphur and the solubility of mercuric sulphide is so low (the solubility product in water is 10^{-53}). It was assumed therefore that virtually none would ever dissolve and appear as a chemically available ion in the water column, particularly in the cold waters of Canadian rivers.

This turns out to be largely true, but not exactly. At any time, a small fraction (one or two per cent) of the mercury in the sediment exists in the form of monomethyl mercury ion, which is of course much more soluble. Just how the conversion to this chemical form takes place has never been made clear; initially (Jensen and Jernelov, 1969) it was thought to be bacterially mediated, and there was even a detailed process suggested in which the methylmercury would be the result of an incorrect biological synthesis of methylcobalamine (Landner, 1971). Recently, the idea of a rate-determined process has fallen into disfavour, and the concept of a quickly re-established equilibrium between the organic and inorganic forms seems to be the consensus opinion (Miller, 1977; Beijer and Jernelov, 1979). Whether the equilibrium is essentially chemical or whether there is a balance between biologically-mediated demethylation as well as methylation steps is not clear. Certainly a demethylation process of some sort is occurring, since mercury released into the atmosphere from a sediment–water system contaminated with a mercuric salt appears to be exclusively in the form of elemental mercury vapour (Miller, D. R., unpublished data).

One thing that is abundantly clear is that the monomethyl mercury does leave the sediment and enter the water column and ultimately the biota. This removal from the sediment allows more to be produced as natural processes tend to re-establish the equilibrium in the sediment. Although the level of the organic form at any time is low (Akagi *et al.*, 1979), as long as it continues to be effectively removed from the sediment–water system by the biota, more will continue to be produced. The end result is that, although the proportion of organic mercury in the sediment is only one or two per cent, and although the partition coefficient for mercury between water and sediment is around 5000:1 in favour of the sediment (less for organic mercury) (Kudo, 1977), nevertheless the large fish in a contaminated river will typically have a tissue concentration of the same order as the sediment concentration of mercury, except that in the fish the mercury is almost all in the highly toxic methylmercury form (Miller, 1977). We might also note that the clearance rate of methylmercury from fish is slow; the half-life of mercury is somewhat more than one year in a relatively large (say, one kilogram) fish (Norstrom *et al.*, 1979).

As was noted above, the release of mercury into the atmosphere from a contaminated aquatic system appears to take place after a demethylation process rather than the reverse, and this may have important environmental consequences. Elemental mercury vapour would not be expected to adhere to a particulate, even one with high organic content, but would be transported in the vapour phase (Miller and Buchanan, 1977). Thus, dry deposition would not be an effective clearance mechanism. This is quite consistent with what is observed; numerous reports have indicated that soil samples taken in the immediate vicinity of a known mercury source contain far less mercury than would be expected if dry deposition were dominant (Miller and Buchanan, 1977). It should be noted that this is quite contrary to the traditional wisdom of the field, in which it was supposed that mercury was methylated twice and released into the atmosphere as the volatile dimethyl form (Jernelov, 1969). The dimethyl form would have a substantial affinity for organic particulates, and would be expected to be found in soil samples. This appears not to be the case.

The more important implication for environmental concerns is, of course, that mercury in the elemental form may be expected to be transported over long distances indeed. Even when removed from the atmosphere by wet deposition through a precipitation event, a resuspension of the substance through volatilization will allow the mercury to re-enter the global atmospheric circulation pattern, contributing again to the long-range transport that seems consistent with environmental observations (Miller, 1979).

This widespread distribution is reflected in several ways, for example in game fish in areas like Ontario, Canada, where fish of large size (a kilogram or so) almost invariably contain substantial body burdens of mercury, even when caught in locations far removed from known sources of mercury pollution.

We might also point out that this recycling of mercury in the atmosphere due to resuspension or volatilization means that if we attempt to estimate the global atmospheric content of mercury by precipitation sampling, we will be counting the same mercury several times. Thus, we would be led to significantly overestimate the amount in general circulation. The conclusion is that a larger fraction of the mercury in general circulation is due to man's operations than is generally thought, and therefore a fractional increase in anthropogenic mercury generation will significantly increase the overall global atmospheric levels (Miller, 1979).

5.8.4 EFFECTS OF MERCURY CONTAMINATION

In spite of the widespread distribution of mercury and the danger of general environmental contamination that might result, so far in Canada the only identifiable problems have been due to point discharges at a small number of sites, either from chloralkali plants or from pulp and paper manufacturing facilities. The most serious took place in the English–Wabigoon river system in Northwestern Ontario (Parks *et al.*, 1984); other locations of concern were Lake Erie and the Bell River in Northwestern Quebec (Miller *et al.*, 1979).

The English–Wabigoon river system lies in a relatively isolated area, the main inhabitants of which are occupants of two reserves of the Ojibwa Indians, called White Dog and Grassy Narrows Reserves. On these lands, the inhabitants follow to a large extent a traditional life-style, and this includes the consumption of locally caught fish as a dietary staple. Employment in the area is scarce, and many of the inhabitants have historically made most of their real income by serving as guides for tourists who visit one of several local lodges for the purpose of sport fishing (or by working in the lodges themselves).

When it was realized that fish in the area were contaminated, fishing was restricted in the entire system. This had the effect not only of cutting off the population from their main source of protein, but also of removing any possibility of income from the sport fishing industry; the fishing restrictions and the accompanying publicity caused such a slump in the tourist industry that the tourist lodges closed completely in the area. Although government agencies attempted to supply the local inhabitants with imported fish, there were serious problems. Maintenance of the relatively large refrigeration facilities required for such an operation was difficult, and there was no immediate way to replace the lost employment opportunities.

The problem has persisted for many years. The river system has been very slow to clear itself, and at time of writing is still noticeably contaminated, although inputs of mercury to the system have long since stopped. (The offending chloralkali plant changed its operation from the mercury-electrode process to the mercury-free membrane process more than a decade ago.)

The result has been described by the leaders of the Indian Bands themselves as producing a 'complete breakdown of the traditional society'. Incidents of abuse of alcohol and other drugs, and also abuse within the family unit, reached frightening proportions. Truly heroic efforts at the local level were required to deal with the problem. These societal efforts are now succeeding; some new employment has been introduced with extensive government support, and social changes, such as the complete banning of alcohol from the Reserves, has had a noticeable effect.

It is interesting to note that what was initially seen as a corresponding problem in Northern Quebec, specifically on the Bell River, was handled in a quite different fashion. Negotiations among the parties involved, namely the Grand Council of the Cree Indians, the industries involved, and the federal and provincial governments, resulted in a curtailment of the mercury input without anything like the social upset that had taken place in Ontario. It is perhaps necessary to add that even to this day the problem has not been resolved to the satisfaction of all concerned (*Globe and Mail* Newspaper, Toronto, July 13, 1986).

5.8.5 THE FUTURE OF MERCURY IN CANADA

In some previously contaminated Canadian waters, the effective mercury levels have decreased rapidly. In Lake Erie, where sediment is constantly being deposited and remains to some extent undisturbed, the sediments actually in contact with the water column are relatively clean today, and mercury content of the water is low; in other words, the mercury appears to have been effectively buried. The time taken for the fish to become sensibly free of mercury should therefore have depended either on the time taken for an individual fish to clear itself of mercury (the half-life is approximately one year) or should depend on the actual lifetime of a fish (rarely more than ten years). Thus, in the decade or so after the problem was identified and abatement procedures introduced, the situation was seen to greatly improve.

In the Ottawa River, the situation is also much better today than ten years ago, but for a different set of reasons. Like many Canadian (and other) rivers, the Ottawa is subject to greatly increased flow rates during spring melt and runoff (higher than summer flow rates by a factor of at least four, and this in spite of the existence of a number of flood control and hydroelectric dams). This flow causes a great deal of scouring of bottom sediment, which as a result moves far downstream. This is especially true for the very fine-grained sediment, which of course is the kind that preferentially binds most of the mercury; a given particle of clay may move 50 km in one year. Thus the sediment is regularly flushed out of the system and replaced by clean sediment from upstream (Miller *et al.*, 1979). The contaminated sediment is either swept out of the system or settles in front of a large dam to be subsequently buried as more sediment follows in its path. Seen from a single geographical location, the effective result is that sediment in a given location decreases in its mercury content by 30–40% each year (of course it is not the same sediment that is being measured). At a typical location near the city of Ottawa, the sediment concentrations decreased to background levels in four years (1969–1973) (Miller *et al.*, 1979). It is perhaps worthy of note that the problem could reappear if large-scale dredging of reservoirs were to be undertaken in the future.

In some cases, as noted above, the situation is not so fortunate. The English–Wabigoon river system is not losing its mercury at anything like the same rate. This is because of geography; the spring flood is not sufficiently strong to scour the sediment significantly, the waterway is not importing new sediment to cover up the old, and the sediment in the first place is of a very fine kind that binds mercury very effectively. Solutions have been proposed, ranging from large-scale dredging and dry-land disposal to the addition of selenium, known to be protective against mercury toxicity, in the hope of providing an antidote at the ecosystem level. So far, attempts to artificially speed up the clearance of the system have met with virtually no success in practical terms (Parks *et al.*, 1984).

In a longer time frame, we should be concerned about mercury as a general environmental contaminant in the future. It is true that all of the recognized problems so far have been due to very local releases. However, levels of mercury sufficient to justify governmental warnings are now found in fish taken from waters far from any known sources, and our new understanding of the atmospheric dynamics of the metal (see above) makes it perfectly understandable that this should be so.

Since one of the significant sources of mercury in the environment is its presence as a contaminant in fossil fuels, and since every prediction is that the consumption of such fuels will markedly increase in the next few years, we should be aware of the potential for mercury to emerge as a more widespread hazard a decade or so from now. The problem of mercury in Canadian waters is by no means over.

5.8.6 INFLUENCE OF A COLD ENVIRONMENT

Geographical and climatological features of Canada have affected the development of mercury problems in the country in several ways. The country has always had an extensive pulp and paper (and hence chloralkali) industry. In many cases the relevant installations have been located in remote areas where pollution control was not thought to be a matter of concern, and the development of such problems went largely unnoticed. (We must always bear in mind that the overall visual impression given by a mercury-contaminated waterway in an isolated northern location is typically that of a rather attractive, unspoiled wilderness environment.) Abatement procedures may have been delayed since the problem had to be not only noticed but also demonstrated from outside; if an outraged local population existed at all, it was often without much political power. Finally, simple considerations of logistics indicate that all kinds of monitoring, abatement, and enforcement procedures are much more difficult in remote areas.

The cold temperature itself is also part of the problem. At depressed temperatures, mercury deposited on the ground does not re-volatilize so rapidly as it does in warmer areas, so we see more ground- and water-level contamination for a given atmospheric burden (Miller and Buchanan, 1977). The generally lower level of biological activity in colder waters, and in particular the lower standing crop of higher plants (which are excellent absorbers of heavy metals in general and mercury in particular, so much so that they have been proposed as one way that mercury can be physically removed from contaminated systems), means that the residence time of mercury in colder waters is longer.

In addition, the significant changes in temperature over the course of the year lead to sudden large concentrations of this and other substances being presented to target organisms at particular times. Over the year, most organisms, at least at the animal level, build up lipid stores which are then called upon and metabolized during short periods of stress, thus effectively presenting a sudden high concentration within the tissues. Similarly, spring runoff brings a sudden load of winter-contaminated snow into the water systems at precisely the time when many aquatic organisms are in the process of reproducing.

The other notable characteristic of cold environments which influences the problem is, of course, the simplicity of the food chain. Animals for which fish constitutes a dietary staple are not only endangered in the vicinity of a point release (it was the appearance of a poisoned domestic animal, specifically a cat, that first alerted people to the problem at Grassy Narrows), but now would seem to be at risk over much larger areas. For example, a population of poisoned wild mink has recently been studied (Wren and Stokes, 1986). Other animals, such as bears and fish-eating birds, may well develop problems in the future. We seem to have been able to prevent human deaths from mercury poisoning (although at the cost of severe social disruption). We will find it much more difficult to protect the rest of the ecosystem.

5.8.7 REFERENCES

Akagi, H., Mortimer, D. C., and Miller, D. R. (1979). Mercury methylation and partition in aquatic systems. *Bull. Environ. Contam. Toxicol.*, **23**, 372–376.

Bakir, F., Dmaluji, S. F., Amin-Zaki, L., Murtadha, M., Khalidi, A., Al-Rawi, N. Y., Tikriti, S., Dhahir, H. I., Clarkson, T. W., Smith, J. C., and Doherty, R. A. (1973). Methylmercury poisoning in Iraq. *Science*, **181**, 230–241.

Beijer, K., and Jernelov, A. (1979). Methylation of mercury in aquatic environments. In: Nriagu, J. (ed.), *The Biogeochemistry of Mercury in the Environment*, pp. 211–230. Elsevier/North Holland Biomedical Press, Amsterdam.

Hocking, M. B. (1979). Uses and emissions of mercury in Canada. In: *Effects of Mercury in the Canadian Environment*, pp. 50–75. Publication NRCC 16739 of the Nat. Res. Council of Canada, Associate Committee on Scientific Criteria for Environ. Quality, Ottawa, Canada.

Jensen, S., and Jernelov, A. (1969). Biological methylation of mercury in aquatic organisms. *Nature*, **223**, 753–754.

Jernelov, A. (1969). Conversion of mercury compounds. In: Miller, W. M., and Berg, G. C. (eds), *Chemical Fallout*, pp. 68–74. Thomas (publishers), Springfield, Illinois.

Kudo, A. (1977). Equilibrium concentrations of methyl mercury in Ottawa River sediments. *Nature*, **270**, 419–420.

Landner, L. (1971). Biochemical model for the biological methylation of mercury suggested from methylation studies *in vitro* with *Neurospora crassa*. *Nature*, **230**, 452–453.

Miller, D. R. (ed.) (1977). Distribution and transport of pollutants in flowing water ecosystems. *Ottawa River Project Final Report*. National Research Council of Canada, Ottawa.

Miller, D. R. (1979). Mercury transport in the environment. In: *Effects of Mercury in the Canadian Environment*, pp. 76–83. Publication NRCC 16739 of the National Research Council of Canada, Associate Committee on Scientific Criteria for Environ. Quality, Ottawa, Canada.

Miller, D. R., and Buchanan, J. (1977). *Atmospheric Transport of Mercury*. Technical report # 14, Monitoring and Assessment Research Centre (MARC), Chelsea College, University of London.

Miller, D. R. *et al.* (Ottawa River Project Group) (1979). Mercury in the Ottawa River. *Environ. Research*, **19**, 231–243.

Norstrom, R. J., McKinnon, A. E., and de Freitas, A. S. W. (1979). A bioenergetics-based model for pollutant accumulation by fish. *J. Fish. Res. Board Canada*, **33**, 248–267.

Parks, J. W., Sutton, J. A., and Hollinger, J. D. (1984). *Mercury Contamination of the Wabigoon/English/Winnipeg River System*. Ontario Ministry of the Environment. Published as document No. Em 37-67/1984E (2 vols), Supplies and Services Canada, Ottawa, Canada.

Stuart, M., and Down, R. F. (1974). Mercury removal from copper concentrate. *Soc. Min. Eng., AIME Trans.*, **256**, 196–204.

Wren, C., and Stokes, P. (1986). Mercury levels in Ontario mink and otter relative to food levels and environmental acidification. *Can. J. Zool.*, **64**, 2854–2859.

Index